Chemical
Process
Design

Other McGraw-Hill Chemical Engineering Books of Interest

BRUNNER · *Hazardous Waste Incineration*

CHOPEY · *Environmental Engineering in the Process Plant*

CHOPEY · *Handbook of Chemical Engineering Calculations*

COOK, DuMONT · *Process Drying Practice*

DEAN · *Lange's Handbook of Chemistry*

DILLON · *Materials Selection for the Chemical Process Industries*

FREEMAN · *Hazardous Waste Minimization*

FREEMAN · *Standard Handbook of Hazardous Waste Treatment and Disposal*

KISTER · *Distillation Design*

KISTER · *Distillation Operation*

KOLLURU · *Environmental Strategies Handbook*

LEVIN, GEALT · *Biotreatment of Industrial Hazardous Waste*

MANSFIELD · *Engineering Design for Process Facilities*

McGEE · *Molecular Engineering*

MILLER · *Flow Measurement Handbook*

PALLUZI · *Pilot Plant and Laboratory Safety*

PALLUZI · *Pilot Plant Design, Construction, and Operation*

PERRY, GREEN · *Perry's Chemical Engineers' Handbook*

POWER · *Steam Jet Ejectors for the Process Industries*

REID ET AL. · *Properties of Gases and Liquids*

REIST · *Introduction to Aerosol Science*

SANDLER, LUCKIEWICZ · *Practical Process Engineering*

SATTERFIELD · *Heterogeneous Catalysis in Practice*

SHINSKEY · *Process Control Systems*

SHINSKEY · *Feedback Controllers for the Process Industries*

SHUGAR, DEAN · *The Chemist's Ready Reference Handbook*

SHUGAR, BALLINGER · *The Chemical Technicians' Ready Reference Handbook*

SMITH, VAN LAAN · *Piping and Pipe Support Systems*

STOCK · *AI in Process Control*

TATTERSON · *Fluid Mixing and Gas Dispersion in Agitated Tanks*

TATTERSON · *Scale-up of Industrial Mixing Processes*

WILLIG · *Environmental TQM*

YOKELL · *A Working Guide to Shell-and-Tube Heat Exchangers*

Chemical Process Design

Robin Smith
Centre of Process Integration
University of Manchester
Institute of Science and Technology

McGraw-Hill, Inc.

New York San Francisco Washington, D.C. Auckland Bogotá
Caracas Lisbon London Madrid Mexico City Milan
Montreal New Delhi San Juan Singapore
Sydney Tokyo Toronto

Library of Congress Cataloging-in-Publication Data

Smith, R. (Robin)
    Chemical process design / by Robin Smith.
        p.   cm.
    Includes bibliographical references.
    ISBN 0-07-059220-9
    1. Chemical processes. I. Title.
    TP155.7.S57   1995
    660'.2812—dc20   94-31182
                                                CIP

1 2 3 4 5 6 7 8 9 0   DOC/DOC   9 0 9 8 7 6 5 4

ISBN 0-07-059220-9

*The sponsoring editor for this book was Gail F. Nalven, the editing
supervisor was Peggy Lamb, and the production supervisor was
Donald F. Schmidt. This book was set in Century Schoolbook by The
Universities Press (Belfast) Ltd.*

Printed and bound by R. R. Donnelley & Sons Company.

This book is printed on acid-free paper.

To Christine, Rachele,
Nicholas, and Anna

Contents

Preface

Chemical process design starts with the selection of a series of processing steps and their interconnection into a flowsheet to transform raw materials into desired products. Whereas a great emphasis in chemical engineering traditionally has been placed on the analysis or simulation of flowsheets, the creation or synthesis of flowsheets has received, by comparison, little attention. Yet the decisions made during the synthesis of the flowsheet are of paramount importance in determining the economic viability, safety, and environmental impact of the final design. This text will concentrate on developing an understanding of the concepts required at each stage of the synthesis of process flowsheets.

Chemical processes will in the future need to be designed as part of a sustainable industrial development which retains the capacity of ecosystems to support industrial activity and life. This book therefore places a high emphasis on waste minimization and energy efficiency in the context of good economic performance and good health and safety practices.

The structure of the text largely follows the order in which the decisions should be made during the development of a process design. Economic evaluation has been included, but as an appendix so as not to interrupt the flow of the text. This book is intended to provide a practical guide to chemical process design for advanced undergraduate and postgraduate students of chemical engineering, practicing process designers, and practicing chemical engineers working in process development.

Robin Smith

Acknowledgments

Chapter 10 on waste minimization features sections of text that were originally published in *The Chemical Engineer* (Smith, R., and Petela, E. A., *The Chemical Engineer,* no. 506, 24–25, 31 Oct. 1991; No. 509/510, 17–23, 12 Dec. 1991; No. 513, 24–28, 13 Feb. 1992; No. 517, 21–23, 9 April 1992; No. 523, 32–35, 16 July 1992). These sections of text are reproduced by kind permission of the Institution of Chemical Engineers.

A number of people have contributed either directly or indirectly to this book. Gratitude is expressed to the following colleagues at UMIST: Geoffrey Clegg, Vikas Dhole, Antonis Kokossis, Janice Kuo, Bodo Linnhoff, Graham Polley, Paul Sharratt, and Gavin Towler. Gratitude is also expressed to the following colleagues outside UMIST: Truls Gunderson, of Telemark Institute of Technology, Norway, Stephen Hall, of M. W. Kellogg, Houston, David Lott, of Zenica, Vic Marshall, of the University of Bradford, Eric Petela, of Linnhoff March, Dinos Triantafyllou, of Esso Chemicals, Yaping Wang, of Linnhoff March, Norman White, of Esso Chemicals, and Duncan Woodcock, of ICI C & P. Finally, gratitude is expressed to Eileen Boocock and Amanda Brown for typing the manuscript.

Nomenclature

a	order of reaction (–) or cost law coefficient (–)
A	heat exchanger area (m^2) or annual cash flow ($)
A_{NETWORK}	heat exchanger network area target (m^2)
b	order of reaction (–) or cost law coefficient (–)
BOD	biological oxygen demand (mg liter^{-1})
c	cost law coefficient (–)
C	number of components in heat exchanger network design (–)
C_{FEED}	molar concentration of FEED (kmol m^{-3})
C_P	specific heat capacity (kJ kg^{-1} K^{-1})
C_{PRODUCT}	molar concentration of PRODUCT (kmol m^{-3})
COD	chemical oxygen demand (mg liter^{-1})
COP_{HP}	coefficients of performance of heat pump (–)
COP_{REF}	coefficients of performance of refrigeration system (–)
CP	heat capacity flowrate (kJ s^{-1} K^{-1})
CP_{EXHAUST}	heat capacity flowrate of heat engine exhaust (kJ kg^{-1} K^{-1})
CW	cooling water
D	distillate flow rate (kg s^{-1}, kmol s^{-1})
EP	economic potential ($ yr^{-1})
f	cost factor to allow for design pressure, material of construction, or installation costs (–)
F	feed flow rate (kg s^{-1}, kmol s^{-1}) or future worth of a sum of money allowing for interest rates ($)
F_T	correction factor for noncountercurrent flow in shell and tube heat exchangers (–)

F_{Tmin} minimum acceptable F_T for noncountercurrent heat exchangers

h_i specific enthalpy (kJ kg^{-1}) or
film heat transfer coefficient ($\text{kJ m}^{-2}\,\text{s}^{-1}\,\text{K}^{-1}$)

H stream enthalpy (kJ s^{-1})

HP   high pressure

i component or stream number (–) or
fractional rate of interest on money (–)

I total number of hot streams (–) or
capital cost investment ($)

j stream number (–)

J total number of cold streams (–)

K total number of enthalpy intervals in heat exchanger networks (–)

K_i ratio of vapor to liquid composition at equilibrium for component i (–)

k reaction-rate constant (units depend on order of reaction) or enthalpy interval number in heat exchanger networks (–)

L liquid flow rate (kg s^{-1}, kmol s^{-1}) or
number of independent loops in a network (–)

LP   low pressure (–)

m mass, molar flow rate (kg s^{-1}, kmol s^{-1})

MP   medium pressure (–)

N number of 1–2 shell-and-tube heat exchangers per match (–)

N_{SHELLS} number of 1–2 shell-and-tube heat exchangers (–)

N_{UNIT} number of units (–)

NC   number of components in a multicomponent mixture (–)

NPV   net present value ($)

P pressure (N m^{-2}) or
thermal effectiveness of 1–2 shell-and-tube heat exchanger (–) or
present worth of a future sum of money ($)

P_{max} maximum thermal effectiveness of 1-2 shell-and-tube heat exchangers (–)

P_{N-2N} thermal effectiveness over N number of 1–2 shell-and-tube heat exchangers in series (–)

P_{1-2} thermal effectiveness over each 1–2 shell-and-tube heat exchanger in series (–)

q thermal condition of the feed in distillation (–)

q_i individual stream heat duty (kJ s^{-1})

Q	heat duty (kJ s^{-1})
Q_C	cooling duty (kJ s^{-1})
$Q_{C\min}$	target for cold utility (kJ s^{-1})
Q_{COND}	condenser heat duty (kJ s^{-1})
Q_{EXHAUST}	heat duty for heat engine exhaust (kJ s^{-1})
Q_{FUEL}	heat from fuel in a furnace, boiler, or gas turbine (kJ s^{-1})
$Q_{H\min}$	target for hot utility (kJ s^{-1})
Q_{HE}	heat engine duty (kJ s^{-1})
Q_{HP}	heat pump duty (kJ s^{-1})
Q_{LOSS}	stack loss from furnace, boiler, or gas turbine (kJ s^{-1})
Q_{REACT}	reactor heating or cooling duty (kJ s^{-1})
Q_{REB}	reboiler heat duty (kJ s^{-1})
Q_{REC}	target for heat recovery (kJ s^{-1})
r	rate of reaction (kmol m^{-3} s^{-1})
R	distillation column reflux ratio (–) or heat capacity ratio of 1–2 shell-and-tube heat exchanger (–)
$R_{\min}$	minimum reflux ratio (–)
s	specific entropy (kJ kg^{-1} K^{-1})
S	reactor selectivity (–) or number of streams in heat exchanger network (–)
S_C	number of cold streams (–)
S_H	number of hot streams (–)
T_{BT}	normal boiling point (°C, K)
T	temperature (°C, K)
T_C	cold-stream temperature (°C, K)
T_{COND}	condenser temperature (°C, K)
T_H	hot-stream temperature (°C, K)
T_{REB}	reboiler temperature (°C, K)
T_S	stream supply temperature (°C, K)
T_T	stream target temperature (°C, K)
T^*	interval temperature; hot streams are represented $\Delta T_{\min}/2$ colder and cold streams $\Delta T_{\min}/2$ hotter than actual temperature (°C, K)
ΔT_{LM}	logarithmic mean temperature difference (°C, K)
$\Delta T_{\min}$	minimum temperature difference (°C, K)
$\Delta T_{\text{THRESHOLD}}$	threshold temperature difference (°C, K)
TOD	total oxygen demand (mg liter^{-1})
U	overall heat transfer coefficient (kJ m^{-2} s^{-1} K^{-1})
V	vapor flow (kg s^{-1}, kmol s^{-1})
$V_{\min}$	minimum vapor flow (kg s^{-1}, kmol s^{-1})

VF	vapor fraction (–)
W	shaftwork (kJ s^{-1})
x	liquid-phase mole fraction (–) or steam wetness fraction (–)
x_F	mole fraction in the feed (–)
x_D	mole fraction in the distillate (–)
X	reactor conversion (–)
X_E	equilibrium reactor conversion (–)
X_{OPT}	optimal reactor conversion (–)
X_P	fraction of maximum thermal effectiveness P_{max} allowed in a 1–2 shell-and-tube heat exchanger (–)
XP	cross-pinch heat transfer in heat exchanger network (kJ s^{-1})
y	vapor-phase mole fraction (–)
z	feed mole fraction (–)

Greek Letters

α	relative volatility (–)
λ	latent heat of vaporization (kJ kg^{-1})
η	Carnot factor (–) or efficiency (–)
η_T	turbine isentropic efficiency (–)
θ	parameter in the Underwood equation
ϕ	cost weighing factor applied to film heat transfer coefficients to allow for mixed materials of construction, pressure rating, and equipment types in heat exchanger networks

The Hierarchy of Chemical Process Design

1.1 Introduction

In a chemical process, the transformation of raw materials into desired products usually cannot be achieved in a single step. Instead, the overall transformation is broken down into a number of steps that provide intermediate transformations. These are carried out through reaction, separation, mixing, heating, cooling, pressure change, particle size reduction and enlargement, etc. Once individual steps have been selected, they must be interconnected to carry out the overall transformation (Fig. 1.1a). Thus the *synthesis* of a chemical process involves two broad activities. First, individual transformation steps are selected. Second, these individual transformations are interconnected to form a complete structure that achieves the required overall transformation. A *flowsheet* is the diagrammatic representation of the process steps with their interconnections.

Once the flowsheet structure has been defined, a *simulation* of the process can be carried out. A simulation is a mathematical model of the process which attempts to predict how the process would behave if it was constructed (see Fig. 1.1b). Having created a model of the process, we assume the flow rates, compositions, temperatures, and pressures of the feeds. The simulation model then predicts the flow rates, compositions, temperatures, and pressures of the products. It also allows the individual items of equipment in the process to be sized and predicts how much raw material is being used, how much energy is being consumed, etc. The performance of the design can then be evaluated.

There are many facets to the evaluation of performance. Good economic performance is an obvious first criterion, but it is certainly not the only one. Chemical processes should be designed as part of a sustainable industrial development which retains the capacity of ecosystems to support both industrial activity and life. In practical terms this means that waste should be minimized and that any waste byproducts which are produced must not be environmentally harmful. Sustainable development also demands that the process should use as little energy as practicable. The process also must meet required health and safety criteria. Start-up, emergency shutdown, and ease of control are other important factors. Flexibility, i.e., the ability to operate under different conditions such as differences in feedstock and product specification, etc., may be important. Availability, i.e., the number of operating hours per year, also may be important. Some of these factors, such as economic performance, can be readily quantified; others, such as safety, often cannot. Evaluation of the factors which are not readily quantifiable, the intangibles, requires the judgment of the designer.

Once the basic performance of the design has been evaluated, changes can be made to improve the performance; in other words, we *optimize*. These changes might involve the synthesis of alternative structures, i.e., *structural optimization*. Thus we simulate and

(a)   Process design starts with the synthesis of a process to convert raw materials into desired products.

(b)   Simulation predicts how a process would behave if it was constructed.

Figure 1.1 Synthesis is the creation of a process to transform feed streams into product streams. Simulation predicts how it would behave if it was constructed.

evaluate again, and so on, optimizing the structure. Alternatively, each structure can be subjected to *parameter optimization* by changing operating conditions within that structure.

We might think that we can find all the structural options by inspection, at least all of the significant ones. The fact that even long-established processes are still being improved bears evidence to just how difficult this is.

This text will attempt to develop an understanding of the concepts required at each stage during the creation of a chemical process design.

1.2 Overall Process Design

Consider the process illustrated in Fig. 1.2.[1] The process requires a reactor to transform the FEED into PRODUCT (Fig. 1.2a). Unfortunately, not all the FEED reacts. Also, part of the FEED reacts to form BYPRODUCT instead of the desired PRODUCT. A

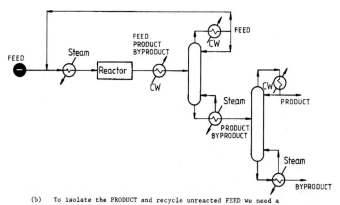

(a)   A reactor transforms FEED into PRODUCT and BYPRODUCT.

(b)   To isolate the PRODUCT and recycle unreacted FEED we need a separation system.

Figure 1.2   Process design starts with the reactor. The reactor design dictates the separation and recycle problem. *(From Smith and Linnhoff, Trans. IChemE, ChERD, 66:195, 1988; reproduced by permission of the Institution of Chemical Engineers.)*

separation system is needed to isolate the PRODUCT at the required purity. Figure 1.2b shows one possible separation system consisting of two distillation columns. The unreacted FEED in Fig. 1.2b is recycled, and the PRODUCT and BYPRODUCT are removed from the process. Figure 1.2b shows a flowsheet in which all heating and cooling is provided by external utilities (steam and cooling water in this case). This flowsheet is probably too inefficient in its use of energy, and we would attempt to recover heat. Thus we *heat integrate* and exchange heat between those streams which need to be cooled and those which need to be heated. Figure 1.3 shows two possible designs for the *heat exchanger network,* but many other heat integration arrangements are possible.

The flowsheets shown in Fig. 1.3 feature the same reactor design. It could be useful to explore changes in reactor design.[1] For example, the size of the reactor could be increased to increase the amount of FEED which reacts (Fig. 1.4a). Now there is not only much less FEED in the reactor effluent but more PRODUCT and BYPRODUCT. However, the increase in BYPRODUCT is larger than the increase in PRODUCT. Thus, although the reactor in Fig. 1.4a has the same three components in its effluent as the reactor in Fig. 1.2a, there is less FEED, more PRODUCT, and significantly more BYPRODUCT. This change in reactor design generates a different task for the separation system, and it is possible that a separation system different from that shown in Figs. 1.2 and 1.3 is now appropriate. Figure 1.4b shows a possible alternative. This also uses two distillation columns, but the separations are carried out in a different order.

Figure 1.4b shows a flowsheet without any heat integration for the different reactor and separation system. As before, this is probably too inefficient in the use of energy, and heat integration schemes can be explored. Figure 1.5 shows two of the many possible flowsheets.

Different complete flowsheets can be evaluated by simulation and costing. On this basis, the flowsheet in Fig. 1.3b might be more promising than the flowsheets in Figs. 1.3a, 1.5a, and 1.5b. However, we cannot be sure that we have the best flowsheet without first optimizing the operating conditions for each. The flowsheet in Fig. 1.5b might have greater scope for improvement than that in Fig. 1.3b.

Thus the complexity of chemical process synthesis is twofold.[1] First, can we identify all possible structures? Second, can we optimize each structure for a valid comparison? When optimizing the structure, there may be many ways in which each individual task can be performed and many ways in which the individual tasks can be interconnected. This means that we must simulate and optimize operating conditions for a multitude of structural options. At first sight this appears to be an overwhelmingly complex problem.

1.3   The Hierarchy of Process Design and the Onion Model

Our attempt to develop a methodology will be helped if we have a clearer picture of the structure of the problem. If the process requires a reactor, this is where the design starts. This is likely to be the only

(a)

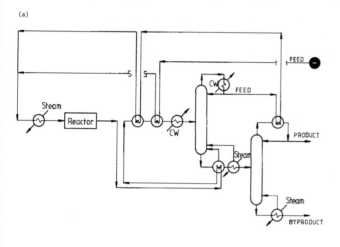

(b)

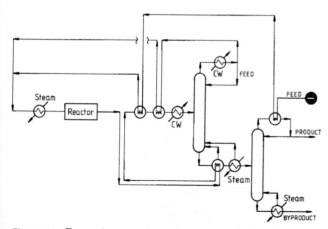

Figure 1.3 For a given reactor and separator design, there are different possibilities for heat integration. *(From Smith and Linnhoff, Trans. IChemE, ChERD, 66:195, 1988; reproduced by permission of the Institution of Chemical Engineers.)*

place in the process where raw materials are converted into products. The chosen reactor design produces a mixture of unreacted feed materials, products, and byproducts that need separating. Unreacted feed material is recycled. The reactor design dictates the separation and recycle problem. Thus design of the separation and recycle system follows reactor design. The reactor and separation and recycle system designs together define the process heating and cooling duties. Thus heat exchanger network design comes third. Those heating and cooling duties which cannot be satisfied by heat recovery dictate the need for external *utilities* (steam, cooling water, etc.). Thus utility selection and design come fourth. This hierarchy can be represented symbolically by the layers of the "onion diagram" shown in Fig. 1.6.[2] The diagram emphasizes the sequential, or hierarchical, nature of process design.

Of course, some processes do not require a reactor, e.g., some oil refinery processes. Here, the design starts with the separation system and moves outward to the heat exchanger network and utilities. However, the basic hierarchy prevails.

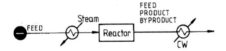

(a)   Changing the reactor design decreases the unreacted FEED, increases the PRODUCT, and significantly increases the BYPRODUCT.

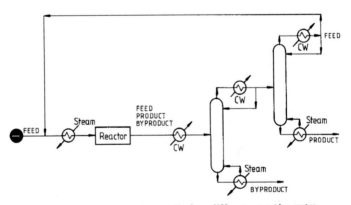

(b)   The alternative reactor design calls for a different separation system.

Figure 1.4 Changing the reactor dictates a different separation and recycle problem. (*From Smith and Linnhoff, Trans. IChemE, ChERD, 66:195, 1988, reproduced by permission of the Institution of Chemical Engineers.*)

The hierarchical nature of process design has been represented in different ways by different authors. A *hierarchy of decisions*[3] and a *process design ladder*[4] also have been suggested.

The synthesis of the correct structure and the optimization of parameters in the design of the reaction and separation systems are often the single most important tasks of process design. Usually there are many options, and it is impossible to fully evaluate them unless a complete design is furnished for the "outer layers" of the onion. For example, it is not possible to assess which is better,

(a)

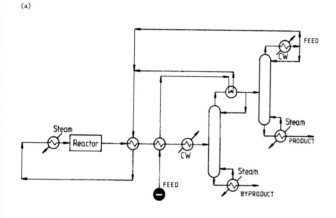

(b)

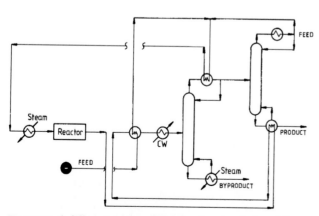

Figure 1.5 A different reactor design leads not only to a different separation system but also to additional possibilities for heat integration. *(From Smith and Linnhoff, Trans. IChemE, ChERD, 66:195, 1988; reproduced by permission of the Institution of Chemical Engineers.)*

the basic scheme from Fig. 1.2*b* or that from Fig. 1.4*b,* without fully evaluating all possible designs such as shown in Figs. 1.3*a* and *b* and 1.5*a* and *b,* etc., all completed, including utilities, etc. Such a complete search is normally too time consuming to be practical.

Later in this text an approach is presented in which some early decisions (i.e., decisions regarding reactor and separator options) can be evaluated without a complete design for the outer layers.[1]

1.4 Approaches to Process Design

In broad terms, there are two approaches to chemical process design:

1. *Building an irreducible structure.* The first approach follows the "onion logic," starting the design by choosing a reactor and then moving outward by adding a separation and recycle system, and so on. At each layer we must make decisions based on the information available at that stage. The ability to look ahead to the completed design might lead to different decisions. Unfortunately, this is not possible, and instead, decisions must be based on an incomplete picture.

This approach to synthesis is one of making a series of best local decisions. Equipment is added only if it can be justified economically on the basis of the information available, albeit an incomplete picture. This keeps the structure irreducible, and features which are technically or economically redundant are not included.

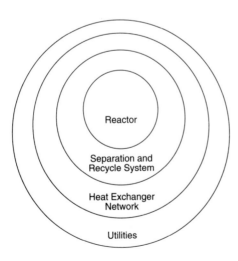

Reactor

Separation and Recycle System

Heat Exchanger Network

Utilities

Figure 1.6 The "onion model" of process design. A reactor design is needed before the separation and recycle system can be designed, and so on. *(From Smith and Linnhoff, Trans. IChemE, ChERD, 66:195, 1988; reproduced by permission of the Institution of Chemical Engineers.)*

There are two drawbacks to this approach:

a. Different decisions are possible at each stage of the design. To be sure we have made the best decisions, the other options must be evaluated. However, each option cannot be evaluated properly without completing the design for that option and optimizing the operating conditions. This means that many designs must be completed and optimized in order to find the best.
b. Even completing and evaluating many options gives no guarantee of ultimately finding the best possible design. Complex interactions can occur between different items of equipment in a flowsheet. The effort to keep the system simple and to not add equipment in the early stages of design may result in missing the benefit of interactions between different items of equipment in a more complex system.

The main advantage of this approach is that the designer can keep control of the basic decisions and interact with the design as it develops. By staying in control of the basic decisions, the intangibles of the design can be included in the decision making.

2. *Creating and optimizing a reducible structure.* In this approach, a structure known as a *superstructure* or *hyperstructure* is first created that has embedded within it all feasible process operations and all feasible interconnections that are candidates for an optimal design.[5] Initially, redundant features are built into the structure. As an example, consider Fig. 1.7. This shows one possible structure of a process for the manufacture of benzene from the reaction between toluene and hydrogen.[6] In Fig. 1.7, the hydrogen enters the process with a small amount of methane as an impurity. Thus in Fig. 1.7 the option is embedded of either purifying the hydrogen feed with a membrane or passing directly to the process. The hydrogen and toluene are mixed and preheated to reaction temperature. Only a furnace has been considered feasible in this case because of the high temperature required. Then two alternative reactor options, isothermal and adiabatic reactors, are embedded, and so on. Redundant features have been included in an effort to ensure that all features that could be part of an optimal solution have been included.

The design problem is next formulated as a mathematical problem with *design equations* and *design variables*. The design equations are the modeling equations of the units and their specification constraints. Design variables are of two types. The first type of design variables describe the operation of each unit (flow rate, composition, temperature, and pressure), its size (volume, heat transfer area, etc.), as well as the costs or profits associated with the units. Since

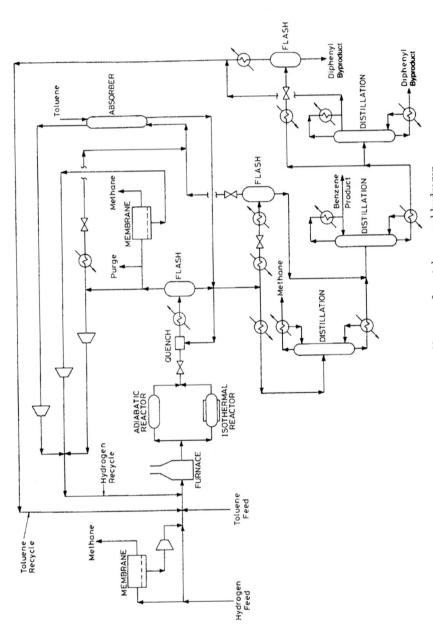

Figure 1.7 An initial structure for the manufacture of benzene from toluene and hydrogen incorporating some redundant features.

the dependence of the design equations on these variables is described in a continuous manner, they are known as *continuous variables*. The second type of design variables, known as *integer variables,* concern decisions on the structure of the flowsheet. These describe the existence of a particular unit or connection. They take on a value of unity if the unit or connection exists and a value of zero otherwise.

Once the problem is formulated mathematically, its solution is carried out through implementation of an optimization algorithm. Economic potential is maximized or cost is minimized (see App. A) in a *structural and parameter optimization*. Should an integer variable be optimized to zero, the corresponding feature is deleted from the structure and the structure is reduced in complexity. In effect, the discrete decision-making aspects of process design are replaced by a discrete/continuous optimization. Thus the initial structure in Fig. 1.7 is optimized to reduce the structure to the final design shown in Fig. 1.8. In Fig. 1.8, the membrane separator on the hydrogen feed has been removed by optimization, as have the isothermal reactor and many other features of the initial structure shown in Fig. 1.7.

There are a number of difficulties associated with this approach:

a. The approach will fail to find the optimal structure if the initial structure does not have the optimal structure embedded somewhere within it. The more options included, the more likely it will be that the optimal structure has been included.

b. If the individual unit operations are represented accurately, the resulting economic potential profile (see App. A) that must be optimized is both extremely large and irregular. The economic-potential profile is rather like the terrain in a range of mountains with many peaks. Each peak in the mountain range represents a *local optimum* in the economic potential. The highest peak represents the point of maximum economic potential and is the *global optimum*. Optimization requires searching around the mountains in a thick fog to find the highest peak, without the benefit of a map and only a compass to tell direction and an altimeter to show height. On reaching the top of any peak, there is no way of knowing whether it is the highest peak because of the fog. All peaks must be searched to find the highest. There are crevasses into which we might fall and not be able to climb out. Such problems can be overcome by changing the model such that the solution space becomes more regular, making the optimization simpler. This most often means simplifying the representation of the operations to make the representation as linear as possible.

c. The most serious drawback is that the design engineer is removed from the decision making. Thus the many intangibles in design,

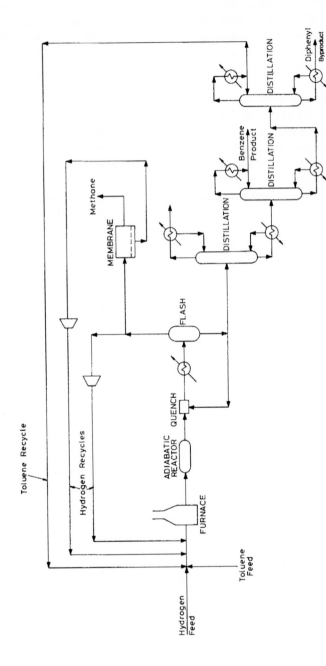

Figure 1.8 Optimization discards many structural features, leaving an optimized structure.

such as safety, layout, etc., which are difficult to include in the mathematical formulation, cannot be taken into account satisfactorily.

On the other hand, this approach has a number of advantages. Many different design options can be considered at the same time. Also, the entire design procedure can be accommodated in a computer program capable of producing designs quickly and efficiently.

In summary, the two general approaches to chemical process design of building an irreducible structure and creating and optimizing a reducible structure both have advantages and disadvantages. Whichever is used in practice, however, there is no substitute for understanding the problem.

This text concentrates on developing an understanding of the concepts required at each stage of the chemical process design. Such understanding is a vital part of process design, whichever approach is followed.

1.5   The Hierarchy of Chemical Process Design—Summary

When developing a chemical process design, there are two basic problems:

- Can all possible structures be identified?
- Can each structure be optimized such that all structures can be compared on a valid basis?

Design starts at the reactor because it is likely to be the only place in the process where raw materials are converted into desired products. The reactor design dictates the separation and recycle problem. The reactor design and separation and recycle problem together dictate the heating and cooling duties for the heat exchanger network. Those duties which cannot be satisfied by heat recovery dictate the need for external utilities. This hierarchy is represented by the layers in the "onion diagram" (see Fig. 1.6).

There are two general approaches to chemical process design:

- Building an irreducible structure
- Creating and optimizing a reducible structure

Both these approaches have advantages and disadvantages.

1.6 References

1. Smith, R., and Linnhoff, B., "The Design of Separators in the Context of Overall Processes," *Trans. IChemE, ChERD,* 66: 195, 1988.
2. Linnhoff, B., Townsend, D. W., Boland, D., et al., *A Users Guide on Process Integration for the Efficient Use of Energy,* IChemE, Rugby, U.K., 1982.
3. Douglas, J. M., "A Hierarchical Decision Procedure for Process Synthesis," *AIChE J.* 31: 353, 1985.
4. Lott, D. H., *Simulation Software as an Aid to Process Synthesis,* IChemE Symposium Series No. 109, IChemE, 1988.
5. Grossmann, I. E., "Mixed Integer Programming Approach for the Synthesis of Integrated Process Flowsheets," *Comp. Chem. Eng.,* 9: 463, 1985.
6. Kocis, G. R., and Grossmann, I. E., "A Modeling/Decomposition Strategy for MINLP Optimization of Process Flowsheets," paper no. 76a, AIChE Meeting, Washington, D.C., 1988.

Choice of Reactor

Since process design starts with the reactor, the first decisions are those which lead to the choice of reactor. These decisions are among the most important in the whole design. Good reactor performance is of paramount importance in determining the economic viability of the overall design and fundamentally important to the environmental impact of the process. In addition to the desired products, reactors produce unwanted byproducts. These unwanted byproducts create environmental problems. As we shall discuss later in Chap. 10, the best solution to environmental problems is not elaborate treatment methods but not to produce waste in the first place.

Most processes are catalyzed where catalysts for the reaction are known. The strategy will be to choose the catalyst, if one is to be used, and the ideal characteristics and operating conditions needed for the reaction system. Decisions must be made in terms of reactor

- Type
- Concentration
- Temperature
- Pressure
- Phase
- Catalyst

Then a practical reactor is selected, approaching as nearly as possible the ideal in order that the design can proceed. However, the reactor design cannot be fixed at this stage, since, as we shall see later, it interacts strongly with the rest of the flowsheet. We shall concentrate here on the choice of reactor and not its detailed sizing, which is outside our scope (for the details of sizing, see Denbigh and Turner,[1] Levenspiel,[2] and Rase[3]).

2.1   Reaction Path

Given that the objective is to manufacture a certain product, there are often a number of alternative *reaction paths* to that product. Reaction paths which use the cheapest raw materials and produce the smallest quantities of byproducts are to be preferred. Reaction paths which produce significant quantities of unwanted byproducts should especially be avoided, since they create significant environmental problems.

However, there are many other factors to be considered in the choice of reaction path. Some are commercial, such as uncertainties regarding future prices of raw materials and byproducts. Others are technical, such as safety and energy consumption.

The lack of suitable catalysts is the most common reason preventing the exploitation of novel reaction paths. At the first stage of design, it is impossible to look ahead and see *all* the consequences of choosing one reaction path or another, but some things are clear even at this stage. Consider the following example.

Example 2.1   Given that the objective is to manufacture vinyl chloride, there are at least three reaction paths which can be readily exploited.[4]

Path 1

$$C_2H_2 \ + \ HCl \ \longrightarrow \ C_2H_3Cl$$

acetylene   hydrogen   vinyl
          chloride   chloride

Path 2

$$C_2H_4 \ + \ Cl_2 \ \longrightarrow \ C_2H_4Cl_2$$

ethylene   chlorine   dichloroethane

$$C_2H_4Cl_2 \ \xrightarrow{\text{heat}} \ C_2H_3Cl + \ HCl$$

dichloroethane   vinyl   hydrogen
             chloride   chloride

Path 3

$$C_2H_4 \ + 1/2O_2 + \ 2HCl \ \longrightarrow \ C_2H_4Cl_2 + H_2O$$

ethylene   oxygen   hydrogen   dichloro   water
             chloride   ethane

$$C_2H_4Cl_2 \ \xrightarrow{\text{heat}} \ C_2H_3Cl + \ HCl$$

dichloroethane   vinyl   hydrogen
             chloride   chloride

The market values and molecular weights of the materials involved are given in Table 2.1. Oxygen is considered to be free at this stage, coming from the

TABLE 2.1   Molecular Weights and Values of Materials in Example 2.1

Material	Molecular weight	Value, $\$\,kg^{-1}$
Acetylene	26	0.94
Chlorine	71	0.21
Ethylene	28	0.53
Hydrogen chloride	36	0.35
Vinyl chloride	62	0.42

atmosphere. Which reaction path makes most sense on the basis of raw material costs and product and byproduct values?

Solution Decisions can be made based on the economic potential of the process (see App. A). At this stage, the best we can do is to define the economic potential (EP) as

$$EP = (\text{value of products}) - (\text{raw materials costs})$$

Path 1

$$EP = (62 \times 0.42) - (26 \times 0.94 + 36 \times 0.35)$$

$$= -\$11.0 \, \text{kmol}^{-1} \text{ vinyl chloride product}$$

Path 2

$$EP = (62 \times 0.42 + 36 \times 0.35) - (28 \times 0.53 + 71 \times 0.21)$$

$$= \$8.89 \, \text{kmol}^{-1} \text{ vinyl chloride product}$$

This assumes the sale of the byproduct HCl. If it cannot be sold, then

$$EP = (62 \times 0.42) - (28 \times 0.53 + 71 \times 0.21)$$

$$= -\$3.71 \, \text{kmol}^{-1} \text{ vinyl chloride product}$$

Path 3

$$EP = (62 \times 0.42) - (28 \times 0.53 + 36 \times 0.35)$$

$$= -\$1.40 \, \text{kmol}^{-1} \text{ vinyl chloride product}$$

Paths 1 and 3 are clearly not viable. Only path 2 shows a positive economic potential when the byproduct HCl can be sold. In practice, this might be quite difficult, since the market for HCl tends to be limited.

The preference is for a process based on ethylene rather than the more expensive acetylene and chlorine rather than the more expensive hydrogen chloride. Electrolytic cells are a much more convenient and cheaper source of chlorine than hydrogen chloride. In addition, we prefer to produce no byproducts.

Example 2.2 Devise a process from the three reaction paths in Example 2.1 which uses ethylene and chlorine as raw materials and produces no byproducts other than water. Does the process look attractive economically?

Solution A study of the stoichiometry of the three paths shows that this can be achieved by combining path 2 and path 3 to obtain a fourth path.

Paths 2 and 3

$$\underset{\text{ethylene}}{C_2H_4} + \underset{\text{chlorine}}{Cl_2} \longrightarrow \underset{\text{dichloroethane}}{C_2H_4Cl_2}$$

$$\underset{\text{ethylene}}{C_2H_4} + \underset{\text{oxygen}}{1/2\,O_2} + \underset{\substack{\text{hydrogen} \\ \text{chloride}}}{2HCl} \longrightarrow \underset{\text{dichloroethane}}{C_2H_4Cl_2} + \underset{\text{water}}{H_2O}$$

$$\underset{\text{dichloroethane}}{2C_2H_4Cl_2} \xrightarrow{\text{heat}} \underset{\substack{\text{vinyl} \\ \text{chloride}}}{2C_2H_3Cl} + \underset{\substack{\text{hydrogen} \\ \text{chloride}}}{2HCl}$$

If we add these three reactions to get the overall stoichiometry, we obtain

Path 4

$$2C_2H_4 + Cl_2 + 1/2O_2 \longrightarrow 2C_2H_3Cl + H_2O$$

ethylene   chlorine   oxygen   vinyl   water
                                                                chloride

or

$$C_2H_4 + 1/2Cl_2 + 1/4O_2 \longrightarrow C_2H_3Cl + 1/2H_2O$$

ethylene   chlorine   oxygen   vinyl   water
                                                            chloride

Now the economic potential is given by

$$EP = (62 \times 0.42) - (28 \times 0.53 + 1/2 \times 71 \times 0.21)$$

$$= \$3.75 \, kmol^{-1} \, vinyl \, chloride \, product$$

In summary, path 2 from Example 2.1 is the most attractive reaction path if there is a large market for hydrogen chloride. In practice, it tends to be difficult to sell the large quantities of hydrogen chloride produced by such processes. Path 4 is the usual commercial route to vinyl chloride.

2.2   Types of Reaction Systems

Having made a choice of the reaction path, we need to choose a reactor type and make some assessment of the conditions in the reactor. This allows assessment of reactor performance for the chosen reaction path in order for the design to proceed.

Before we can proceed with the choice of reactor and operating conditions, some general classifications must be made regarding the types of reaction systems likely to be encountered. We can classify reaction systems into five broad types:

1. *Single reactions.* Most reaction systems involve multiple reactions. In practice, the *secondary* reactions can sometimes be neglected, leaving a single *primary* reaction to consider. Single reactions are of the type

$$\text{FEED} \longrightarrow \text{PRODUCT} \tag{2.1}$$

or

$$\text{FEED} \longrightarrow \text{PRODUCT} + \text{BYPRODUCT} \tag{2.2}$$

or

$$\text{FEED 1} + \text{FEED 2} \longrightarrow \text{PRODUCT} \tag{2.3}$$

etc.

An example of this type of reaction which does not produce a byproduct is the production of allyl alcohol from propylene oxide:

$$CH_3HC\!\!-\!\!CH_2 \longrightarrow CH_2\!\!=\!\!CHCH_2OH$$
$$\diagdown \diagup$$
$$O$$

propylene oxide   allyl alcohol

An example of a reaction which does produce a byproduct is the production of acetone from isopropyl alcohol, which produces a hydrogen byproduct:

$$(CH_3)_2CHOH \longrightarrow CH_3COCH_3 + H_2$$
isopropyl   acetone
alcohol

2. *Multiple reactions in parallel producing byproducts.* Rather than a single reaction, a system may involve secondary reactions producing (additional) byproducts in *parallel* with the primary reaction. Multiple reactions in parallel are of the type

$$\text{FEED} \longrightarrow \text{PRODUCT}$$
$$\text{FEED} \longrightarrow \text{BYPRODUCT} \tag{2.4}$$

or

$$\text{FEED} \longrightarrow \text{PRODUCT} + \text{BYPRODUCT 1}$$
$$\text{FEED} \longrightarrow \text{BYPRODUCT 2} + \text{BYPRODUCT 3} \tag{2.5}$$

or

$$\text{FEED 1} + \text{FEED 2} \longrightarrow \text{PRODUCT}$$
$$\text{FEED 1} + \text{FEED 2} \longrightarrow \text{BYPRODUCT} \tag{2.6}$$

etc.

An example of a parallel reaction system occurs in the production of ethylene oxide:

$$CH_2\!\!=\!\!CH_2 + 1/2O_2 \longrightarrow H_2C\!\!-\!\!CH_2$$
$$\diagdown \diagup$$
$$O$$

ethylene   oxygen   ethylene oxide

with the parallel reaction

$$CH_2\!\!=\!\!CH_2 + 3O_2 \longrightarrow 2CO_2 + 2H_2O$$
ethylene   oxygen   carbon   water
                           dioxide

3. *Multiple reactions in series producing byproducts.* Rather than

the primary and secondary reactions being in parallel, they can be in *series*. Multiple reactions in series are of the type

$$\text{FEED} \longrightarrow \text{PRODUCT}$$
$$\text{PRODUCT} \longrightarrow \text{BYPRODUCT}$$

(2.7)

or

$$\text{FEED} \longrightarrow \text{PRODUCT} + \text{BYPRODUCT 1}$$
$$\text{PRODUCT} \longrightarrow \text{BYPRODUCT 2} + \text{BYPRODUCT 3}$$

(2.8)

or

$$\text{FEED 1} + \text{FEED 2} \longrightarrow \text{PRODUCT}$$
$$\text{PRODUCT} \longrightarrow \text{BYPRODUCT 1} + \text{BYPRODUCT 2}$$

(2.9)

etc.

An example of a series reaction system is the production of formaldehyde from methanol:

$$\underset{\text{methanol}}{CH_3OH} + \underset{\text{oxygen}}{1/2O_2} \longrightarrow \underset{\text{formaldehyde}}{HCHO} + \underset{\text{water}}{H_2O}$$

A series reaction of the formaldehyde occurs:

$$\underset{\text{formaldehyde}}{HCHO} \longrightarrow \underset{\text{carbon monoxide}}{CO} + \underset{\text{hydrogen}}{H_2}$$

4. *Mixed parallel and series reactions producing byproducts.* In more complex reaction systems, both parallel and series reactions can occur together. *Mixed parallel and series* reactions are of the type

$$\text{FEED} \longrightarrow \text{PRODUCT}$$
$$\text{FEED} \longrightarrow \text{BYPRODUCT}$$
$$\text{PRODUCT} \longrightarrow \text{BYPRODUCT}$$

(2.10)

or

$$\text{FEED} \longrightarrow \text{PRODUCT}$$
$$2\,\text{FEED} \longrightarrow \text{BYPRODUCT 1}$$
$$\text{PRODUCT} \longrightarrow \text{BYPRODUCT 2}$$

(2.11)

or

$$\text{FEED 1} + \text{FEED 2} \longrightarrow \text{PRODUCT}$$
$$\text{FEED 1} + \text{FEED 2} \longrightarrow \text{BYPRODUCT 1}$$
$$\text{PRODUCT} \longrightarrow \text{BYPRODUCT 2} + \text{BYPRODUCT 3}$$

(2.12)

etc.

An example of mixed parallel and series reactions is the production of ethanolamines by reaction between ethylene oxide and ammonia:[5]

$$H_2C\overset{\diagdown}{\underset{O}{\diagup}}CH_2 + NH_3 \longrightarrow NH_2CH_2CH_2OH$$

ethylene oxide   ammonia   monoethanolamine

$$NH_2CH_2CH_2OH + H_2C\overset{\diagdown}{\underset{O}{\diagup}}CH_2 \longrightarrow NH(CH_2CH_2OH)_2$$

monoethanolamine   ethylene oxide   diethanolamine

$$NH(CH_2CH_2OH)_2 + H_2C\overset{\diagdown}{\underset{O}{\diagup}}CH_2 \longrightarrow N(CH_2CH_2OH)_3$$

diethanolamine   ethylene oxide   triethanolamine

Here the ethylene oxide undergoes parallel reactions, whereas the monoethanolamine undergoes a series reaction to diethanolamine and triethanolamine.

5. *Polymerization reactions.* There are two broad types of polymerization reactions, those which involve a *termination step* and those which do not.[1] An example that involves a termination step is free-radical polymerization of an alkene molecule. The polymerization requires a free radical from an initiator compound such as a peroxide. The initiator breaks down to form a free radical (e.g., $\dot{C}H_3$ or $\dot{O}H$), which attaches to a molecule of alkene and in so doing generates another free radical. Consider the polymerization of vinyl chloride from a free-radical initiator $\dot{R}$. An *initiation step* first occurs:

$$\dot{R} + CH_2{=}CHCl \longrightarrow RCH_2{-}\dot{C}HCl$$
initiator   vinyl   vinyl chloride
              chloride   free radical

A *propagation step* involving growth around an *active center* follows:

$$RCH_2{-}\dot{C}HCl + CH_2{=}CHCl \longrightarrow RCH_2{-}CHCl{-}CH_2{-}\dot{C}HCl$$

and so on, leading to molecules of the structure

$$R{-}(CH_2{-}CHCl)_n{-}CH_2{-}\dot{C}HCl$$

Eventually, the chain is terminated by steps such as the union of two radicals that consumes but does not generate radicals:

$$R—(CH_2—CHCl)_n—CH_2—\dot{C}HCl + \dot{C}HCl—CH_2—(CHCl—CH_2)_m—R$$

$$\longrightarrow R—(CH_2—CHCl)_n—CH_2—CHCl—CHCl—CH_2—(CHCl—CH_2)_m—R$$

This *termination step* stops the subsequent growth of the polymer chain. The period during which the chain length grows, i.e., before termination, is known as the *active life* of the polymer. Other termination steps are possible.

An example of a polymerization without a termination step is *polycondensation*:[1]

$$HO—(CH_2)_n—COOH + HO—(CH_2)_n—COOH$$

$$\longrightarrow HO—(CH_2)_n—COO—(CH_2)_n—COOH + H_2O \qquad \text{etc.}$$

Here the polymer grows by successive esterification with elimination of water and no termination step.

2.3 Reactor Performance

Before we can explore how reactor conditions can be chosen, we require some measure of reactor performance. For polymerization reactors, the most important measure of performance is the distribution of molecular weights in the polymer product. The distribution of molecular weights dictates the mechanical properties of the polymer. For other types of reactors, three important parameters are used to describe their performance:[6]

$$\text{Conversion} = \frac{\text{(reactant consumed in the reactor)}}{\text{(reactant } \textit{fed} \text{ to the reactor)}} \qquad (2.13)$$

$$\text{Selectivity} = \frac{\text{(desired product produced)}}{\text{(reactant } \textit{consumed} \text{ in the reactor)}}$$

$$\times \text{ stoichiometric factor} \qquad (2.14)$$

$$\text{Reactor yield} = \frac{\text{(desired product produced)}}{\text{(reactant } \textit{fed} \text{ to the reactor)}}$$

$$\times \text{ stoichiometric factor} \qquad (2.15)$$

where the stoichiometric factor is the stoichiometric moles of reactant required per mole of product. When more than one reactant is required (or more than one desired product produced) Eqs. (2.13) to (2.15) can be applied to each reactant (or product).

The following example will help to clarify the distinctions between these three parameters.

Example 2.3 Benzene is to be produced from toluene according to the reaction:[7]

$$C_6H_5CH_3 + H_2 \longrightarrow C_6H_6 + CH_4$$
$$\text{toluene} \quad \text{hydrogen} \quad\quad \text{benzene} \quad \text{methane}$$

Some of the benzene formed undergoes a secondary reaction in series to an unwanted byproduct, diphenyl, according to the reaction

$$2C_6H_6 \rightleftharpoons C_{12}H_{10} + H_2$$
$$\text{benzene} \quad\quad \text{diphenyl} \quad \text{hydrogen}$$

Table 2.2 gives the compositions of the reactor feed and effluent streams. Calculate the conversion, selectivity, and reactor yield with respect to (a) the toluene feed and (b) the hydrogen feed.

Solution

(a)
$$\text{Toluene conversion} = \frac{\text{(toluene consumed in the reactor)}}{\text{(toluene fed to the reactor)}}$$

$$= \frac{372 - 93}{372}$$

$$= 0.75$$

$$\text{Stoichiometric factor} = \text{stoichiometric moles of toluene required}$$
$$\text{per mole of benzene produced}$$

$$= 1$$

TABLE 2.2 Reactor Feed and Effluent streams in Example 2.3

Component	Inlet flow rate, kmol h^{-1}	Outlet flow rate, kmol h^{-1}
H_2	1858	1583
CH_4	804	1083
C_6H_6	13	282
$C_6H_5CH_3$	372	93
$C_{12}H_{10}$	0	4

$$\text{Benzene selectivity from toluene} = \frac{\text{(benzene produced in the reactor)}}{\text{(toluene consumed in the reactor)}}$$

$$\times \text{ stoichiometric factor}$$

$$= \frac{282-13}{372-93} \times 1$$

$$= 0.96$$

$$\text{Reactor yield of benzene from toluene} = \frac{\text{(benzene produced in the reactor)}}{\text{(toluene fed to the reactor)}}$$

$$\times \text{ stoichiometric factor}$$

$$= \frac{282-13}{372} \times 1$$

$$= 0.72$$

(b) $$\text{Hydrogen conversion} = \frac{\text{(hydrogen consumed in the reactor)}}{\text{(hydrogen fed to the reactor)}}$$

$$= \frac{1858-1583}{1858}$$

$$= 0.15$$

Stoichiometric factor = stoichiometric moles of hydrogen required per mole of benzene produced

$$= 1$$

$$\text{Benzene selectivity from hydrogen} = \frac{\text{(benzene produced in the reactor)}}{\text{(hydrogen consumed in the reactor)}}$$

$$\times \text{ stoichiometric factor}$$

$$= \frac{282-13}{1858-1583} \times 1$$

$$= 0.98$$

$$\text{Reactor yield of benzene from hydrogen} = \frac{\text{(benzene produced in the reactor)}}{\text{(hydrogen fed to the reactor)}}$$

$$\times \text{ stoichiometric factor}$$

$$= \frac{282-13}{1858} \times 1$$

$$= 0.14$$

Because there are two feeds to this process, the reactor performance can be calculated with respect to both feeds. However, the principal concern is performance with respect to toluene, since it is much more expensive than hydrogen.

If a reaction is reversible, there is a maximum conversion that can be achieved, the equilibrium conversion, which is less than 1.0. Fixing the mole ratio of reactants, temperature, and pressure fixes the equilibrium conversion.[1-3]

In describing reactor performance, selectivity is usually a more meaningful parameter than reactor yield. Reactor yield is based on the reactant fed to the reactor rather than on that which is consumed. Clearly, part of the reactant fed might be material that has been recycled rather than fresh feed. Because of this, reactor yield takes no account of the ability to separate and recycle unconverted raw materials. Reactor yield is only a meaningful parameter when it is not possible for one reason or another to recycle unconverted raw material to the reactor inlet. By constrast, the yield of the overall process is an extremely important parameter when describing the performance of the overall plant, as will be discussed later.

It is now possible to define the goals for the selection of the reactor at this stage in the design. Unconverted material usually can be separated and recycled later. Because of this, the reactor conversion cannot be fixed finally until the design has progressed much further than just choosing the reactor. As we shall see later, the choice of reactor conversion has a major influence on the rest of the process. Nevertheless, some decisions must be made regarding the reactor for the design to proceed. Thus we must make some guess for the reactor conversion in the knowledge that this is likely to change once more detail is added to the total system.

Unwanted byproducts usually cannot be converted back to useful products or raw materials. The reaction to unwanted byproducts creates both raw materials costs due to the raw materials which are wasted in their formation and environmental costs for their disposal. Thus maximum selectivity is wanted for the chosen reactor conversion. The objectives at this stage can be summarized as follows:

1. *Single reactions.* In a single reaction such as Eq. (2.2) which produces a byproduct, there can be no influence on the relative amount of product and byproduct formed. Thus, with single reactions such as Eqs. (2.1) to (2.3), the goal is to minimize the reactor capital cost (which usually means minimizing reactor volume) for a given reactor conversion. Increasing the reactor conversion increases size and hence cost of the reactor but, as we shall see later,

decreases the cost of many other parts of the flowsheet. Because of this, the initial setting for reactor conversion for single irreversible reactions is around 95 percent and that for a single reversible reaction is around 95 percent of the equilibrium conversion.[8]

For batch reactors, account has to be taken of the time required to achieve a given conversion. Batch cycle time is addressed later.

2. *Multiple reactions in parallel producing byproducts.* Raw materials costs usually will dominate the economics of the process. Because of this, when dealing with multiple reactions, whether parallel, series, or mixed, the goal is usually to minimize byproduct formation (maximize selectivity) for a given reactor conversion. Choice of reactor conditions should exploit differences between the kinetics and equilibrium effects in the primary and secondary reactions to favor the formation of the desired product rather than the byproduct, i.e., improve selectivity. Making an initial guess for conversion is more difficult than with single reactions, since the factors that affect conversion also can have a significant effect on selectivity.

Consider the system of parallel reactions from Eq. (2.4) with the corresponding rate equations.[1–3]

$$\text{FEED} \longrightarrow \text{PRODUCT } r_1 = k_1 C_{\text{FEED}}^{a_1}$$

$$\text{FEED} \longrightarrow \text{BYPRODUCT } r_2 = k_2 C_{\text{FEED}}^{a_2}$$

(2.16)

where r_1, r_2 = rates of reaction for primary and secondary reactions
k_1, k_2 = reaction-rate constants for primary and secondary reactions
C_{FEED} = molar concentration of FEED in the reactor
a_1, a_2 = constants (order of reaction) for primary and secondary reactions

The ratio of the rates of the secondary and primary reactions gives[1–3]

$$\frac{r_2}{r_1} = \frac{k_2}{k_1} C_{\text{FEED}}^{a_2 - a_1}$$

(2.17)

Maximum selectivity requires a minimum ratio r_2/r_1 in Eq. (2.17). A high conversion in the reactor tends to decrease C_{FEED}. Thus

- $a_2 > a_1$ selectivity increases as conversion increases.
- $a_2 < a_1$ selectivity decreases as conversion increases.

If selectivity increases as conversion increases, the initial setting for reactor conversion should be on the order of 95 percent, and that for reversible reactions should be on the order of 95 percent of the equilibrium conversion. If selectivity decreases with increasing conversion, then it is much more difficult to give guidance. An initial setting of 50 percent for the conversion for irreversible reactions or 50 percent of the equilibrium conversion for reversible reactions is as reasonable as can be guessed at this stage. However, these are only initial guesses and will almost certainly be changed later.

3. *Multiple reactions in series producing byproducts.* Consider the system of series reactions from Eq. (2.7):

$$\text{FEED} \longrightarrow \text{PRODUCT } r_1 = k_1 C_{\text{FEED}}^{a_1}$$

$$\text{PRODUCT} \longrightarrow \text{BYPRODUCT } r_2 = k_2 C_{\text{PRODUCT}}^{a_2}$$

(2.18)

where r_1, r_2 = rates of reaction for primary and secondary reactions
k_1, k_2 = reaction-rate constants
C_{FEED} = molar concentration of FEED
C_{PRODUCT} = molar concentration of PRODUCT
a_1, a_2 = constants (order of reaction) for primary and secondary reactions

Selectivity for series reactions of the types given in Eqs. (2.7) to (2.9) is increased by low concentrations of reactants involved in the secondary reactions. In the preceding example, this means reactor operation with a low concentration of PRODUCT—in other words, with low conversion. For series reactions, a significant reduction in selectivity is likely as the conversion increases.

Again, it is difficult to select the initial setting of the reactor conversion with systems of reactions in series. A conversion of 50 percent for irreversible reactions or 50 percent of the equilibrium conversion for reversible reactions is as reasonable as can be guessed at this stage.

Multiple reactions also can occur with impurities that enter with the feed and undergo reaction. Again, such reactions should be minimized, but the most effective means of dealing with byproduct reactions caused by feed impurities is not to alter reactor conditions but to introduce feed purification.

The next requirement is to achieve the initial setting for the reactor conversion while maximizing selectivity for that conversion. The first consideration is reactor type.

2.4 Idealized Reactor Models

Three idealized models are used for the design of reactors.[1-3] In the first (Fig. 2.1a), the *ideal batch model,* the reactants are charged at the beginning of the operation. The contents are subjected to perfect mixing for a certain period, after which the products are discharged. Concentration changes with time, but the perfect mixing ensures that at any instant the composition and temperature throughout the reactor are both uniform.

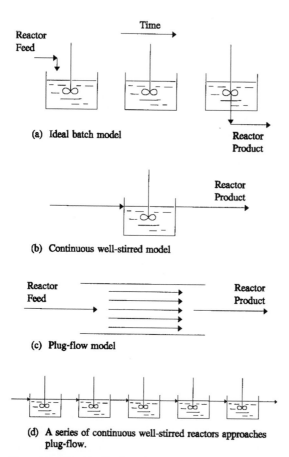

(a) Ideal batch model

(b) Continuous well-stirred model

(c) Plug-flow model

(d) A series of continuous well-stirred reactors approaches plug-flow.

Figure 2.1 The idealized models used for reactor design. *(From Smith and Petela, The Chemical Engineer, Dec. 17, 1991; reproduced by permission of the Institution of Chemical Engineers.)*

In the second model (Fig. 2.1b) the *continuous well-stirred model,* feed and product takeoff are continuous, and the reactor contents are assumed to be perfectly mixed. This leads to uniform composition and temperature throughout. Because of the perfect mixing, a fluid element can leave at the instant it enters the reactor or stay for an extended period. The residence time of individual fluid elements in the reactor varies.

In the third model (Fig. 2.1c), the *plug-flow model,* a steady uniform movement of the reactants is assumed, with no attempt to induce mixing along the direction of flow. Like the ideal batch reactor, the residence time in a plug-flow reactor is the same for all fluid elements. Plug-flow operation can be approached by using a number of continuous well-mixed reactors in series (Fig. 2.1d). The greater the number of well-mixed reactors in series, the closer is the approach to plug-flow operation.

Consider now which of the idealized models is preferred for the five categories of reaction systems introduced in Sec. 2.2.

1. *Single reactions.* Consider the single reaction from Eq. (2.1):

$$\text{FEED} \longrightarrow \text{PRODUCT} \quad r = kC_{\text{FEED}}^a \qquad (2.19)$$

where r = rate of reaction
  k = reaction rate constant
 C_{FEED} = molar concentration of FEED
  a = order of reaction

Clearly, the highest rate of reaction is maintained by the highest concentration of feed (C_{FEED}, kmol m^{-3}). In the continuous well-mixed reactor, the incoming feed is instantly diluted by the product that has already been formed. The rate of reaction is thus lower in the continuous well-stirred reactor than in the ideal batch and plug-flow reactors, since it operates at the low reaction rate corresponding with the outlet concentration of feed. Thus the continuous well-stirred model requires a greater volume than the ideal batch and plug-flow reactors. Consequently, for single reactions, the ideal batch or plug-flow reactors are preferred.

2. *Multiple reactions in parallel producing byproducts.* Consider again the system of parallel reactions from Eqs. (2.16) and (2.17). A batch or plug-flow reactor maintains higher average concentrations of feed (C_{FEED}) than a continuous well-mixed reactor, in which the incoming feed is instantly diluted by the PRODUCT and

BYPRODUCT. If $a_1 > a_2$ in Eqs. (2.16) and (2.17), the primary reaction to PRODUCT is favored by a high concentration of FEED. If $a_1 < a_2$, the primary reaction to PRODUCT is favored by a low concentration of FEED. Thus, if

- $a_2 < a_1$, use a batch or plug-flow reactor.
- $a_2 > a_1$, use a continuous well-mixed reactor.

In general terms, if the reaction to the desired product has a higher order than the byproduct reaction, use a batch or plug-flow reactor. If the reaction to the desired product has a lower order than the byproduct reaction, use a continuous well-mixed reactor.

If the reaction involves more than one feed, the picture becomes more complex. Consider the reaction system from Eq. (2.6) with the corresponding rate equations:

$$\text{FEED 1} + \text{FEED 2} \longrightarrow \text{PRODUCT } r_1 = k_1 C_{\text{FEED1}}^{a_1} C_{\text{FEED2}}^{b_1}$$

$$\text{FEED 1} + \text{FEED 2} \longrightarrow \text{BYPRODUCT } r_2 = k_2 C_{\text{FEED1}}^{a_2} C_{\text{FEED2}}^{b_2} \qquad (2.20)$$

where $C_{\text{FEED1}}, C_{\text{FEED2}}$ = molar concentrations of FEED 1 and FEED 2 in the reactor

a_1, b_1 = constants (order of reaction) for the primary reaction

a_2, b_2 = constants (order of reaction) for the secondary reaction

Now the ratio we wish to minimize is given by[1-3]

$$\frac{r_2}{r_1} = \frac{k_2}{k_1} C_{\text{FEED1}}^{a_2-a_1} C_{\text{FEED2}}^{b_2-b_1} \qquad (2.21)$$

Given this reaction system the options are,

- Keep both C_{FEED1} and C_{FEED2} low (i.e., use a continuous well-mixed reactor).
- Keep both C_{FEED1} and C_{FEED2} high (i.e., use a batch or plug-flow reactor).
- Keep one of the concentrations high while maintaining the other low (this is achieved by charging one of the feeds as the reaction progresses).

Figure 2.2 summarizes these arguments to choose a reactor for systems of multiple reactions in parallel.

3. *Multiple reactions in series producing byproducts.* Consider the series reaction system from Eq. (2.18). For a certain reactor conversion, the FEED should have a corresponding residence time in the reactor. In the continuous well-mixed reactor, FEED can leave the instant it enters or remains for an extended period. Similarly, PRODUCT can remain for an extended period or leave immediately. Substantial fractions of both FEED and PRODUCT leave before and after what should be the specific residence time for a given conversion. Thus the continuous well-mixed model would be expected to give a poorer selectivity than a batch or plug-flow reactor for a given conversion. A batch or plug-flow reactor should be used for multiple reactions in series.

4. *Mixed parallel and series reactions producing byproducts.* Consider the mixed parallel and series reaction system from Eq. (2.10) with the corresponding kinetic equations:

$$\text{FEED} \longrightarrow \text{PRODUCT } r_1 = k_1 C_{\text{FEED}}^{a_1}$$

$$\text{FEED} \longrightarrow \text{BYPRODUCT } r_2 = k_2 C_{\text{FEED}}^{a_2} \qquad (2.22)$$

$$\text{PRODUCT} \longrightarrow \text{BYPRODUCT } r_3 = k_3 C_{\text{PRODUCT}}^{a_3}$$

As far as the parallel byproduct reaction is concerned, for high selectivity, if

- $a_1 > a_2$, use a batch or plug-flow reactor.
- $a_1 < a_2$, use a continuous well-mixed reactor.

The series byproduct reaction requires a plug-flow reactor. Thus, for the mixed parallel and series system above, if

- $a_1 > a_2$, use a batch or plug-flow reactor.

But what is the correct choice if $a_1 < a_2$? Now the parallel byproduct reaction calls for a continuous well-mixed reactor. On the other hand, the byproduct series reaction calls for a plug-flow reactor. It would seem that, given this situation, some level of mixing between a plug-flow and a continuous well-mixed reactor will give the best

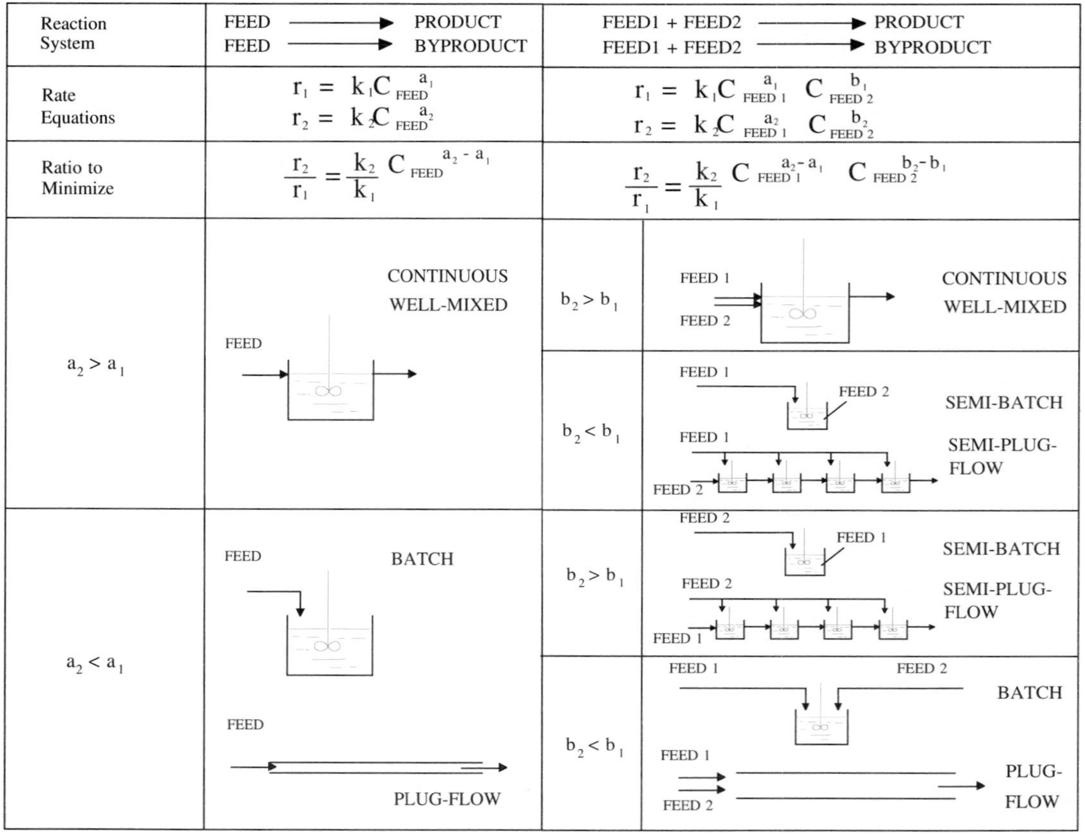

Single Feed System

Reaction System:

FEED ⟶ PRODUCT
FEED ⟶ BYPRODUCT

Rate Equations:

$$r_1 = k_1 C_{FEED}^{a_1}$$
$$r_2 = k_2 C_{FEED}^{a_2}$$

Ratio to Minimize:

$$\frac{r_2}{r_1} = \frac{k_2}{k_1} C_{FEED}^{a_2 - a_1}$$

- $a_2 > a_1$: CONTINUOUS WELL-MIXED
- $a_2 < a_1$: BATCH / PLUG-FLOW

Two Feed System

Reaction System:

FEED1 + FEED2 ⟶ PRODUCT
FEED1 + FEED2 ⟶ BYPRODUCT

Rate Equations:

$$r_1 = k_1 C_{FEED1}^{a_1} \, C_{FEED2}^{b_1}$$
$$r_2 = k_2 C_{FEED1}^{a_2} \, C_{FEED2}^{b_2}$$

Ratio to Minimize:

$$\frac{r_2}{r_1} = \frac{k_2}{k_1} C_{FEED1}^{a_2 - a_1} \, C_{FEED2}^{b_2 - b_1}$$

- $a_2 > a_1$, $b_2 > b_1$: CONTINUOUS WELL-MIXED
- $a_2 > a_1$, $b_2 < b_1$: SEMI-BATCH / SEMI-PLUG-FLOW
- $a_2 < a_1$, $b_2 > b_1$: SEMI-BATCH / SEMI-PLUG-FLOW
- $a_2 < a_1$, $b_2 < b_1$: BATCH / PLUG-FLOW

Figure 2.2 Reactor choice for parallel reaction systems. (From Smith and Petela. The

overall selectivity.[10] This could be obtained by

- A series of continuous well-mixed reactors (Fig. 2.3a)
- A plug-flow reactor with a recycle (Fig. 2.3b)
- A series combination of plug-flow and continuous well-mixed reactors (Fig. 2.3c and d)

The arrangement which gives the highest overall selectivity can only be deduced by detailed sizing and costing calculations specific to the reaction system.

5. *Polymerization reactions.* Polymers are characterized by the distribution of molecular weight about the mean as well as by the mean itself. The breadth of this distribution depends on whether a batch or plug-flow reactor is used on the one hand or a continuous well-mixed reactor on the other. The breadth has an important influence on the mechanical and other properties of the polymer, and this is an important factor in the choice of reactor.

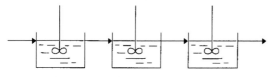

(a) Series of continuous well-mixed reactors

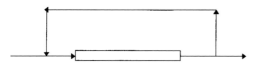

(b) Plug-flow reactor with recycle

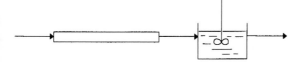

(c) Plug-flow followed by continuous well-mixed reactor

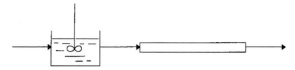

(d) Continuous well-mixed followed by plug-flow reactor

Figure 2.3 Choice of reactor type for mixed parallel and series reactions when the parallel reaction has a higher order than the primary reaction.

Two broad classes of polymerization reactions can be identified:[1]

a. *No termination reaction or the active polymer life is long compared with the average residence time in the reactor.* In a batch or plug-flow reactor, all molecules have the same residence time, and without the effect of termination (see Sec. 2.1), all will grow to approximately equal lengths, producing a narrow distribution of molecular weights. By contrast, a continuous well-mixed reactor will cause a wide distribution because of the distribution of residence times in the reactor.

b. *Active polymer life is short compared with average residence time in the reactor.* When polymerization takes place by mechanisms involving free radicals, the life of these actively growing centers may be extremely short due to termination processes such as the union of two free radicals (see Sec. 2.2). These termination processes are influenced by free-radical concentration, which in turn is proportional to monomer concentration. In batch or plug-flow reactors, the monomer and free-radical concentrations decline. This produces increasing chain lengths with increasing residence time and thus a broad distribution of molecular weights. The continuous well-mixed reactor maintains a uniform concentration of monomer and thus a constant chain-termination rate. This results in a narrow distribution of molecular weights. Because the active life of the polymer is short, the variation in residence time does not have a significant effect.

2.5   Reactor Concentration

In the preceding section, the choice of reactor type was made on the basis of which gave the most appropriate concentration profile as the reaction progressed in order to minimize volume for single reactions or maximize selectivity for multiple reactions for a given conversion. However, after making the decision to choose one type of reactor or another, there are still important concentration effects to be considered.

When more than one reactant is used, it is often desirable to use an excess of one of the reactants. It is sometimes desirable to feed an inert material to the reactor or to separate the product partway through the reaction before carrying out further reaction. Sometimes it is desirable to recycle unwanted byproducts to the reactor. Let us now examine these cases.

1. *Single irreversible reactions.* An excess of one feed component can force another component toward complete conversion. As an

example, consider the reaction between ethylene and chlorine to dichloroethane:

$$\underset{\text{ethylene}}{C_2H_4} + \underset{\text{chlorine}}{Cl_2} \longrightarrow \underset{\text{dichloroethane}}{C_2H_4Cl_2}$$

An excess of ethylene is used to ensure essentially complete conversion of the chlorine, which is thereby eliminated as a problem for the downstream separation system.

In a single reaction (where selectivity is not a problem), the usual choice of excess reactant is to eliminate the component which is more difficult to separate in the downstream separation system. Alternatively, if one of the components is more hazardous (as is chlorine in this example), again we try to ensure complete conversion.

2. *Single reversible reactions.* The maximum conversion in reversible reactions is limited by the equilibrium conversion, and conditions in the reactor are usually chosen to increase the equilibrium conversion. *Le Châtelier's principle* dictates the changes required to increase equilibrium conversion:

> If any change in the conditions of a system in equilibrium causes the equilibrium to be displaced, the displacement will be in such a direction as to oppose the effect of the change.

We shall see later how temperature and pressure affect equilibrium conversion. For now, let us consider how concentration affects equilibrium conversion.

a. *Feed ratio.* If to a system in equilibrium an excess of one of the feeds is added, then the effect is to shift the equilibrium to decrease the feed concentration. In other words, an excess of one feed can be used to increase the equilibrium conversion. As an example, ethyl acetate can be produced by the esterification of ethyl alcohol with acetic acid in the presence of a catalyst such as sulfuric acid according to the reaction

$$\underset{\text{ethyl alcohol}}{C_2H_5OH} + \underset{\text{acetic acid}}{CH_3COOH} \rightleftharpoons \underset{\text{ethyl acetate}}{CH_3COOC_2H_5} + H_2O$$

The equilibrium conversion can be increased by employing one reactant in excess (or removing the water formed, or both).

b. *Inerts' concentration.* Sometimes, an inert material is present in the reactor. This might be a solvent in a liquid-phase reaction or an inert gas in a gas-phase reaction. Consider the reaction system

$$\text{FEED} \rightleftharpoons \text{PRODUCT 1} + \text{PRODUCT 2} \qquad (2.23)$$

This reaction brings about an increase in the number of moles per unit volume. Adding inert material causes the number of moles per unit volume to be decreased, and the equilibrium will be displaced to oppose this by shifting to a higher conversion.

If the reaction involves a decrease in the number of moles, e.g.,

$$\text{FEED 1} + \text{FEED 2} \rightleftharpoons \text{PRODUCT} \qquad (2.24)$$

inert material should be decreased, if any inerts are present. Removing inert material causes the number of moles per unit volume to be increased, and the equilibrium will be displaced to oppose this by shifting to a higher conversion.

If inert material is to be added, then ease of separation is an important consideration. For example, steam is added as an inert to hydrocarbon cracking reactions and is an attractive material in this respect because it is easily separated from the hydrocarbon components by condensation. If the reaction does not involve any change in the number of moles, inert material has no effect on equilibrium conversion.

c. *Product removal during reaction.* Sometimes the equilibrium conversion can be increased by removing the product (or one of the products) continuously from the reactor as the reaction progresses, e.g., by allowing it to vaporize from a liquid-phase reactor. Another way is to carry out the reaction in stages with intermediate separation of the products. As an example of intermediate separation, consider the production of sulfuric acid as illustrated in Fig. 2.4. Sulfur dioxide is oxidized to sulfur trioxide:

$$\underset{\text{sulfur dioxide}}{2SO_2} + O_2 \rightleftharpoons \underset{\text{sulfur trioxide}}{2SO_3}$$

This reaction can be forced to effective complete conversion by

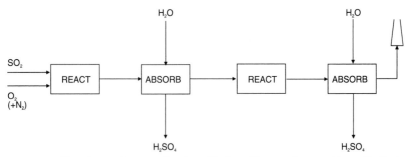

Figure 2.4 Reaction and separation in sulfuric acid manufacture allow effectively complete conversion despite reversibility of the reaction.

first carrying out the reaction to approach equilibrium. The sulfur trioxide is then separated (by absorption). Removal of sulfur trioxide shifts the equilibrium, and further reaction of the remaining sulfur dioxide and oxygen allows effective complete conversion of the sulfur dioxide (Fig. 2.4).

Intermediate separation followed by further reaction is clearly most appropriate when the intermediate separation is straightforward, as in the case of sulfuric acid production.

3. *Multiple reactions in parallel producing byproducts.* Once the reactor type is chosen to maximize selectivity, we are in a position to alter selectivity further in parallel reaction systems. Consider the parallel reaction system from Eq. (2.20). To maximize selectivity for this system, we minimize the ratio given by Eq. (2.21):

$$\frac{r_2}{r_1} = \frac{k_2}{k_1} C_{\text{FEED1}}^{a_2-a_1} C_{\text{FEED2}}^{b_2-b_1}$$

Even after the type of reactor is chosen, excess of FEED 1 or FEED 2 can be used:

- If $(a_2 - a_1) > (b_2 - b_1)$, use excess FEED 2.
- If $(a_2 - a_1) < (b_2 - b_1)$, use excess FEED 1.

If the secondary reaction is reversible and involves a decrease in the number of moles, such as

$$\text{FEED 1} + \text{FEED 2} \longrightarrow \text{PRODUCT}$$

$$\text{FEED 1} + \text{FEED 2} \rightleftharpoons \text{BYPRODUCT}$$

$$(2.25)$$

then, if inerts are present, increasing the concentration of inerts will decrease byproduct formation. If the secondary reaction is reversible and involves an increase in the number of moles, such as

$$\text{FEED 1} + \text{FEED 2} \longrightarrow \text{PRODUCT}$$

$$\text{FEED 1} \rightleftharpoons \text{BYPRODUCT 1} + \text{BYPRODUCT 2}$$

$$(2.26)$$

then, if inerts are present, decreasing the concentration of inerts will decrease byproduct formation. If the secondary reaction has no change in the number of moles, then concentration of inerts does not affect it.

For all reversible secondary reactions, deliberately feeding BYPRODUCT to the reactor inhibits its formation at the source by shifting the equilibrium of the secondary reaction. This is achieved in practice by separating and recycling BYPRODUCT rather than separating and disposing of it directly.

An example of such recycling in a parallel reaction system is in the "Oxo" process for the production of C_4 alcohols. Propylene and synthesis gas (a mixture of carbon monoxide and hydrogen) are first reacted to n- and isobutyraldehydes using a cobalt-based catalyst. Two parallel reactions occur:

$$\underset{\text{propylene}}{C_3H_6} + \underset{\substack{\text{carbon} \\ \text{monoxide}}}{CO} + \underset{\text{hydrogen}}{H_2} \; \rightleftharpoons \; \underset{\text{n-butyraldehyde}}{CH_3CH_2CH_2CHO}$$

$$\underset{\text{propylene}}{C_3H_6} + \underset{\substack{\text{carbon} \\ \text{monoxide}}}{CO} + \underset{\text{hydrogen}}{H_2} \; \rightleftharpoons \; \underset{\text{iso-butyraldehyde}}{CH_3CH(CH_3)CHO}$$

The n-isomer is more valuable. Recycling the iso-isomer can be used as a means of suppressing its formation.[5]

4. *Multiple reactions in series producing byproducts.* For the series reaction system in Eq. (2.18), the series reaction is inhibited by low concentrations of PRODUCT. It has been noted already that this can be achieved by operating with a low conversion.

If the reaction involves more than one feed, it is not necessary to operate with the same low conversion on all the feeds. Using an excess of one of the feeds enables operation with a relatively high conversion of other feed material, and still inhibits series reactions. Consider again the series reaction system from Example 2.3:

$$\underset{\text{toluene}}{C_6H_5CH_3} + \underset{\text{hydrogen}}{H_2} \; \longrightarrow \; \underset{\text{benzene}}{C_6H_6} + \underset{\text{methane}}{CH_4}$$

$$\underset{\text{benzene}}{2C_6H_6} \; \rightleftharpoons \; \underset{\text{diphenyl}}{C_{12}H_{10}} + \underset{\text{hydrogen}}{H_2}$$

It is usual to operate this reactor with a large excess of hydrogen.[7] The mole ratio of hydrogen to toluene entering the reactor is on the order of $5:1$. The excess of hydrogen encourages the primary reaction directly and discourages the secondary reaction by reducing the concentration of the benzene product. Also, in this case, because hydrogen is a byproduct of the secondary reversible reaction, an excess of hydrogen favors the reverse reaction to benzene. In fact, unless a large excess of hydrogen is used, series reactions that decompose the benzene all the way to carbon become significant. This is known as *coke formation*.

Another way to keep the concentration of PRODUCT low is to remove the product as the reaction progresses, e.g., by intermediate separation followed by further reaction. For example, in a reaction system such as Eq. (2.18), intermediate separation of the PRODUCT followed by further reaction maintains a low concentration of PRODUCT as the reaction progresses. Such intermediate separation is most appropriate when separation of the product from the reactants is straightforward.

If the series reaction is also reversible, such as

$$FEED \longrightarrow PRODUCT$$

$$PRODUCT \rightleftharpoons BYPRODUCT$$

(2.27)

then, again, removal of the PRODUCT as the reaction progresses, e.g., by intermediate separation of the PRODUCT, maintains a low concentration of PRODUCT and at the same time shifts the equilibrium for the secondary reaction toward PRODUCT rather than BYPRODUCT formation.

If the secondary reaction is reversible and inerts are present, then we should

- Increase the concentration of inerts if the BYPRODUCT reaction involves a decrease in the number of moles.

- Decrease the concentration of inerts if the BYPRODUCT reaction involves an increase in the number of moles.

An alternative way to improve selectivity for the reaction system in Eq. (2.27) is again to deliberately feed BYPRODUCT to the reactor to shift the equilibrium of the secondary reaction away from BYPRODUCT formation.

An example of where recycling can be effective in improving selectivity is in the production of benzene from toluene. The series reaction is reversible. Hence recycling diphenyl to the reactor can be used to suppress its formation at the source.

5. *Mixed parallel and series reactions producing byproducts.* As with parallel and series reactions, use of an excess of one of the feeds can be effective in improving selectivity with mixed reactions. As an

example, consider the chlorination of methane to produce chloromethane.[5] The primary reaction is

$$\underset{\text{methane}}{CH_4} + \underset{\text{chlorine}}{Cl_2} \longrightarrow \underset{\text{chloromethane}}{CH_3Cl} + \underset{\text{hydrogen chloride}}{HCl}$$

Secondary reactions can occur to higher chlorinated compounds:

$$\underset{\text{chloromethane}}{CH_3Cl} + Cl_2 \longrightarrow \underset{\text{dichloromethane}}{CH_2Cl_2} + HCl$$

$$\underset{\text{dichloromethane}}{CH_2Cl_2} + Cl_2 \longrightarrow \underset{\text{chloroform}}{CHCl_3} + HCl$$

$$\underset{\text{chloroform}}{CHCl_3} + Cl_2 \longrightarrow \underset{\substack{\text{carbon} \\ \text{tetrachloride}}}{CCl_4} + HCl$$

The secondary reactions are series with respect to the chloromethane but parallel with respect to chlorine. A very large excess of methane (mole ratio of methane to chlorine on the order of 10:1) is used to suppress selectivity losses.[5] The excess of methane has two effects. First, because it is only involved in the primary reaction, it encourages the primary reaction. Second, by diluting the product, chloromethane, it discourages the secondary reactions, which prefer a high concentration of chloromethane.

Removal of the product as the reaction progresses is also effective in suppressing the series element of the byproduct reactions, providing the separation is straightforward.

If the byproduct reaction is reversible and inerts are present, then changing the concentration of inerts if there is a change in the number of moles should be considered, as discussed above. Whether or not there is a change in the number of moles, recycling byproducts can suppress their formation if the byproduct-forming reaction is reversible. An example is in the production of ethylbenzene from benzene and ethylene:[5]

$$\underset{\text{benzene}}{C_6H_6} + \underset{\text{ethylene}}{C_2H_4} \rightleftharpoons \underset{\text{ethylbenzene}}{C_6H_5CH_2CH_3}$$

Polyethylbenzenes (diethylbenzene, triethylbenzene, etc.) are also formed as unwanted byproducts through reversible reactions in series with respect to ethylbenzene but parallel with respect to ethylene. For example,

$$\underset{\text{ethylbenzene}}{C_6H_5CH_2CH_3} + \underset{\text{ethylene}}{C_2H_4} \rightleftharpoons \underset{\text{diethylbenzene}}{C_6H_4(C_2H_5)_2}$$

These polyethylbenzenes are recycled to the reactor to inhibit formation of fresh polyethylbenzenes.[5]

2.6 Reactor Temperature

The choice of reactor temperature depends on many factors. Generally, the higher the rate of reaction, the smaller the reactor volume. Practical upper limits are set by safety considerations, materials-of-construction limitations, or maximum operating temperature for the catalyst. Whether the reaction system involves single or multiple reactions, and whether the reactions are reversible, also affects the choice of reactor temperature, as we shall now discuss.

1. *Single reactions*

a. *Endothermic reactions.* If an endothermic reaction is reversible, then Le Châtelier's principle dictates that operation at a high temperature increases the maximum conversion. Also, operation at high temperature increases the rate of reaction, allowing reduction of reactor volume. Thus, for endothermic reactions, the temperature should be set as high as possible, consistent with safety, materials-of-construction limitations, and catalyst life.

b. *Exothermic reactions.* For exothermic irreversible reactions, the temperature should be set as high as possible, consistent with materials-of-construction, catalyst life, and safety, in order to minimize reactor volume. If an exothermic reaction is reversible, then Le Châtelier's principle dictates that operation at a low temperature increases maximum conversion. However, operation at a low temperature decreases the rate of reaction, thereby increasing the reactor volume. Thus, initially, when far from equilibrium, it is advantageous to use a high temperature to increase the rate of reaction. As equilibrium is approached, the temperature should be lowered to increase the maximum conversion. Thus, for reversible exothermic reactions, the ideal temperature is continuously decreasing as conversion increases.

2. *Multiple reactions.* The arguments presented for minimizing reactor volume for single reactions can be used for the primary reaction when dealing with multiple reactions. However, the goal at this stage of the design, when dealing with multiple reactions, is to maximize selectivity rather than to minimize volume for a given conversion.

Consider Eqs. (2.16), (2.18), and (2.20). Both rates of reaction change with temperature, since the reaction-rate constants k_1 and k_2 both increase with increasing temperature. The rate of change with temperature might be significantly different for the primary and secondary reactions.

- If k_1 increases faster than k_2, operate at high temperature (but beware of safety and materials-of-construction constraints).

- If k_2 increases faster than k_1, operate at low temperature (but beware of capital cost, since low temperature, although increasing selectivity, also increases reactor size). Here there is an economic tradeoff between decreasing byproduct formation and increasing capital cost.

3. *Temperature control.* Let us now consider temperature control of the reactor. In the first instance, adiabatic operation of the reactor should be considered, since this leads to the simplest and cheapest reactor design. If adiabatic operation produces an unacceptable rise in temperature for exothermic reactions or an unacceptable fall in temperature for endothermic reactions, this can be dealt with in a number of ways:

a. *Indirect heat transfer with the reactor.* If adiabatic operation is not feasible, indirect heating or cooling should be considered. This might be by heat transfer surface internal or external to the reactor. Different possible arrangements are considered later for practical reactors.

b. *Cold shot and hot shot.* The injection of cold fresh feed directly into the reactor at intermediate points, called *cold shot,* can be extremely effective for control of temperature in exothermic reactions. If the reaction is endothermic, then fresh feed which has been preheated can be injected at intermediate points, known as *hot shot.*

c. *Heat carrier.* An inert material can be introduced with the reactor feed to increase its heat capacity flow rate [i.e., the product of mass flow rate $(kg\,s^{-1})$ and specific heat capacity $(J\,kg^{-1}\,{}^{\circ}C^{-1})$] and reduce the temperature rise for exothermic reactions or reduce temperature fall for endothermic reactions. Where possible, we should try to use one of the existing process fluids as heat carrier.

Even if the reactor temperature is controlled within acceptable limits, the reactor effluent may need to be cooled rapidly, or *quenched,* to stop the reaction quickly to prevent excessive byproduct formation. This quench can be accomplished by indirect heat transfer using conventional heat transfer equipment or by direct heat transfer by mixing with another fluid. A commonly encountered situation is one in which gaseous products from a reactor need rapid cooling and this is accomplished by mixing with a liquid that evaporates. The heat required to evaporate the liquid causes the gaseous products to cool rapidly. The quench liquid can be recycled, cooled product or an inert material such as water.

In fact, cooling of the reactor effluent by direct heat transfer can be used for a variety of reasons:

- The reaction is very rapid and must be stopped quickly to prevent excessive byproduct formation.

- The reactor products are so hot or corrosive that if passed directly to a heat exchanger, special materials-of-construction or an expensive mechanical design would be required.

- The reactor product cooling would cause excessive fouling in a conventional exchanger.

The liquid used for the direct heat transfer should be chosen such that it can be separated easily from the reactor product and so recycled with the minimum expense. Use of extraneous materials, i.e., materials that do not already exist in the process, should be avoided because it is often difficult to separate and recycle them with high efficiency. Extraneous material not recycled becomes an effluent problem. As we shall discuss later, the best way to deal with effluent problems is not to create them in the first place.

2.7 Reactor Pressure

Increasing the pressure of irreversible vapor-phase reactions increases the rate of reaction and hence decreases reactor volume both by decreasing the residence time required for a given reactor conversion and increasing the vapor density. In general, pressure has little effect on the rate of liquid-phase reactions.

The selection of reactor pressure for vapor-phase reversible reactions depends on whether there is a decrease or increase in the number of moles and whether there is a system of single or multiple reactions.

1. *Single reactions*
a. *Decrease in the number of moles.* A decrease in the number of moles for vapor-phase reactions decreases the volume as reactants are converted to products. For a fixed reactor volume, this means a decrease in pressure as reactants are converted to products. The

effect of an increase in pressure of the system is to cause a shift of the composition of the gaseous mixture toward one occupying a smaller volume. Increasing the reactor pressure increases the equilibrium conversion. Increasing the pressure increases the rate of reaction and also reduces reactor volume. Thus, if the reaction involves a decrease in the number of moles, the pressure should be set as high as practicable, bearing in mind that the high pressure might be costly to obtain through compressor power, mechanical construction might be expensive, and high pressure brings safety problems.

b. *Increase in the number of moles.* An increase in the number of moles for vapor-phase reactions increases the volume as reactants are converted to products. Le Châtelier's principle dictates that a decrease in reactor pressure increases equilibrium conversion. However, operation at a low pressure decreases the rate of reaction in vapor-phase reactions and increases the reactor volume. Thus, initially, when far from equilibrium, it is advantageous to use high pressure to increase the rate of reaction. As equilibrium is approached, the pressure should be lowered to increase the conversion. The ideal pressure would continuously decrease as conversion increases to the desired value. The low pressure required can be obtained by operating the system at reduced absolute pressure or by introducing a diluent to decrease the partial pressure. The diluent is an inert material (e.g., steam) and is simply used to lower the partial pressure in the vapor phase. For example, ethylbenzene can be dehydrogenated to styrene according to the reaction[5]

$$C_6H_5CH_2CH_3 \rightleftharpoons C_6H_5CH=CH_2 + H_2$$
$$\text{ethylbenzene} \qquad\qquad \text{styrene}$$

This is an endothermic reaction accompanied by an increase in the number of moles. High conversion is favored by high temperature and low pressure. The reduction in pressure is achieved in practice by the use of superheated steam as a diluent and by operating the reactor below atmospheric pressure. The steam in this case fulfills a dual purpose by also providing heat for the reaction.

2. *Multiple reactions producing byproducts.* The arguments presented for the effect of pressure on single vapor-phase reactions can be used for the primary reaction when dealing with multiple reactions. Again, selectivity is likely to be more important than reactor volume for a given conversion.

If there is a significant difference between the effect of pressure on the primary and secondary reactions, the pressure should be chosen to reduce as much as possible the rate of the secondary reaction relative to the primary reaction. Improving the selectivity in this way may require changing the system pressure or perhaps introducing a diluent.

For liquid-phase reactions, the effect of pressure on the selectivity and reactor volume is less pronounced, and the pressure is likely to be chosen to

- Prevent vaporization of the products
- Allow vaporization of liquid in the reactor so that it can be condensed and refluxed back to the reactor as a means of removing the heat of reaction
- Allow vaporization of one of the components in a reversible reaction in order that removal increases maximum conversion.

2.8   Reactor Phase

Having considered reactor temperature and pressure, we are now in a position to judge whether the reactor phase will be gas, liquid, or multiphase. Given a free choice between gas- and liquid-phase reactions, operation in the liquid phase is usually preferred. Consider the single reaction system from Eq. (2.19):

$$\text{FEED} \longrightarrow \text{PRODUCT} \quad r = kC_{\text{FEED}}^{a}$$

Clearly, in the liquid phase much higher concentrations of C_{FEED} (kmol m^{-3}) can be maintained than in the gas phase. This makes liquid-phase reactions in general more rapid and hence leads to smaller reactor volumes for liquid-phase reactors.

However, a note of caution should be added. In many multiphase reaction systems, rates of mass transfer between different phases can be just as important or more important than reaction kinetics in determining the reactor volume. Mass transfer rates are generally higher in gas-phase than liquid-phase systems. In such situations, it is not so easy to judge whether gas or liquid phase is preferred.

Very often the choice is not available. For example, if reactor temperature is above the critical temperature of the chemical species, then the reactor must be gas phase. Even if the temperature can be lowered below critical, an extremely high pressure may be required to operate in the liquid phase.

The choice of reactor temperature, pressure, and hence phase must, in the first instance, take account of the desired equilibrium and selectivity effects. If there is still freedom to choose between gas and liquid phase, operation in the liquid phase is preferred.

2.9 Catalysts

Most processes are catalyzed where catalysts for the reaction are known. The choice of catalyst is crucially important. Catalysts increase the rate of reaction but are unchanged in quantity and chemical composition at the end of the reaction. If the catalyst is used to accelerate a reversible reaction, it does not by itself alter the position of the equilibrium. When systems of multiple reactions are involved, the catalyst may have different effects on the rates of the different reactions. This allows catalysts to be developed which increase the rate of the desired reactions relative to the undesired reactions. Hence the choice of catalyst can have a major influence on selectivity.

Unfortunately, despite much research into the fundamentals of catalysis, the choice of catalyst is still largely empirical. The catalytic process can be homogeneous or heterogeneous.

1. *Homogeneous catalysts.* With a homogeneous catalyst, the reaction proceeds entirely in the vapor or liquid phase. The catalyst may modify the reaction mechanism by participation in the reaction but is regenerated in a subsequent step. The catalyst is then free to promote further reaction. An example of such a homogeneous catalytic reaction is the production of acetic anhydride. In the first stage of the process, acetic acid is pyrolyzed to ketene in the gas phase at 700°C:

$$\underset{\text{acetic acid}}{CH_3COOH} \longrightarrow \underset{\text{ketene}}{CH_2{=}C{=}O} + \underset{\text{water}}{H_2O}$$

The reaction uses triethyl phosphate as a homogeneous catalyst.[5]

In general, heterogeneous catalysts are preferred to homogeneous catalysts because the separation and recycling of homogeneous catalysts often can be very difficult. Loss of homogeneous catalyst not only creates a direct expense through loss of material but also creates an environmental problem.

2. *Heterogeneous catalysts.* In heterogeneous catalysis, the catalyst is in a different phase from the reacting species. Most often, the

heterogeneous catalyst is a solid, acting on species in the liquid or gas phase. In this case, the reactants diffuse to the catalyst surface and are adsorbed onto the surface, where the reaction takes place. After reaction, the products desorb and diffuse back to the bulk gas or liquid phase. The solid catalyst can be either

- *Bulk catalytic materials,* in which the gross composition does not change significantly through the material, such as platinum wire mesh.

- *Supported catalysts,* in which the active catalytic material is dispersed over the surface of a porous solid.

Catalytic gas-phase reactions play an important role in many bulk chemical processes, such as in the production of methanol, ammonia, sulfuric acid, and nitric acid. In most processes, the effective area of the catalyst is critically important. Since these reactions take place at surfaces through processes of adsorption and desorption, any alteration of surface area naturally causes a change in the rate of reaction. Industrial catalysts are usually supported on porous materials, since this results in a much larger active area per unit of reactor volume.

As well as depending on catalyst porosity, the reaction rate is some function of the reactant concentrations, temperature, and pressure. However, this function may not be as simple as in the case of uncatalyzed reactions. Before a reaction can take place, the reactants must diffuse through the pores to the solid surface. This results in a situation where either reaction or diffusion can be the rate-limiting process. Alternatively, it may be that reaction speed and diffusion have an almost equal effect. If reaction is rate limiting, as tends to occur in a lower temperature range, the effects of concentration and temperature are those typical of chemical reaction. On the other hand, if diffusion is rate limiting, as tends to occur in a higher temperature range, the effects of concentration and temperature are those characteristic of diffusion. In the transitional region, where both reaction and diffusion affect the overall rate, the effects of temperature and concentration are often rather complex.

More often than not, solid-catalyzed reactions are multiple reactions. For reactions in parallel, the key to high selectivity is to maintain the appropriate high or low concentration levels of reactants at the catalyst surface, to encourage the desired reaction, and to discourage the byproduct reactions. For reactions in series, the key is to avoid the mixing of fluids of different compositions. These arguments for the gross flow pattern of fluid through any reactor have already been developed.

However, before extrapolating the arguments from the gross patterns through the reactor for homogeneous reactions to solid-catalyzed reactions, it must be recognized that in catalytic reactions the fluid in the interior of catalyst pellets may differ from the main body of fluid. The local inhomogeneities caused by lowered reactant concentration within the catalyst pellets result in a product distribution different from that which would otherwise be observed.

a. *Surface reaction controls.* When surface reaction is rate controlling, the concentrations of reactant within the pellets and in the main gas stream are essentially the same. In this situation, the considerations for the gross flow pattern of fluid through the reactor apply.

b. *Diffusion controls.* When diffusional resistance across the gas film around the particles controls, then the concentration of reactant at the catalyst surface is lower than in the main gas stream. Referring back to Eq. (2.16), for example, lowered reactant concentration favors the reaction of lower order. Hence, if the desired reaction is of lower order, operating under conditions of diffusion control increases selectivity. If the desired reaction is of higher order, the opposite holds.

The choice of catalyst and the conditions of reaction can be critical in the performance of the process because of the resulting influence on selectivity.

3. *Catalytic degradation.* The performance of most catalysts deteriorates with time. The rate at which the deterioration takes place is another important factor in the choice of catalyst and the choice of reactor conditions. Deterioration in performance lowers the rate of reaction, which, for a given reactor design, manifests itself as a lowering of the conversion. This often can be compensated by increasing the temperature of the reactor. However, significant increases in temperature can degrade selectivity considerably and often accelerate the mechanisms that cause catalyst degradation.

Loss of catalyst performance can occur in a number of ways:

a. *Physical loss.* Physical loss is particularly important with homogeneous catalysts, which need to be separated from reaction products and recycled. Unless this can be done with high efficiency, it leads to physical loss (and subsequent environmental problems). However, physical loss as a problem is not restricted to homogeneous catalysts. It also can be a problem with heterogeneous catalysts. This is particularly the case when catalytic fluidized-bed reactors are employed. Attrition of the particles causes the catalyst particles to be broken down in size. Particles which are carried over from the fluidized bed are normally separated from the reactor effluent and recycled to the bed. However, the finest particles are not separated and recycled and are lost.

b. *Surface deposits.* The formation of deposits on the surface of solid catalysts introduces a physical barrier to the reacting species. The deposits are most often insoluble (in liquid-phase reactions) or involatile (in gas-phase reactions) byproducts of reaction. An example of this is the formation of carbon deposits (known as *coke*) on the surface of catalysts involved in hydrocarbon reactions. Such coke formation can sometimes be suppressed by suitable adjustment of the feed composition. If coke formation occurs, the catalyst often can be regenerated by air oxidation of the carbon deposits at elevated temperatures.

c. *Sintering.* With high-temperature gas-phase reactions which use solid catalysts, *sintering* of the support or the active material can occur. Sintering is a molecular rearrangement that occurs below the melting point of the material and causes a reduction in the effective surface area of the catalyst. This problem is accelerated if poor heat transfer or poor mixing of reactants leads to local "hot spots" in the catalyst bed. Sintering can begin at temperatures equal to about half the melting point of the catalyst.

d. *Poisoning.* Poisons are materials which chemically react with or form strong chemical bonds with the catalyst. Such reactions degrade the catalyst and reduce its activity. Poisons are usually impurities in the raw materials or products of corrosion.

e. *Chemical change.* In theory, a catalyst should not undergo chemical change. However, some catalysts can slowly change chemically with a consequent reduction in activity.

The rate at which the catalyst is lost or degrades has a major influence on the design. If degradation is rapid, the catalyst needs to be regenerated or replaced on a continuous basis. In addition to the cost implications, there are also environmental implications, since the lost or degraded catalyst represents waste. While it is often possible to recover useful materials from degraded catalyst and to recycle those materials in the manufacture of new catalyst, this still inevitably creates waste, since the recovery of material can never be complete.

Having discussed the choice of reactor type and operating conditions at length, let us try two examples.

Example 2.4 Monoethanolamine is required as a product. This can be produced from the reaction between ethylene oxide and ammonia:[5]

$$H_2C\text{---}CH_2 + NH_3 \longrightarrow NH_2CH_2CH_2OH$$
$$\diagdown_O\diagup$$

ethylene oxide   ammonia   monoethanolamine

Two principal secondary reactions occur, to diethanolamine and triethanolamine:

$$NH_2CH_2CH_2OH + H_2C\overset{\diagdown}{\underset{O}{}}\diagup CH_2 \longrightarrow NH(CH_2CH_2OH)_2$$

monoethanolamine ethylene oxide   diethanolamine

$$NH(CH_2CH_2OH)_2 + H_2C\overset{\diagdown}{\underset{O}{}}\diagup CH_2 \longrightarrow N(CH_2CH_2OH)_3$$

diethanolamine   ethylene oxide   triethanolamine

The secondary reactions are parallel with respect to ethylene oxide but series with respect to monoethanolamine. Monoethanolamine is more valuable than both the di- and triethanolamine. As a first step in the flowsheet synthesis, make an initial choice of reactor which will maximize the production of monoethanolamine relative to di- and triethanolamine.

Solution We wish to avoid as much as possible the production of di- and triethanolamine, which are formed by series reactions with respect to mo-noethanolamine. In a continuous well-mixed reactor, part of the monoethanol-amine formed in the primary reaction could stay for extended periods, thus increasing its chances of being converted to di- and triethanolamine. The ideal batch or plug-flow arrangement is preferred, to carefully control the residence time in the reactor.

Further consideration of the reaction system reveals that the ammonia feed takes part only in the primary reaction and in neither of the secondary reactions. Consider the rate equation for the primary reaction:

$$r_1 = k_1 C_{EO}^{a_1} C_{NH_3}^{b_1}$$

where r_1 = reaction rate of primary reaction
   k_1 = reaction-rate constant for the primary reaction
   C_{EO} = molar concentration of ethylene oxide in the reactor
   C_{NH_3} = molar concentration of ammonia in the reactor
   a_1, b_1 = order of primary reaction

Operation with an excess of ammonia in the reactor has the effect of increasing the rate due to the $C_{NH_3}^{b_1}$ term. However, operation with excess ammonia decreases the concentration of ethylene oxide, and the effect is to decrease the rate due to the $C_{EO}^{a_1}$. Whether the overall effect is a slight increase or decrease in reaction rate depends on the relative magnitude of a_1 and b_1. Consider now the rate equations for the byproduct reactions:

$$r_2 = k_2 C_{MEA}^{a_2} C_{EO}^{b_2}$$

$$r_3 = k_3 C_{DEA}^{a_3} C_{EO}^{b_3}$$

where r_2, r_3 = rates of reaction to di- and triethanolamine, respectively
   k_2, k_3 = reaction-rate constants for the di- and triethanolamine reactions, respectively
   C_{MEA} = molar concentration of monoethanolamine
   C_{DEA} = molar concentration of diethanolamine
   a_2, b_2 = order of reaction for the diethanolamine reaction
   a_3, b_3 = order of reaction for the triethanolamine reaction

An excess of ammonia in the reactor decreases the concentrations of monoethanolamine, diethanolamine, and ethylene oxide and decreases the rates of reaction for both secondary reactions.

Thus an excess of ammonia in the reactor has a marginal effect on the primary reaction but significantly decreases the rate of the secondary reactions. Using excess ammonia also can be thought of as operating the reactor with a low conversion with respect to ammonia.

The use of an excess of ammonia is borne out in practice.[5] A mole ratio of ammonia to ethylene oxide of 10:1 yields 75 percent monoethanolamine, 21 percent diethanolamine, and 4 percent triethanolamine. Using equimolar proportions under the same reaction conditions, the respective proportions become 12, 23, and 65 percent.

Another possibility to improve selectivity is to reduce the concentration of monoethanolamine in the reactor by using more than one reactor with intermediate separation of the monoethanolamine. Considering the boiling points of the components given in Table 2.3, then separation by distillation is apparently possible. Unfortunately, repeated distillation operations are likely to be very expensive. Also, there is a market to sell both di- and triethanolamine, even though their value is lower than that of monoethanolamine. Thus, in this case, repeated reaction and separation are probably not justified, and the choice is a single plug-flow reactor.

An initial guess for the reactor conversion is very difficult to make. A high conversion increases the concentration of monoethanolamine and increases the rates of the secondary reactions. As we shall see later, a low conversion has the effect of decreasing the reactor capital cost but increasing the capital cost of many other items of equipment in the flowsheet. Thus an initial value of 50 percent conversion is probably as good as a guess as can be made at this stage.

Example 2.5   *tert*-Butyl hydrogen sulfate is required as an intermediate in a reaction sequence. This can be produced by the reaction between isobutylene and moderately concentrated sulfuric acid:

$$
\begin{array}{ccc}
\underset{\substack{\text{isobutylene}}}{CH_3-\overset{\displaystyle \overset{CH_3}{|}}{C}=CH_2} + \underset{\substack{\text{sulfuric}\\\text{acid}}}{H_2SO_4} & \longrightarrow & \underset{\substack{tert\text{-butyl}\\\text{hydrogen sulfate}}}{CH_3-\overset{\displaystyle \overset{CH_3}{|}}{\underset{\displaystyle \underset{OSO_3H}{|}}{C}}-CH_3}
\end{array}
$$

TABLE 2.3 Normal Boiling Points of the Components

Component	Normal boiling point (K)
Ammonia	240
Ethylene oxide	284
Monoethanolamine	444
Diethanolamine	542
Triethanolamine	609

Series reactions occur in which the *tert*-butyl hydrogen sulfate reacts to unwanted *tert*-butyl alcohol:

$$\underset{\substack{\text{\textit{tert}-butyl} \\ \text{hydrogen sulfate}}}{\underset{\begin{array}{c}\text{CH}_3 \\ | \\ \text{CH}_3\text{—C—CH}_3 \\ | \\ \text{OSO}_3\text{H}\end{array}}{}} \; + \; \underset{\text{water}}{\text{H}_2\text{O}} \; \xrightarrow{\text{heat}} \; \underset{\text{\textit{tert}-butyl alcohol}}{\underset{\begin{array}{c}\text{CH}_3 \\ | \\ \text{CH}_3\text{—C—CH}_3 \\ | \\ \text{OH}\end{array}}{}} \; + \; \underset{\text{sulfuric acid}}{\text{H}_2\text{SO}_4}$$

Other series reactions form unwanted polymeric material. Further information on the reaction is

- The primary reaction is rapid and exothermic.
- Laboratory studies indicate that the reactor yield is a maximum when the concentration of sulfuric acid is maintained at 63 percent.[11]
- The temperature should be maintained around 0°C or excessive byproduct formation occurs.[11,12]

Make an initial choice of reactor.

Solution   The byproduct reactions to avoid are all series in nature. This suggests that we should not use a continuous well-mixed reactor but rather use either a batch or plug-flow reactor.

However, the laboratory data seem to indicate that a constant concentration in the reactor to maintain 63 percent sulfuric acid would be beneficial. Careful temperature control is also important. These two factors would suggest that a continuous well-mixed reactor is appropriate. There is a conflict. How can a well-defined residence time be maintained and simultaneously a constant concentration of sulfuric acid be maintained?

Using a batch reactor, a constant concentration of sulfuric acid can be maintained by adding concentrated sulfuric acid as the reaction progresses, i.e., semi-batch operation. Good temperature control of such systems can be maintained, as we shall discuss later.

Choosing to use a continuous rather than a batch reactor, plug-flow behavior can be approached using a series of continuous well-mixed reactors. This again allows concentrated sulfuric acid to be added as the reaction progresses, in a similar way as suggested for some parallel systems in Fig. 2.2. Breaking the reactor down into a series of well-mixed reactors also allows good temperature control, as we shall discuss later.

To make an initial guess for the reactor conversion is again difficult. The series nature of the byproduct reactions suggests that a value of 50 percent is probably as good as can be suggested at this stage.

2.10   Practical Reactors

By contrast with ideal models, practical reactors must consider many factors other than variations in temperature, concentration, and residence time. Practical reactors deviate from the three idealized models but can be classified into a number of common types.

1. *Stirred-tank reactors.* Stirred-tank reactors consist simply of an agitated tank and are used for reactions involving a liquid. Applications include

- Homogeneous liquid-phase reactions
- Heterogeneous gas-liquid reactions
- Heterogeneous liquid-liquid reactions
- Heterogeneous solid-liquid reactions
- Heterogeneous gas-solid-liquid reactions

Stirred-tank reactors can be operated in batch, semi-batch, or continuous mode. In batch operation,

- Operation is more flexible for variable production rates or manufacture of a variety of similar products in the same equipment.
- Labor costs tend to be higher (although this can be overcome to some extent by use of computer control).

In continuous operation:

- Automatic control tends to be more straightforward (leading to lower labor costs and greater consistency of operation).

In fact, it is often possible with stirred-tank reactors to come close to the idealized well-stirred model in practice, providing the fluid phase is not too viscous. Such reactors should be avoided for some types of parallel reaction systems (see Fig. 2.2) and for all systems in which byproduct formation is via series reactions.

Stirred-tank reactors become unfavorable if the reaction must take place at high pressure. Under high-pressure conditions, a small-diameter cylinder requires a thinner wall than a large-diameter cylinder. Under high-pressure conditions, use of a tubular reactor is preferred, as described in the next section, although mixing problems with heterogeneous reactions and other factors may prevent this. Another important factor to the disadvantage of the continuous stirred-tank reactor is that for a given conversion it requires a large inventory of material relative to, say, a tubular reactor. This is not desirable for safety reasons if the reactants or products are particularly hazardous.

Heat can be added to or removed from stirred-tank reactors via external jackets (Fig. 2.5a), internal coils (Fig. 2.5b), or separate heat exchangers by means of a flow loop (Fig. 2.5c). Figure 2.5d shows vaporization of the contents being condensed and refluxed to remove heat.

If plug-flow is required, but the volume of the reactor is large, then plug-flow operation can be approached by using stirred tanks in series, since large volumes are often more economically arranged in stirred tanks than in tubular devices. This also can offer the advantage of better temperature control than the equivalent tubular reactor arrangement.

2. *Tubular reactors.* Although tubular reactors often take the actual form of a tube, they can be any reactor in which there is steady movement in one direction only. The tubes may be arranged in parallel, in a construction similar to a shell-and-tube heat exchanger. This design is used when external heating or cooling is required. In high-temperature reactions, the tubes are constructed inside a furnace.

Because the characteristic of tubular reactors approximates plug-flow, they are used if careful control of residence time is important, as in the case where there are multiple reactions in series. High surface area to volume ratios are possible, which is an advantage if high rates of heat transfer are required. It is sometimes possible to approach isothermal conditions or a predetermined temperature profile by careful design of the heat transfer arrangements.

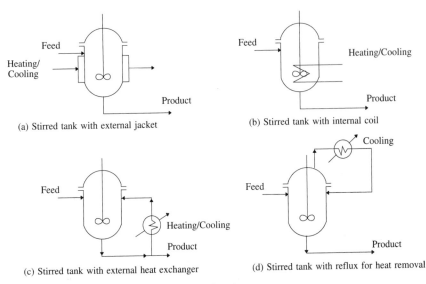

Figure 2.5 Heat transfer to and from stirred tanks.

Tubular reactors, as previously stated, are also advantageous for high-pressure reactions where smaller-diameter cylindrical vessels can be used to allow thinner vessel walls. Tubular reactors should be avoided when carrying out multiphase reactions, since it is often difficult to achieve good mixing between phases.

3. *Fixed-bed catalytic reactors.* Tubular reactors are also used extensively for catalytic reactions. Here the reactor is packed with particles of solid catalyst. Most designs approximate to plug-flow behavior. Figure 2.6 shows four possible arrangements for fixed-bed reactors. The first (Fig. 2.6a) is similar to a shell-and-tube exchanger in which the tubes are packed with catalyst. The second (Fig. 2.6b) has the tubes constructed inside a furnace for high temperatures. The third (Fig. 2.6c) is a series of adiabatic beds with intermediate cooling or heating to maintain temperature control. The heating or cooling can be effected by internal or external exchangers. The fourth (Fig. 2.6d) uses direct injection of a fluid to perform heat transfer. The injected fluid might typically be cold fresh feed or cooled recycled product to control the temperature rise in an exothermic reaction. This is known as *cold-shot cooling*. Many other arrangements are possible.

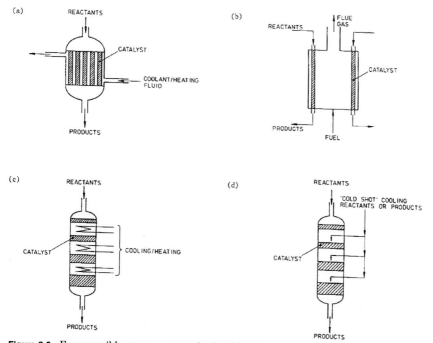

Figure 2.6 Four possible arrangements for fixed-bed reactors.

Generally speaking, temperature control in fixed beds is difficult because heat loads vary through the bed. Also, in exothermic reactors, the temperature in the catalyst can become locally excessive. Such "hot spots" can cause the onset of undesired reactions or catalyst degradation. In tubular devices such as shown in Fig. 2.6a and b, the smaller the diameter of tube, the better is the temperature control. Temperature-control problems also can be overcome by using a mixture of catalyst and inert solid to effectively "dilute" the catalyst. Varying this mixture allows the rate of reaction in different parts of the bed to be controlled more easily.

However, if high rates of heat transfer are required or the catalyst requires frequent regeneration, then fixed beds are not suitable, and under these circumstances, a fluidized bed is preferred, as we shall discuss later.

As an example of the application of a fixed-bed tubular reactor, consider the production of methanol. Synthesis gas (a mixture of hydrogen, carbon monoxide, and carbon dioxide) is reacted over a copper-based catalyst.[5] The main reactions are

$$CO + 2H_2 \rightleftharpoons \underset{\text{methanol}}{CH_3OH}$$

$$CO_2 + H_2 \longrightarrow CO + H_2O$$

The first reaction is exothermic, and the second is endothermic. Overall, the reaction evolves considerable heat. Figure 2.7 shows two alternative reactor designs.[13] Figure 2.7a shows a shell and tube type of device which generates steam on the shell side. The temperature profile through the reactor in Fig. 2.7a is seen to be relatively smooth. Figure 2.7b shows an alternative reactor design that uses cold-shot cooling. By contrast with the tubular reactor, the cold-shot reactor shown in Fig. 2.7b experiences significant temperature fluctuations. Such fluctuations can cause accidental catalyst overheating and shorten catalyst life.

Gas-liquid mixtures are sometimes reacted in packed beds. The gas and the liquid usually flow cocurrently. Such *trickle-bed reactors* have the advantage that residence times of the liquid are shorter than in countercurrent operation. This can be useful in avoiding unwanted side reactions.

4. *Fixed-bed noncatalytic reactors.* Fixed-bed reactors can be used to react a gas and a solid. For example, hydrogen sulfide can be removed from fuel gases by reaction with ferric oxide:

$$\underset{\text{ferric oxide}}{Fe_2O_3} + \underset{\substack{\text{hydrogen}\\\text{sulfide}}}{3H_2S} \longrightarrow \underset{\substack{\text{ferric}\\\text{sulfide}}}{Fe_2S_3} + 3H_2O$$

The ferric oxide is regenerated using air:

$$2Fe_2S_3 + 3O_2 \longrightarrow 2Fe_2O_3 + 6S$$

Two fixed-bed reactors can be used in parallel, one reacting and the

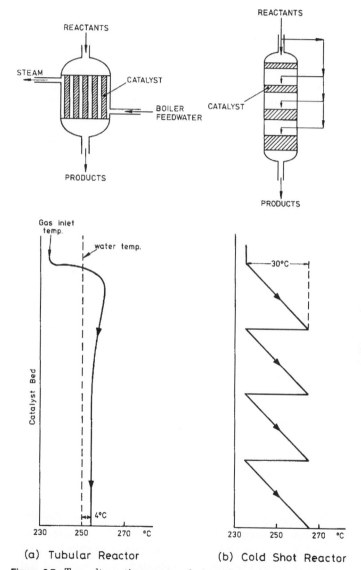

(a) Tubular Reactor   (b) Cold Shot Reactor

Figure 2.7 Two alternative reactor designs for methanol production give quite different thermal profiles.

other regenerating. However, there are many disadvantages to carrying out this type of reaction in a packed bed. The operation is not under steady-state conditions, and this can present control problems. Eventually, the bed must be taken off-line to replace the solid. Fluidized beds are usually preferred for gas-solid noncatalytic reactions.

Fixed-bed reactors in the form of gas absorption equipment are used commonly for noncatalytic gas-liquid reactions. Here the packed bed serves only to give good contact between the gas and liquid. Both cocurrent and countercurrent operations are used. Countercurrent operation gives the highest reaction rates. Cocurrent operation is preferred if a short liquid residence time is required.

For example, hydrogen sulfide and carbon dioxide can be removed from natural gas by reaction with monoethanolamine in an absorber according to the following reactions:[14]

$$\underset{\text{monoethanolamine}}{HOCH_2CH_2NH_2} + \underset{\substack{\text{hydrogen}\\\text{sulfide}}}{H_2S} \;\rightleftharpoons\; \underset{\substack{\text{monoethanolamine}\\\text{hydrogen}\\\text{sulfide}}}{HOCH_2CH_2NH_3HS}$$

$$\underset{\text{monoethanolamine}}{HOCH_2CH_2NH_2} + \underset{\substack{\text{carbon}\\\text{dioxide}}}{CO_2} + H_2O \;\rightleftharpoons\; \underset{\substack{\text{monoethanolamine}\\\text{hydrogen}\\\text{carbonate}}}{HOCH_2CH_2NH_3HCO_3}$$

These reactions can be reversed in a distillation column. This releases the hydrogen sulfide and carbon dioxide for further processing. The monoethanolamine can then be recycled.

5. *Fluidized-bed catalytic reactors.* In fluidized-bed reactors, solid material in the form of fine particles is held in suspension by the upward flow of the reacting fluid. The effect of the rapid motion of the particles is good heat transfer and temperature uniformity. This prevents the formation of the hot spots that can occur with fixed-bed reactors.

The performance of fluidized-bed reactors is not approximated by either the well-stirred or plug-flow idealized models. The solid phase tends to be well-mixed, but the bubbles lead to the gas phase having a poorer performance than well mixed. Overall, the performance of a fluidized-bed reactor often lies somewhere between the well-stirred and plug-flow models.

In addition to the advantage of high heat transfer rates, fluidized beds are also useful in situations where catalyst particles need frequent regeneration. Under these circumstances, particles can be removed continuously from the bed, regenerated, and recycled back to the bed. In exothermic reactions, the recycling of catalyst can be used to remove heat from the reactor, or in endothermic reactions, it can be used to add heat.

One disadvantage of fluidized beds is that attrition of the catalyst can cause the generation of catalyst fines, which are then carried over from the bed and lost from the system. This carryover of catalyst fines sometimes necessitates cooling the reactor effluent through direct-contact heat transfer by mixing with a cold fluid, since the fines tend to foul conventional heat exchangers.

Figure 2.8 shows the essential features of a refinery catalytic cracker. This particular reaction is accompanied by the deposition of carbon on the surface of the catalyst. The fluidized-bed reactor allows the catalyst to be withdrawn continuously and circulated to a fluidized regenerator, where the carbon is burnt off in an air stream, allowing regenerated catalyst to be returned to the cracker.

6. *Fluidized bed noncatalytic reactors.* Fluidized beds are also suited to gas-solid noncatalytic reactions. All the advantages described earlier for gas-solid catalytic reactions apply. As an example,

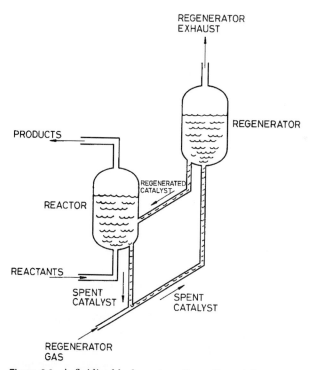

Figure 2.8 A fluidized-bed reactor allows the catalyst to be continuously withdrawn and regenerated as with the refinery catalytic cracker.

limestone (principally calcium carbonate) can be heated to produce calcium oxide in a fluidized-bed reactor according to the reaction

$$CaCO_3 \xrightarrow{\text{heat}} CaO + CO_2$$

The solid particles are fluidized by air and fuel, which are fed to the bed and burnt to produce the high temperatures necessary for the reaction.

7. *Kilns.* Reactions involving free-flowing solid, paste, and slurry materials can be carried out in kilns. In a rotary kiln, a cylindrical shell is mounted with its axis making a small angle to horizontal and rotated slowly. The material to be reacted is fed to the elevated end of the kiln and tumbles down the kiln as a result of the rotation. The behavior of the reactor usually approximates plug flow. High-temperature reactions demand refractory lined steel shells and are usually heated by direct firing. An example of a reaction carried out in such a device is the production of hydrogen fluoride:

$$\underset{\substack{\text{calcium} \\ \text{fluoride}}}{CaF_2} + \underset{\substack{\text{sulfuric} \\ \text{acid}}}{H_2SO_4} \longrightarrow \underset{\substack{\text{hydrogen} \\ \text{fluoride}}}{2HF} + \underset{\substack{\text{calcium} \\ \text{sulfate}}}{CaSO_4}$$

Other designs of kilns use static shells rather than rotating shells and rely on mechanical rakes to move solid material through the reactor.

2.11 Choice of Reactor—Summary

1. *Raw materials efficiency.* In choosing the reactor, the overriding consideration is usually raw materials efficiency (bearing in mind materials of construction, safety, etc.). Raw material costs are usually the most important costs in the whole process. Also, any inefficiency in raw materials use is likely to create waste streams that become an environmental problem. The reactor creates inefficiency in the use of raw materials in the following ways:

a. If low conversion is obtained and unreacted feed material is difficult to separate and recycle.

b. Through the formation of unwanted byproducts. Sometimes the byproduct has value as a product in its own right; sometimes it simply has value as fuel. Sometimes it is a liability and requires disposal in expensive waste-treatment processes.

c. Impurities in the feed can undergo reaction to form additional byproducts. This is best avoided by purification of the feed before reaction.

Some guess for the reactor conversion must be made in order that

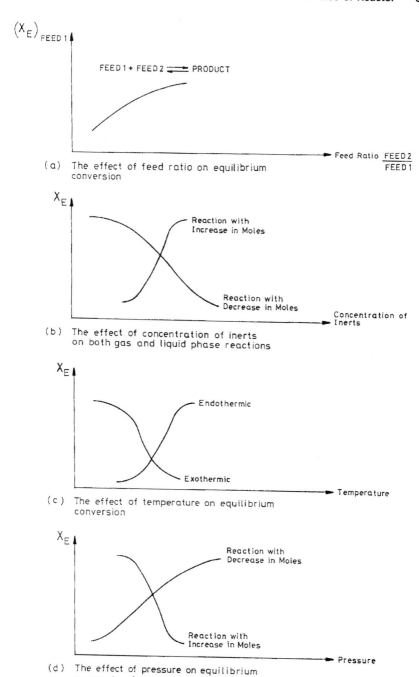

(a) The effect of feed ratio on equilibrium
     conversion

(b) The effect of concentration of inerts
     on both gas and liquid phase reactions

(c) The effect of temperature on equilibrium
     conversion

(d) The effect of pressure on equilibrium
     conversion for gas phase reactions

Figure 2.9 Various measures can be taken to increase equilibrium conversion in reversible reactions. *(From Smith and Petela, The Chemical Engineer, Dec. 17, 1991; reproduced by permission of the Institution of Chemical Engineers.)*

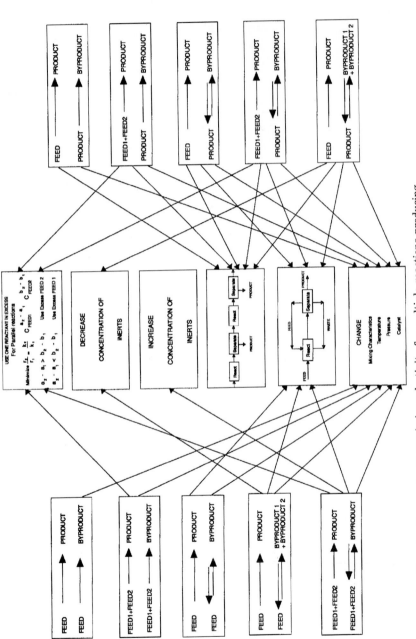

Figure 2.10 Choosing the reactor to maximize selectivity for multiple reactions producing byproducts.

the design can proceed. This is likely to change later in the design because, as will be seen later, there is a strong interaction between the reactor conversion and the rest of the flowsheet.

2. *Single reactions.* For single reactions, a good initial setting is 95 percent conversion for irreversible reactions and 95 percent of the equilibrium conversion for reversible reactions. Figure 2.9 summarizes the influence of feed mole ratio, inert concentration, temperature, and pressure on equilibrium conversion.[2]

3. *Multiple reactions.* For multiple reactions in which the byproduct is formed in parallel, the selectivity may increase or decrease as conversion increases. If the byproduct reaction is a higher order than the primary reaction, selectivity increases for increasing reactor conversion. In this case, the same initial setting as single reactions should be used. If the byproduct reaction of the parallel system is a

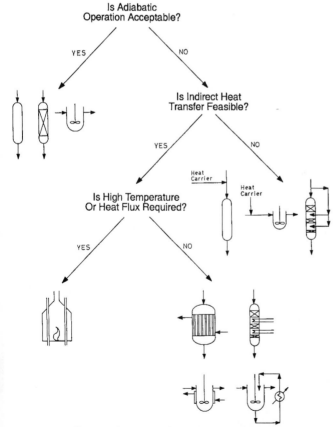

Figure 2.11 Choosing heat transfer in the reactor.

lower order than the primary reaction, a lower conversion than that for single reactions is expected to be appropriate. The best guess at this stage is to set the conversion of 50 percent for irreversible reactions or to 50 percent of the equilibrium conversion for reversible reactions.

For multiple reactions in which the byproduct is formed in series, the selectivity decreases as conversion increases. In this case, lower conversion than that for single reactions is expected to be appropriate. Again, the best guess at this stage is to set the conversion to 50 percent for irreversible reactions or to 50 percent of the equilibrium conversion for reversible reactions.

It should be emphasized that these recommendations for the initial settings of the reactor conversion will almost certainly change at a later stage, since reactor conversion is an extremely important optimization variable. When dealing with multiple reactions, selectivity is maximized for the chosen conversion. Thus a reactor type, temperature, pressure, and catalyst are chosen to this end. Figure 2.10 summarizes the basic decisions which must be made to maximize selectivity.[15]

4. *Heat transfer.* Once the basic reactor type and conditions have been chosen, heat transfer can be a major problem. Figure 2.11 summarizes the basic decisions which must be made regarding heat transfer. If the reactor product is to be cooled by direct contact with a cold fluid, then use of extraneous materials should be avoided.

5. *Reactors in the overall process.* It should be emphasized that many considerations other than those represented in Figs. 2.9, 2.10, and 2.11 also influence the decision on the choice of reactor. Safety considerations, operating pressure, materials of construction, etc. have a considerable effect on the outcome.

The decisions made in the reactor design are often the most important in the whole flowsheet. The design of the reactor usually interacts strongly with the rest of the flowsheet. Hence a return to the decisions made for the reactor must be made when the process design has progressed further and we have fully understood the consequences of those decisions. For the detailed sizing of the reactor, the reader is referred to the many excellent texts on reactor design.[1-3]

2.12   References

1. Denbigh, K. G., and Turner, J. C. R., *Chemical Reactor Theory,* 3d ed., Cambridge University Press, Cambridge, England, 1984.
2. Levenspiel, O., *Chemical Reaction Engineering,* 2d ed., Wiley, New York, 1972.

3. Rase, H. F., *Chemical Reactor Design for Process Plants,* vol. 1, Wiley, New York, 1977.
4. Rudd, D. F., Powers, G. J., and Siirola, J. J., *Process Synthesis,* Prentice-Hall, Englewood Cliffs, N.J., 1973.
5. Waddams, A. L., *Chemicals from Petroleum,* John Murray, London, 1978.
6. Wells, G. L., and Rose, L. M., *The Art of Chemical Process Design,* Elsevier, New York, 1986.
7. Douglas, J. M., "A Hierarchical Decision Procedure for Process Synthesis," *AIChEJ,* 31: 353, 1985.
8. Douglas, J. M., *Conceptual Design of Chemical Processes,* McGraw-Hill, New York, 1988.
9. Smith, R., and Petela, E. A., "Waste Minimisation in the Process Industries: Part 2—Reactors," *The Chemical Engineer,* 12: 509/510, Dec. 17, 1992.
10. Gillespie, B. M., and Carberry, J. J., "Reactor Yield at Intermediate Mixing Levels: An Extension of Van de Vusse's Analysis," *Chem. Eng. Sci.,* 21: 472, 1966.
11. Morrison, R. T., and Boyd, R. N., *Organic Chemistry,* 6th ed. Prentice-Hall, Englewood Cliffs, N.J., 1992.
12. Albright, L. F., and Goldsby, A. R., *Industrial and Laboratory Alkylations,* ACS Symposium Series No. 55, ACS, Washington, DC, 1977.
13. Supp. E., "Technology of Lurgi's Low Pressure Methanol Process," *Chem. Tech.,* 3: 430, 1973.
14. Kohl, A. L., and Riesenfeld, F. C., *Gas Purification,* Gulf Publishing Company, Houston, Texas 1979.
15. Smith, R., and Petela, E. A., "Waste Minimisation in the Process Industries," paper presented at the IChemE Symposium on Integrated Pollution Control Through Clean Technology, Wilmslow, UK, May 20–21, 1992.

3

Choice of Separator

Having made an initial specification for the reactor, attention is turned to separation of the reactor effluent. In addition, it might be necessary to carry out separation before the reactor to purify the feed. Whether before or after the reactor, the overall separation task normally must be broken down into a number of intermediate separation tasks. The first consideration is the choice of separator for the intermediate separation tasks. Later we shall consider how these separation tasks should be connected to the reactor. As with reactors, we shall concentrate on the choice of separator and not its detailed sizing.

If the mixture to be separated is homogeneous, a separation can only be performed by the addition or creation of another phase within the system. For example, if a gaseous mixture is leaving the reactor, another phase could be created by partial condensation. The vapor resulting from the partial condensation will be rich in the more volatile components and the liquid will be rich in the less volatile components, achieving a separation. Alternatively, rather than creating another phase, one can be added to the gaseous mixture, such as a solvent which would preferentially dissolve one or more of the components from the mixture. Further separation is required to separate the solvent from the process materials allowing recycle of the solvent, etc. A number of physical properties can be exploited to achieve the separation of homogeneous mixtures.[1,2] If a heterogeneous or multiphase mixture leaves the reactor, then separation can be done physically by exploiting differences in density between the phases.

Separation of the different phases of a heterogeneous mixture should be carried out before homogeneous separation, taking advantage of what already exists. Phase separation tends to be easier and should be done first. The phase separations likely to be carried out are

- Vapor-liquid
- Liquid-liquid (immiscible)
- Solid-liquid
- Solid-vapor
- Solid-solid

Now we consider some of the methods by which such separations can be achieved. A comprehensive survey is beyond the scope of this text, and many good surveys are already available.[1-5]

3.1 Separation of Heterogeneous Mixtures

The four principal methods for the separation of heterogeneous mixtures are

- Settling and sedimentation
- Flotation
- Centrifugal separation
- Filtration

1. *Settling and sedimentation.* In *settling* processes, particles are separated from a fluid by gravitational forces acting on the particles. The particles can be solid particles or liquid drops. The fluid can be a liquid or a gas.

Figure 3.1a shows a "flash" drum used to separate by gravity a vapor-liquid mixture. The velocity of the vapor through the flash drum must be less than the settling velocity of the liquid drops.

Figure 3.1b shows a simple gravity settler for removing a

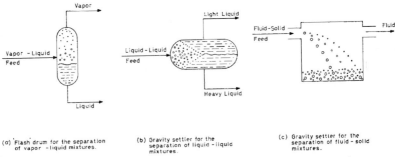

Figure 3.1 Settling processes used for the separation of heterogeneous mixtures.

dispersed liquid phase from another liquid phase. The horizontal velocity must be low enough to allow the low-density droplets to rise from the bottom of the vessel to the interface and for the high-density droplets to settle down to the interface and coalesce. An empty vessel may be employed, but horizontal baffles can be used to reduce turbulence and assist the coalescence. More elaborate methods to assist the coalescence include the use of mesh pads in the vessel or the use of an electric field to promote coalescence.

Figure 3.1c is a schematic diagram of gravity settling chamber. A mixture of vapor or liquid and solid particles enters at one end of a large chamber. Particles settle toward the floor. The vertical height of the chamber divided by the settling velocity of the particles must give a time less than the residence time of the air.

The separation of suspended solid particles from a liquid by gravity settling into a clear fluid and a slurry of higher solids content is called *sedimentation.* Figure 3.2 shows a sedimentation device known as a *thickener,* the prime function of which is to produce a more concentrated slurry. The feed slurry in Fig. 3.2 is fed at the center of the tank below the surface of the liquid. Clear liquid overflows from the top edge of the tank. A slowly revolving rake removes the thickened slurry or sludge and serves to scrape the sludge toward the center of the base for removal. It is common in such operations to add a *flocculating agent* to the mixture to assist the settling process. This agent has the effect of neutralizing electric charges on the particles that cause them to repel each other and remain dispersed. The effect is to form *aggregates* or *flocs* which, because they are larger in size, settle more rapidly. When the prime function of the sedimentation is to remove solids from a liquid rather than to produce a more concentrated solid-liquid mixture, the device is known as a *clarifier.* Clarifiers are often similar in design to thickeners.

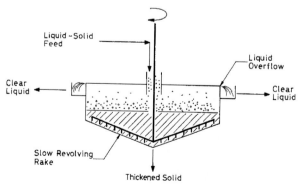

Figure 3.2 A thickener for liquid-solid separation.

Figure 3.3 shows a simple type of *classifier*. In this device, a large tank is subdivided into several sections. A size range of solid particles suspended in vapor or liquid enters the tank. The larger, faster-settling particles settle to the bottom close to the entrance, and the slower-settling particles settle to the bottom close to the exit. The vertical baffles in the tank allow the collection of several fractions.

This type of classification device can be used to carry out solid-solid separation in mixtures of different solids. The mixture of particles is first suspended in a fluid and then separated into fractions of different size or density in a device similar to that in Fig. 3.3.

2. *Flotation. Flotation* is a gravity separation process which exploits differences in the surface properties of particles. Gas bubbles are generated in a liquid and become attached to solid particles or immiscible liquid droplets, causing the particles or droplets to rise to the surface. This is used to separate mixtures of solid-solid particles and liquid-liquid mixtures of finely divided immiscible droplets. It is an important technique in mineral processing, where it is used to separate different types of ore.

When used to separate solid-solid mixtures, the material is ground to a particle size small enough to *liberate* particles of the chemical species to be recovered. The mixture of solid particles is then dispersed in the flotation medium, which is usually water. Gas bubbles become attached to the solid particles, thereby allowing them to float to the surface of the liquid. The solid partices are collected from the surface by an overflow weir or mechanical scraper. The separation of the solid particles depends on the different species having different surface properties such that one species is preferentially attached to the bubbles. A number of chemicals are added to the flotation medium to meet the various requirements of the flotation process:

a. *Modifiers* are added to control the pH of the separation. These could be acids, lime, sodium hydroxide, etc.

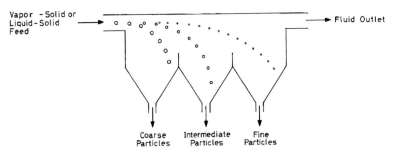

Figure 3.3 Simple gravity settling classifier.

b. *Collectors* are water-repellent reagents which are added to preferentially adsorb onto the surface of one of the solids. Coating or partially coating the surface of one of the solids renders that solid more hydrophobic and increases its tendency to attach to the gas bubbles.

c. *Activators* are used to "activate" the mineral surface for the collector.

d. *Depressants* are used to preferentially attach to one of the solids to make it less hydrophobic and decrease its tendency to attach to the gas bubbles.

e. *Frothers* are surface-active agents added to the flotation medium to create a stable froth and assist the separation.

Flotation is also used in applications such as the separation of oil droplets from oil-water mixtures. In this application it is not necessary to add reagents to make the particles hydrophobic, since the oil is naturally so.

The bubbles of gas can be generated by three methods:

a. Dispersion, in which the bubbles are injected directly by some form of sparging system.

b. Dissolution in the liquid under pressure and then liberation in the flotation cell by reducing the pressure.

c. Electrolysis of the liquid.

3. *Centrifugal separation.* In the preceding processes, the particles were separated from the fluid by gravitational forces acting on the particles. Sometimes gravity separation may be too slow because of the closeness of the densities of the particles and the fluid, because of small particle size leading to low settling velocity, or because of the formation of a stable emulsion.

Centrifugal separators make use of the common principle that an object whirled about an axis at a constant radial distance from the point is acted on by a force. Use of centrifugal forces increases the force acting on the particles. Particles that do not settle readily in gravity settlers often can be separated from fluids by centrifugal force.

The simplest type of centrifugal device is the cyclone separator (Fig. 3.4), which consists of a vertical cylinder with a conical bottom. The centrifugal force is generated by the fluid motion. The mixture enters in a tangential inlet near the top, and the rotating motion so created develops centrifugal force which throws the particles radially toward the wall.

The entering fluid flows downward in a spiral adjacent to the wall. When the fluid reaches the bottom of the cone, it spirals upward in a smaller spiral at the center of the cone and cylinder. The downward

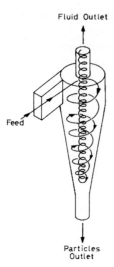

Fluid Outlet

Feed

Particles
Outlet

Figure 3.4 A cyclone generates
centrifugal force by the fluid
motion.

and upward spirals are in the same direction. The particles are
thrown toward the wall and fall downward, leaving at the bottom of
the cone.

Figure 3.5 shows centrifuges in which a cylindrical bowl is rotated
to produce the centrifugal force. In Fig. 3.5*a,* the cylindrical bowl is
shown rotating with a feed consisting of a solid-liquid mixture
admitted at the center. The feed is immediately thrown outward
toward the walls of the container. The particles settle horizontally

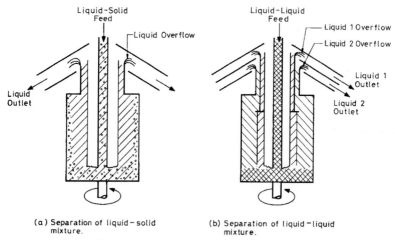

Liquid-Solid
Feed

Liquid Overflow

Liquid
Outlet

(a) Separation of liquid-solid
mixture.

Liquid-Liquid
Feed

Liquid 1 Overflow

Liquid 2 Overflow

Liquid 1
Outlet

Liquid 2
Outlet

(b) Separation of liquid-liquid
mixture.

Figure 3.5 A centrifuge uses a rotating bowl to produce centrifugal force.

outward and press against the vertical bowl wall. Different arrangements are possible to remove the solids from the bowl. In Fig. 3.5*b,* two liquids having different densities are being separated by the centrifuge. The more dense fluid occupies the outer periphery, since the centrifugal force is greater on the more dense fluid.

4. *Filtration.* In filtration, suspended solid particles in a liquid or gas are removed by passing the mixture through a porous medium that retains the particles and passes the fluid. The solid can be retained on the surface of the filter medium, which is *cake filtration,* or captured within the filter medium, which is *depth filtration.* The filter medium can be arranged in many ways.

Figure 3.6 shows four examples of cake filtration in which the filter medium is a *cloth* of natural or artificial fibers or even metal. Figure 3.6*a* shows the filter cloth arranged between plates in an enclosure. Figure 3.6*b* shows the cloth arranged as a thimble. This arrangement is common for the separation of solid particles from vapor and is known as a *bag filter.* Figure 3.6*c* shows a rotating belt for the separation of a slurry of solid particles in a liquid, and Fig. 3.6*d* shows a rotating drum in which the drum rotates through the slurry. When filtering solids from liquids, if the purity of the filter cake is not important, *filter aids,* which are particles of porous solid, can be

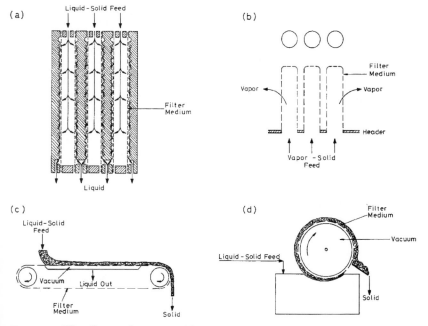

Figure 3.6 Filtration can be arranged in many ways.

added to the mixture to aid the filtration process. When filtering solids from a liquid, a thickener is often used upstream of filtration to concentrate the mixture prior to filtration.

Rather than use a cloth, a *granular* medium consisting of layers of particulate solids on a support grid can be used. Downward flow of the mixture causes the solid particles to be captured within the medium. Such *deep-bed filters* are used to remove small quantities of solids from large quantities of liquids. To release the solid particles captured within the bed, the flow is periodically reversed, causing the bed to expand and release the particles which have been captured. Around 3 percent of the throughput is needed for this backwashing.

In situations where a low concentration of suspended solids needs to be separated from a liquid, then *cross-flow filtration* can be used. The most common design uses a porous tube. The suspension is passed through the tube at high velocity and is concentrated as the liquid flows through the porous medium. The turbulent flow prevents the formation of a filter cake, and the solids are removed as a more concentrated slurry.

3.2 Separation of Homogeneous Fluid Mixtures

As pointed out previously, the separation of homogeneous fluid mixtures requires the creation or addition of another phase. The most common method is by repeated vaporization and condensation—distillation. The three principal advantages of distillation are

1. The ability to handle a wide range of throughput. Many of the alternatives to distillation can only handle low throughput.

2. The ability to handle a wide range of feed concentrations. Many of the alternatives to distillation can only handle relatively pure feeds.

3. The ability to produce high product purity. Many of the alternatives to distillation only carry out a partial separation and cannot produce pure products.

The principal cases which are not suited to distillation include the following:

1. *Separation of low-molecular-weight materials.* Low-molecular-weight materials are distilled at high pressure to increase their condensing temperature and to allow, if possible, the use of cooling water or air cooling in the column condenser. Very low

molecular weight materials require refrigeration in the condenser. This significantly increases the cost of the separation, since refrigeration is very expensive (see App. A). Absorption, adsorption, and membrane gas separators are the most commonly used alternatives to distillation for the separation of low-molecular-weight materials.

2. *Separation of high-molecular-weight heat-sensitive materials.* High-molecular-weight materials are often heat sensitive and as such are usually distilled under vacuum to reduce their boiling temperature.

3. *Separation of components with a low concentration.* Distillation is not well suited to the separation of products which form a low concentration in the feed mixture. Adsorption and absorption are both effective alternative means.

4. *Separation of classes of components.* If a class of components is to be separated (e.g., a mixture of aromatics from a mixture of aliphatics), then distillation can only separate according to boiling points, irrespective of the class of component. In a complex mixture where classes of components need to be separated, this might mean isolating many components unnecessarily. Liquid-liquid extraction can be applied to the separation of classes of components.

5. *Mixtures with low relative volatility or which exhibit azeotropic behavior.* The most common means of dealing with the separation of low-relative-volatility and azeotropic mixtures is to use extractive or azeotropic distillation. These processes are considered in detail later. Crystallization and liquid-liquid extraction also can be used.

6. *Separation of a volatile liquid from an involatile component.* This is a common operation carried out by evaporation and drying. These processes are considered in some detail later.

7. *Separation of mixtures of condensable and non-condensable components.* If a fluid mixture contains both condensable and noncondensable components, then a partial condensation followed by a simple phase separator often can give a good separation. This is essentially a single-stage distillation operation. It is a special case that deserves attention in some detail later.

In summary, distillation is not well suited for separating either low-molecular-weight materials or high-molecular-weight heat-sensitive materials. However, distillation might still be the best method for these cases, since the basic advantages of distillation (potential for high throughput, any feed concentration, and high purity) still prevail.

Even though choices of separators must be made at this stage in the design, it must be borne in mind that the assessment of separation processes ideally should be done in the context of the total system. As is discussed later, separators which use an input of heat to carry out the separation often can be run at effectively zero energy cost if they are appropriately heat integrated with the rest of the process. This includes the three most common types of separators, i.e., distillation columns, evaporators, and dryers. Although they are energy intensive, they also can be energy efficient in terms of the overall process if they are properly heat integrated (see Chaps. 14 and 15).

Despite these problems, a choice of separation system must be made and the design progressed further before it can be properly assessed.

3.3   Distillation

If distillation is the choice of separator, then some preliminary selection of the major design variables must be made to allow the design to proceed. The first decision is operating pressure. As pressure is raised,

- Separation becomes more difficult (relative volatility decreases); i.e., more plates or reflux are required.
- Latent heat of vaporization decreases; i.e., reboiler and condenser duties become lower.
- Vapor density increases, giving a smaller column diameter.
- Reboiler temperature increases with a limit often set by thermal decomposition of the material being vaporized, causing excessive fouling.
- Condenser temperature increases.

As pressure is lowered, these effects obviously reverse. The lower limit is often set by the desire to avoid

- Vacuum operation
- Refrigeration in the condenser

Both vacuum operation and the use of refrigeration incur capital and operating cost penalties and increase the complexity of the design. They should be avoided if possible. For a first pass through the design, it is usually adequate, if process constraints permit, to set distillation pressure to as low a pressure above ambient as allows cooling water or air cooling to be used in the condenser. The pressure should be fixed such that the bubble point of the overhead product is perhaps 10°C above the summer cooling water temperature or to atmospheric pressure if this would mean vacuum operation. Of course, process constraints, especially when distilling high-molecular-weight material, often dictate that vacuum operation must be used in order to reduce the boiling temperature of the material to below a value at which product decomposition occurs.

Another variable that needs to be set for distillation is reflux ratio. For a stand-alone distillation column, there is a capital-energy tradeoff, as illustrated in Fig. 3.7. As the reflux ratio is increased from its minimum, the capital cost decreases initially as the number of plates reduces from infinity, but the utility costs increase as more reboiling and condensation are required (see Fig. 3.7). If the capital

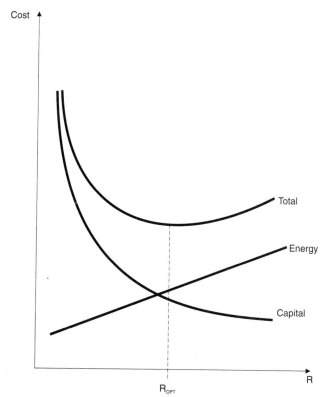

Figure 3.7 The capital-energy tradeoff for stand-alone distillation columns.

costs of the column, reboiler, and condenser are annualized (see App. A) and combined with the annual cost of utilities (see App. A), the optimal reflux ratio is obtained. The optimal ratio of actual to minimum reflux is often less than 1.1. However, most designers are reluctant to design columns closer to minimum reflux than 1.1, except in special circumstances, since a small error in design data or a small change in operating conditions might lead to an infeasible design. No attempt should be made to do the optimization illustrated in Fig. 3.7 at this stage, since, as will be seen later, if the column is heat integrated with the rest of the process, the nature of the tradeoffs changes, and the optimal reflux ratio for the heat-integrated column can be very different from that for a stand-alone column. Any reasonable assumption is adequate at this stage, say, a ratio of actual to minimum reflux of 1.1.

The last design variable which needs to be fixed before the design can proceed, but which is of lesser importance, is feed condition. Subcooled feed (i.e., below bubble point)

- Decreases trays in the rectifying section but increases trays in the stripping section
- Requires more heat in the reboiler but less cooling in the condenser

Partially vaporized feed reverses these effects. For a given separation, the feed conditions can be optimized. No attempt should be made to do this at this stage in the design, since heat integration is likely to change the optimal setting later in the design. It is usually adequate to set the feed to saturated liquid conditions. This tends to equalize the vapor rate below and above the feed.

It must be emphasized that it is not worth expending any effort optimizing pressure, feed condition, or reflux ratio until the overall heat-integration picture has been established. These parameters very often change later in the design.

3.4   Distillation of Mixtures Which Exhibit Azeotropic Behavior or Have Low Relative Volatility

If the light and heavy key components form an azeotrope, then something more sophisticated than simple distillation is required. The first option to consider when separating an azeotrope is exploiting change in azeotropic composition with pressure. If the composition of the azeotrope is sensitive to pressure and it is possible to operate the distillation over a range of pressures without any material decomposition occurring, then this property can be used to carry out a separation. A change in azeotropic composition of at least 5 percent with a change in pressure is usually required.[6]

Figure 3.8a shows the temperature-composition diagram for a minimum-boiling azeotrope that is sensitive to changes in pressure.[6] This azeotrope can be separated using two columns operating at different pressures, as shown in Fig. 3.8b.[6] Feed with mole fraction of A (x_{FA}), of, say, 0.3 is fed to the high-pressure column. The bottom product from this high-pressure column is relatively pure B, whereas the overhead is an azeotrope with $x_{DA} = 0.8$, $x_{DB} = 0.2$. This azeotrope is fed to the low-pressure column, which produces relatively pure A in the bottom and in the overhead an azeotrope with $x_{DA} = 0.6$, $x_{DB} = 0.4$. This azeotrope is added to the feed of the high-pressure column.

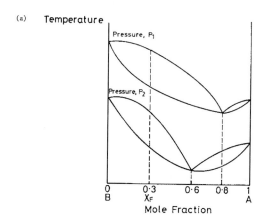

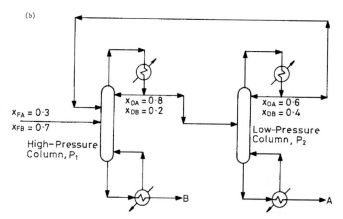

Figure 3.8 Separation of a minimum boiling azeotrope by pressure change. *(From Holland, Gallun, and Lockett, Chemical Engineering, March 23, 1981, 88: 185–200; reproduced by permission.)*

Figure 3.9*a* shows the temperature-composition diagram for a maximum-boiling azeotrope that is sensitive to changes in pressure.[6] Again, this can be separated using two columns operating at different pressures, as shown in Fig. 3.9*b*.[6] Feed with, say, $x_{FA} = 0.8$ is fed to the high-pressure column. This produces relatively pure A in the overheads and an azeotrope with $x_{BA} = 0.2$, $x_{BB} = 0.8$ in the bottoms. This azeotrope is then fed to a low-pressure column, which produces relatively pure B in the overhead and an azeotrope with $x_{BA} = 0.5$, $x_{BB} = 0.5$ in the bottoms. This azeotrope is added to the feed to the high-pressure column.

The problem with using a pressure change is that the smaller the change in azeotropic composition, the larger is the recycle in Figs. 3.8*b* and 3.9*b*. If the azeotrope is not sensitive to changes in pressure, then an extraneous material can be added to the distilla-

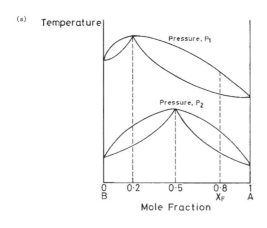

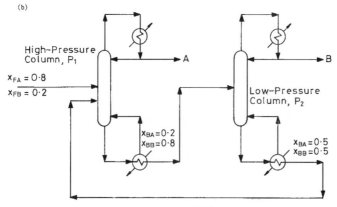

Figure 3.9 Separation of a maximum boiling azeotrope by pressure change. *(From Holland, Gallun, and Lockett, Chemical Engineering, March 23, 1981, 88: 185–200; reproduced by permission.)*

tion to alter in a favorable way the relative volatilty of the key components.

This technique is useful not only when the mixture is impossible to separate by conventional distillation because of an azeotrope but also when the mixture is difficult to separate because of a particularly low relative volatility. Such distillation operations in which an extraneous mass-separating agent is used can be divided into two broad classes.

In the first class, *azeotropic distillation,* the extraneous mass-separating agent is relatively volatile and is known as an *entrainer.* This entrainer forms either a low-boiling binary azeotrope with one of the keys or, more often, a ternary azeotrope containing both keys. The latter kind of operation is feasible only if condensation of the overhead vapor results in two liquid phases, one of which contains the bulk of one of the key components and the other contains the bulk of the entrainer. A typical scheme is shown in Fig. 3.10. The mixture (A + B) is fed to the column, and relatively pure A is taken from the column bottoms. A ternary azeotrope distilled overhead is condensed and separated into two liquid layers in the decanter. One layer contains a mixture of A + entrainer which is returned as reflux. The other layer contains relatively pure B. If the B layer contains a significant amount of entrainer, then this layer may need to be fed to an additional column to separate and recycle the entrainer and produce pure B.

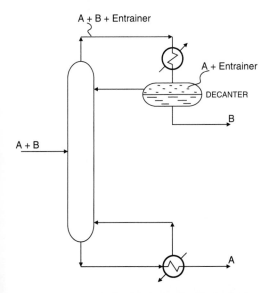

Figure 3.10 A typical azeotropic distillation using an entrainer.

The second class of distillation operation using an extraneous mass-separating agent is *extractive distillation.* Here, the extraneous mass-separating agent is relatively involatile and is known as a *solvent.* This operation is quite different from azeotropic distillation in that the solvent is withdrawn from the column bottoms and does not form an azeotrope with any of the components. A typical extractive distillation process is shown in Fig. 3.11.[7]

As with azeotropic distillation, the separation is possible in extractive distillation because the extraneous mass-separating agent interacts more strongly with one of the components than the other. This in turn alters in a favorable way the relative volatility between the key components.

In principle, extractive distillation is more useful than azeotropic distillation because the process does not depend on the accident of azeotrope formation, and thus a greater choice of mass-separating agent is, in principle, possible. In general, the solvent should have a chemical structure similar to that of the less volatile of the two components. It will then tend to form a near-ideal mixture with the less volatile component and a nonideal mixture with the more volatile component. This has the effect of increasing the volatility of the more volatile component.

The solvent flow rate to the distillation usually can be varied

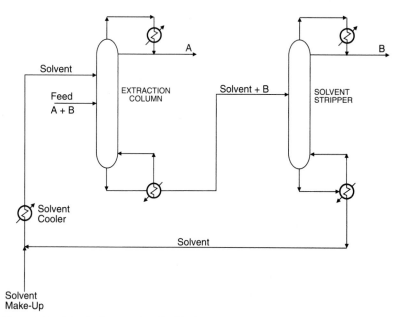

Figure 3.11 A typical extractive distillation.

within limits. A greater solvent flow rate generally yields a better separation. However, the solvent must be reboiled in both extraction and stripping columns, leading to a greater energy demand at greater flow rates. Also, higher solvent flow rates increase distillation temperatures. The solvent flow rate can be optimized on a stand-alone basis. However, no attempt should be made to do this until the overall process context has been established, since heat integration is likely to have a significant effect on the optimization.

When separating azeotropic mixtures, if possible, changes in the azeotropic composition with pressure should be exploited rather than using an extraneous mass-separating agent. When using an extraneous mass-separating agent, there are inevitably losses from the process. Even if these losses are not significant in terms of the cost of the material, they create environmental problems somewhere later in the design. As discussed in detail in Chap. 10, the best way to solve effluent problems is to deal with them at the source. The best way to solve the effluent problems caused by loss of the extraneous mass-separating agent is to eliminate it from the design. However, clearly in many instances practical difficulties and excessive cost might force its use. Occasionally, a component that already exists in the process can be used as the entrainer or solvent, thus avoiding the introduction of extraneous materials for azeotropic and extractive distillation.

3.5 Absorption

The most common alternative to distillation for the separation of low-molecular-weight materials is absorption. In absorption, a gas mixture is contacted with a liquid solvent which preferentially dissolves one or more components of the gas. Absorption processes often require an extraneous material to be introduced into the process to act as liquid solvent. If it is possible to use the materials already in the process, this should be done in preference to introducing an extraneous material for reasons already discussed. Liquid flow rate, temperature, and pressure are important variables to be set.

1. *Liquid flow rate.* The absorption factor for component i, $L/K_i V$, determines how readily component i will absorb into the liquid phase, where L and V are the liquid and vapor flow rates and K_i is the vapor-liquid equilibrium K value for component i. When this factor is large, component i will be absorbed more readily into the liquid phase. The absorption factor must be greater than 1 for a high degree of solute removal; otherwise, removal will be limited to a low value by liquid flow rate. Since $L/K_i V$ is increased by increasing the liquid flow rate, the number of plates required to achieve a given separation decreases. However, at high values of $L/K_i V$, the increase in liquid

flow rate brings diminishing returns. This leads to an economic optimum in the range $1.2 < L/K_iV < 2.0$. An absorption factor around 1.4 is often used.[8]

2. *Temperature.* Decreasing temperature increases the solubility of the solute. In an absorber, the transfer of solute from gas to liquid brings about a heating effect. This usually will lead to temperature increasing down the column. If the component being separated is dilute, the heat of absorption will be small, and the temperature rise down the column also will be small. Otherwise, the temperature rise down the column will be large, which is undesirable, since solubility decreases with increasing temperature. To counteract the temperature rise in absorbers, the liquid is sometimes cooled at intermediate points as it passes down the column. The cooling is usually to temperatures which can be achieved with cooling water, except in special circumstances where refrigeration is used.

If the solvent is volatile, there will be some loss with the vapor. This should be avoided if the solvent is expensive and/or environmentally harmful by using a condenser (refrigerated if necessary) on the vapor leaving the absorber.

3. *Pressure.* High pressure gives greater solubility of solute in the liquid. However, high pressure tends to be expensive to create, since this can require a gas compressor. Thus there is an optimal pressure.

As with distillation, no attempt should be made to carry out any optimization of liquid flow rate, temperature, or pressure at this stage in the design. The separation in absorption is sometimes enhanced by adding a component to the liquid which reacts with the solute.

Having dissolved the solute in the liquid, it is often necessary to then separate the solute from the liquid in a stripping operation so as to recycle to the absorber. Now the stripping factor for component i, K_iV/L, should be large to concentrate it in the vapor phase and thus be stripped out of the liquid phase. For a stripping column, the stripping factor should be in the range $1.2 < K_iV/L < 2.0$ and is often around 1.4.[8] As with absorbers, there can be a significant change in temperature through the column. This time, however, the liquid decreases in temperature down the column, for reasons analogous to those developed for absorbers. Increasing temperature and decreasing pressure will enhance the stripping.

3.6   Evaporators

Single-stage evaporators tend only to be used when the capacity needed is small. It is more usual to employ multistage systems which recover and reuse the latent heat of the vaporized material. Three different arrangements for a three-stage evaporator are illustrated in Fig. 3.12.

1. *Forward-feed operation* is shown in Fig. 3.12a. The fresh feed is added to the first stage and flows to the next stage in the same direction as the vapor flow. The boiling temperature decreases from stage to stage, and this arrangement is thus used when the

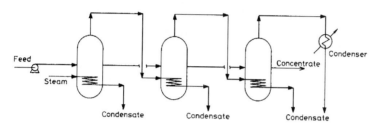

(a) Forward feed operation

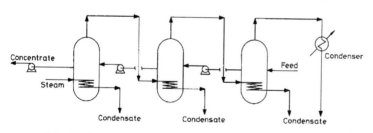

(b) Backward feed operation

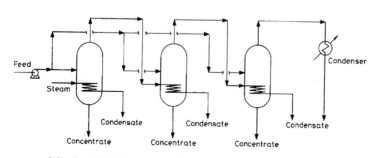

(c) Parallel feed operation

Figure 3.12 Three possible arrangements for a three-stage evaporator.

concentrated product is subjected to decomposition at higher temperatures. It also has the advantage that it is possible to design the system without pumps to transfer the solutions from one stage to the next.

2. *Backward-feed operation* is shown in Fig. 3.12*b*. Here, the fresh feed enters the last and coldest stage and leaves the first stage as concentrated product. This method is used when the concentrated product is highly viscous. The high temperatures in the early stages reduce viscosity and give higher heat transfer coefficients. Because the solutions flow against the pressure gradient between stages, pumps must be used to transfer solutions between stages.

3. *Parallel-feed operation* is illustrated in Fig. 3.12*c*. Fresh feed is added to each stage, and product is withdrawn from each stage. The vapor from each stage is still used to heat the next stage. This arrangement is used mainly when the feed is almost saturated, particularly when solid crystals are the product.

Many other *mixed-feed arrangements* are possible which combine the individual advantages of each type of arrangement. Figure 3.13 shows a three-stage evaporator in temperature-enthalpy terms, assuming that inlet and outlet solutions are at saturated conditions

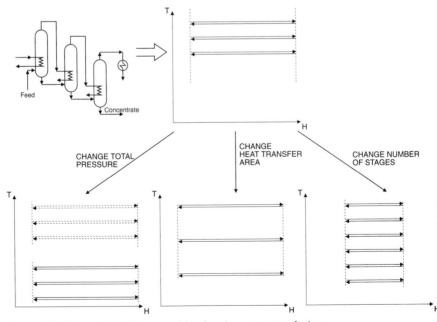

Figure 3.13 The principal degrees of freedom in evaporator design.

and that all evaporation and condensation duties are at constant temperature.

The three principal degrees of freedom in the design of stand-alone evaporators are

1. Temperature levels can be changed by manipulating the operating pressure. Figure 3.13a shows the effect of a decrease in pressure.

2. The temperature difference between stages can be manipulated by changing the heat transfer area. Figure 3.13b shows the effect of a decrease in heat transfer area.

3. The heat flow through the system can be manipulated by changing the number of stages. Figure 3.13c shows the effect of an increase from three to six stages.

Given these degrees of freedom, how can an initialization be made for the design? The most significant degree of freedom is the choice of number of stages. If the evaporator is operated using hot and cold utility, as the number of stages is increased, a tradeoff might be expected, as shown in Fig. 3.14. Here, starting with a single stage, it has a low capital cost but requires a large energy cost. Increasing the stages to two decreases the energy cost in return for a small increase in capital cost, and the total cost decreases. However, as the stages are increased, the increase in capital cost at some point no longer compensates for the corresponding decrease in energy cost, and the total cost increases. Hence there is an optimal number of stages. However, no attempt should be made to carry out this optimization at this point, since the design is almost certain to change significantly when heat integration is considered later (see Chap. 15).

All that can be done is to make a reasonable initial assessment of the number of stages. Having made a decision for the number of stages, the heat flow through the system is temporarily fixed so that the design can proceed. Generally, the maximum temperature in evaporators is set by product decomposition and fouling. Therefore, the highest-pressure stage is operated at a pressure low enough to be below this maximum temperature. The pressure of the lowest-pressure stage is normally chosen to allow heat rejection to cooling water or air cooling. If decomposition and fouling are not a problem, then the stage pressures should be chosen such that the highest-pressure stage is below steam temperature and the lowest-pressure stage above cooling water or air cooling temperature.

For a given number of stages, if

- all heat transfer coefficients are equal,

- all evaporation and condensation duties are at constant temperature,
- boiling point rise of the evaporating mixture is negligible,
- latent heat is constant through the system,

then minimum capital cost is given when all temperature differences are equal.[9] If evaporator pressure is not limited by the steam temperature but by product decomposition and fouling, then the ΔTs should be spread out equally between the upper practical temperature limit and the cold utility. This is usually a good enough initialization for most purposes given that the design might change drastically later when heat integration is considered.

Another factor that can be important in the design of evaporators is the condition of the feed. If the feed is cold, then the backward-feed arrangement has the advantage that a smaller amount of liquid must be heated to the higher temperatures of the second and first stages.

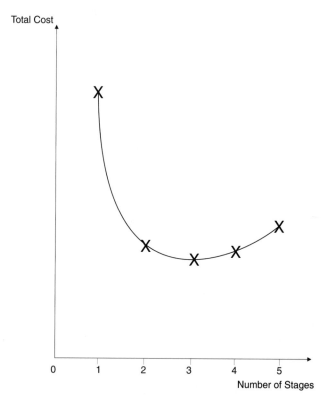

Figure 3.14 Variation of total cost with number of stages indicates that three stages is the optimal number for a stand-alone system in this case.

However, factors such as this should not be allowed to dictate design options at the early stages of flowsheet design because preheating the cold feed by heat integration with the rest of the process might be possible.

If the evaporator design is considered against a background process and heat integration with the background process is possible, then very different designs can emerge. When making an initial choice of separator, a simple, low-capital-cost design of evaporator should be chosen.

3.7 Dryers

Drying refers to the removal of water from a substance through a whole range of processes, including distillation, evaporation, and even physical separations such as with centrifuges. Here, consideration is restricted to the removal of moisture from solids and liquids into a gas stream (usually air) by heat, namely, *thermal drying.* Some of the types of equipment for removal of water also can be used for removal of organic liquids from solids.

Four of the more common types of thermal dryers used in the process industries are illustrated in Fig. 3.15.

1. *Tunnel dryers* are shown in Fig. 3.15*a*. Wet material on trays or a conveyor belt is passed through a tunnel, and drying takes place by hot air. The airflow can be countercurrent, cocurrent, or a mixture of both. This method is usually used when the product is not free flowing.

2. *Rotary dryers* are shown in Fig. 3.15*b*. Here, a cylindrical shell mounted at a small angle to the horizontal is rotated at low speed. Wet material is fed at the higher end and flows under gravity. Drying takes place from a flow of air, which can be countercurrent or cocurrent. The heating may be direct to the dryer gas or indirect through the dryer shell. This method is usually used when the material is free flowing. Rotary dryers are not well suited to materials which are particularly heat sensitive because of the long residence time in the dryer.

3. *Drum dryers* are shown in Fig. 3.15*c*. This consists of a heated metal roll. As the roll rotates, a layer of liquid or slurry is dried. The final dry solid is scraped off the roll. The product comes off in flaked form. Drum dryers are suitable for handling slurries or pastes of solids in fine suspension and are limited to low and moderate throughput.

4. *Spray dryers* are shown in Fig. 3.15*d*. Here, a liquid or slurry

product is exposed to the hot gas for a short period. Also, the solution is sprayed as fine droplets into a hot gas stream. The feed to the dryer must be pumpable to obtain the high pressures required by the atomizer. The product tends to be light, porous particles. An important advantage of the spray dryer is that the

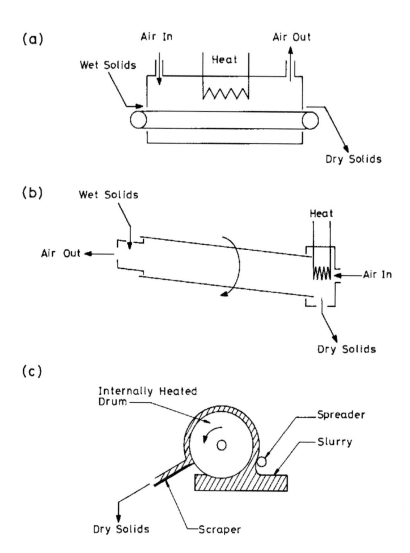

Figure 3.15 Four of the more common types of thermal dryers.

(d)

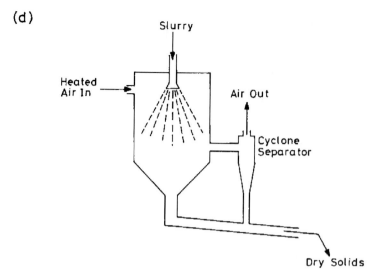

Figure 3.15 *(Continued)*

evaporation of the liquid from the spray keeps the product temperature low even in the presence of hot gases. Spray dryers are thus particularly suited to products which are sensitive to thermal decomposition.

Another important class of dryer is the *fluidized-bed dryers*. Some designs combine spray and fluidized-bed dryers. Choice between dryers is usually based on practicalities such as the materials' handling characteristics, product decomposition, product physical form (e.g., if a porous granular material is required), etc. Also, dryer efficiency can be used to compare the performance of different dryer designs. This is usually defined as follows:

$$\text{Dryer efficiency} = \frac{\text{heat of vaporization}}{\text{total heat consumed}} \tag{3.1}$$

If the total heat consumed is from an external utility (e.g., mains steam), then a high efficiency is desirable, even perhaps at the expense of a high capital cost. However, if the heat consumed is by recovery from elsewhere in the process, as is discussed in Chap. 15, then comparison on the basis of dryer efficiency becomes less meaningful.

3.8 Choice of Separator—Summary

For a heterogeneous or multiphase mixture, separation usually can be achieved by phase separation. Such phase separation should be carried out before any homogeneous separation. Phase separation tends to be easier and should be done first.

Distillation is by far the most commonly used method for the separation of homogeneous fluid mixtures. The cost of distillation varies with operating pressure, which, in turn, is mainly determined by the molecular weight of the materials being separated. Its widespread use can be attributed to its ability to

- Handle a wide range of throughput

- Handle a wide range of feed concentrations

- Produce high-purity products

No attempt should be made to optimize pressure, reflux ratio, or feed condition of distillation in the early stages of design. The optimal values almost certainly will change later once heat integration with the overall process is considered.

If an azeotropic mixture is to be separated by distillation, then use of pressure change to alter the azeotropic composition should be considered before use of an extraneous mass-separating agent. Avoiding the use of extraneous materials often can prevent environmental problems later in the design.

The most common alternative to distillation for the separation of low-molecular-weight materials is absorption. Liquid flow rate, temperature, and pressure are important variables to be set, but no attempts should be made to carry out any optimization at this stage.

As with distillation and absorption, when evaporators and dryers are chosen, no attempt should be made to carry out any optimization at this stage in the design.

3.9 References

1. King, C. J., *Separation Processes,* 2d ed., McGraw-Hill, New York, 1980.
2. Rousseau, R. W., *Handbook of Separation Process Technology,* Wiley, New York, 1987.
3. Foust, A. S., Wenzel, L. A., Clump, C. W., et al., *Principles of Unit Operations,* Wiley, New York, 1980.
4. Svarovsky, L., *Solid-Gas Separation,* Elsevier Scientific, New York, 1981.
5. Walas, S. M., *Chemical Process Equipment Selection and Design,* Butterworth, Reading, Mass., 1988.

6. Holland, C. D., Gallun, S. E. and Lockett, M. J., "Modeling Azeotropic and Extractive Distillations," *Chem. Engg.,* 88: 185, *March* 23, 1981.

7. Sucksmith, I., "Extractive Distillation Saves Energy," *Chem. Engg.,* 88: 185, June 28, 91, 1982.

8. Douglas, J. M., *Conceptual Design of Chemical Processes,* McGraw-Hill, New York, 1988.

9. Smith, R., and Jones, P. S., "The Optimal Design of Integrated Evaporation Systems," *Heat Recovery Systems and CHP,* 10: 341, 1990.

4

Synthesis of Reaction-Separation Systems

4.1 The Function of Process Recycles

The recycling of material is an essential feature of most chemical processes. Therefore, it is necessary to consider the main factors which dictate the recycle structure of a process. We shall start by considering the function of process recycles and restrict consideration to continuous processes. Later the scope will be extended to include batch processes.

1. *Reactor conversion.* In Chap. 2 an initial choice was made of reactor type, operating conditions, and conversion. Only in extreme cases would the reactor be operated close to complete conversion. The initial setting for the conversion varies according to whether there are single reactions or multiple reactions producing byproducts and whether reactions are reversible.

Consider the simple reaction

$$FEED \longrightarrow PRODUCT \tag{4.1}$$

Achieving complete conversion of FEED to PRODUCT in the reactor usually requires an extremely long residence time, which is normally uneconomic (at least in continuous processes). Thus, if there is no byproduct formation, the initial reactor conversion is set to be around 95 percent, as discussed in Chap. 2. The reactor effluent thus contains unreacted FEED and PRODUCT (Fig. 4.1a).

Because we require a pure product, a separator is needed. The unreacted FEED is usually too valuable to be disposed of and is therefore recycled to the reactor inlet via a pump or compressor (see Fig. 4.1*b*). In addition, disposal of unreacted FEED rather than recycling creates an environmental problem.

2. *Byproduct formation.* Consider now the case where a byproduct is formed either by the primary reaction, such as

$$\text{FEED} \longrightarrow \text{PRODUCT} + \text{BYPRODUCT} \qquad (4.2)$$

or via a secondary reaction, such as

$$\text{FEED} \longrightarrow \text{PRODUCT}$$
$$\text{PRODUCT} \longrightarrow \text{BYPRODUCT} \qquad (4.3)$$

An additional separator is now required (Fig. 4.2*a*). Again, the unreacted FEED is normally recycled, but the BYPRODUCT must be removed to maintain the overall material balance. An additional complication now arises with two separators because the *separation sequence* can be changed[1] (see Fig. 4.2*b*). We shall consider separation sequencing in detail in the next chapter.

Also, instead of using two separators, a purge can be used (see Fig. 4.2*c*). Using a purge saves the cost of a separator but incurs raw materials losses and possibly waste treatment and disposal costs.

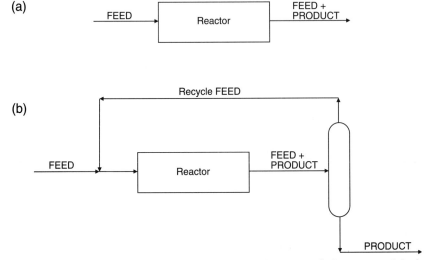

Figure 4.1 Incomplete conversion in the reactor requires a recycle for unreacted feed material.

This might be worthwhile if the FEED-BYPRODUCT separation is expensive. To use a purge, the FEED and BYPRODUCT must be adjacent to each other in order of volatility (assuming distillation is used as the means of separation). Of course, care should be taken to ensure that the resulting increase in concentration of BYPRODUCT in the reactor does not have an adverse effect on reactor performance. Too much BYPRODUCT might, for example, cause a deterioration in the performance of the catalyst.

Clearly, the separation configurations shown in Fig. 4.2 change between different processes as the order of volatility between the components changes.[2]

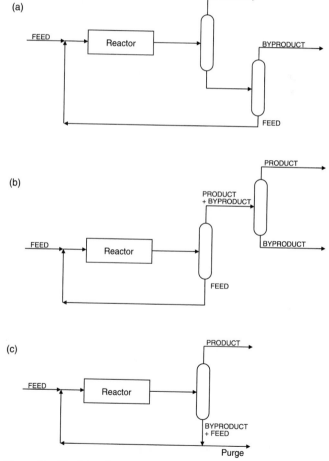

Figure 4.2 If a byproduct is formed in the reactor, then different recycle structures are possible.

3. *Recycling byproducts for improved selectivity.* In systems of multiple reactions, byproducts are sometimes formed in secondary reactions which are reversible, such as

$$\text{FEED} \longrightarrow \text{PRODUCT}$$
$$\text{FEED} \rightleftharpoons \text{BYPRODUCT}$$

(4.4)

The three recycle structures shown in Fig. 4.2 also can be used with this case. Because the BYPRODUCT is now being formed by a secondary reaction which is reversible, its formation can be inhibited by recycling BYPRODUCT as shown in Fig. 4.3a. In Fig. 4.3a, the BYPRODUCT formation is inhibited to the extent that it is effectively stopped. In Fig. 4.3b it is only reduced and the net BYPRODUCT formation removed. Again, the separation configuration will change between different processes as the order of volatility between the components changes.

4. *Recycling byproducts or contaminants which damage the reactor.* When recycling unconverted feed material, it is possible that

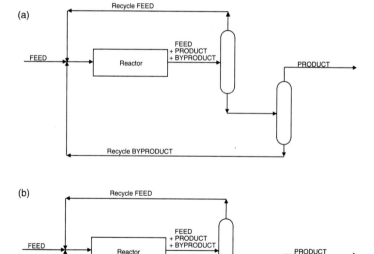

Figure 4.3 If a byproduct is formed via a reversible secondary reaction, then recycling the byproduct can inhibit its formation at the source.

some byproducts or contaminants such as products of corrosion can poison the catalyst in the reactor. Even trace quantities can sometimes be very damaging to the catalyst. It is clearly desirable to remove such damaging components from the recycle in arrangements similar to those shown in Fig. 4.2a and b.

5. *Feed impurities.* So far only cases in which the feed is pure have been considered. An impurity in the feed opens up further options for recycle structures. The first option in Fig. 4.4a shows the impurity

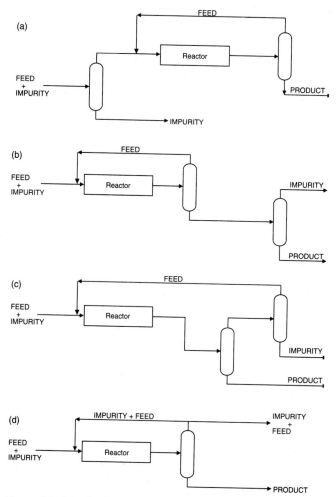

Figure 4.4 Introduction of an impurity with the feed creates further options for recycle structures. (*From Smith and Linnhoff, Trans. IChemE, ChERD, 66: 195, 1988; reproduced by permission of the Institution of Chemical Engineers.*)

being separated before entering the reactor. If the impurity has an adverse effect on the reaction or poisons the catalyst, this is the obvious solution. However, if the impurity does not have a significant effect on the reaction, then it could perhaps be passed through the reactor and be separated as shown in Fig. 4.4b. Alternatively, the separation sequence is changed as shown in Fig. 4.4c.[1]

The fourth option, shown in Fig. 4.4d, uses a purge.[1] As with its use to separate byproducts, the purge saves the cost of a separation but incurs raw materials losses. This might be worthwhile if the FEED-IMPURITY separation is expensive. To use a purge, the FEED and IMPURITY must be adjacent to each other in order of volatility (again, assuming distillation as the means of separation). Care should be taken to ensure that the resulting increase in concentration of IMPURITY in the reactor does not have an adverse effect on reactor performance.

The separation configuration again changes between different processes as the order of volatility changes.

6. *Reactor diluents and solvents.* As pointed out in Sec. 2.5, an inert diluent such as steam is sometimes needed in the reactor to lower the partial pressure of reactants in the vapor phase. Diluents are normally recycled. An example is shown in Fig. 4.5. The actual configuration used depends on the order of volatilities.

Some reactions are carried out in the liquid phase in a solvent. If this is the case, then the solvent is separated and recycled in arrangements similar to that shown in Fig. 4.5.

7. *Reactor heat carrier.* Also as pointed out in Sec. 2.6, if adiabatic operation is not possible and it is not possible to control temperature by direct heat transfer, then an "inert" material can be introduced to the reactor to increase its heat capacity flow rate (i.e., product of mass flow rate and specific heat capacity) and to reduce

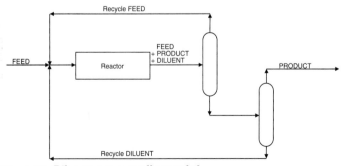

Figure 4.5 Diluents are normally recycled.

temperature rise for exothermic reactions or reduce temperature decrease for endothermic reactions.

The introduction of an extraneous component as a heat carrier affects the recycle structure of the flowsheet. Figure 4.6a presents an example of the recycle structure for just such a process.

Where possible, introducing extraneous materials into the process should be avoided, and a material already present in the process should be used. Figure 4.6b illustrates use of the product as the heat carrier. This simplifies the recycle structure of the flowsheet and removes the need for one of the separators (see Fig. 4.6b). Use of the product as a heat carrier is obviously restricted to situations where the product does not undergo secondary reactions to unwanted byproducts. Note that the unconverted feed which is recycled also acts as a heat carrier itself. Thus, rather than relying on recycled product to limit the temperature rise (or fall), simply opt for a low conversion, a high recycle of feed, and a resulting small temperature change.

Whether an extraneous component, product, or feed is used as a heat carrier, the actual configuration, as before, depends on the order of the volatilities (again assuming distillation as the means of separation).

Having considered the main factors which determine the need for

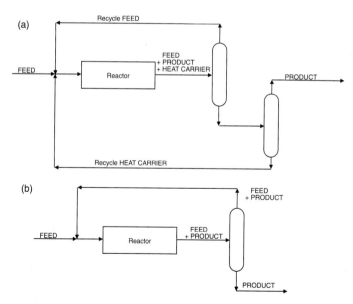

Figure 4.6 Heat carriers are normally recycled.

recycles, care should be taken if a flowsheet requires multiple recycles. It is clearly counterproductive to separate two components adjacent in volatility which are to be recycled, since they would only be remixed at some point before entering the reactor. The designer should always be on guard to avoid unnecessary separation and unnecessary mixing.

Example 4.1 Monochlorodecane (MCD) is to be produced from decane (DEC) and chlorine via the reaction

$$\underset{\text{DEC}}{C_{10}H_{22}} + \underset{\text{chlorine}}{Cl_2} \longrightarrow \underset{\text{MCD}}{C_{10}H_{21}Cl} + \underset{\text{hydrogen chloride}}{HCl}$$

A side reaction occurs in which dichlorodecane (DCD) is produced:

$$\underset{\text{MCD}}{C_{10}H_{21}Cl} + \underset{\text{chlorine}}{Cl_2} \longrightarrow \underset{\text{DCD}}{C_{10}H_{20}Cl_2} + \underset{\text{hydrogen chloride}}{HCl}$$

The byproduct, DCD, is not required for this project. Hydrogen chloride can be sold to a neighboring plant. Assume at this stage that all separations can be carried out by distillation. The normal boiling points are given in Table 4.1.

1. Determine alternative recycle structures for the process by assuming different levels of conversion of raw materials and different excesses of reactants.
2. Which structure is most effective in suppressing the side reaction?
3. What is the minimum selectivity of decane which must be achieved for profitable operation? The values of the materials involved together with their molecular weights are given in Table 4.1.

Solution

1. Four possible arrangements can be considered:
a. *Complete conversion of both feeds.* Figure 4.7a shows the most desirable arrangement: complete conversion of the decane and chlorine in the reactor. The absence of reactants in the reactor effluent means that no recycles are needed.

Although the flowsheet shown in Fig. 4.7a is very attractive, it is not practical. This would require careful control of the stoichiometric ratio of decane to chlorine, taking into account both the requirements of the primary and byproduct reactions. Even if it was possible to balance out the

TABLE 4.1 Data for Process Materials

Material	Molecular weight	Normal boiling point (K)	Value ($ kg^{-1})
Hydrogen chloride	36	188	0.35
Chlorine	71	239	0.21
Decane	142	447	0.27
Monochlorodecane	176	488	0.45
Dichlorodecane	211	514	0

reactants exactly, a small upset in process conditions would create an excess of either decane or chlorine, and these would then appear as components in the reactor effluent. If these components appear in the reactor effluent of the

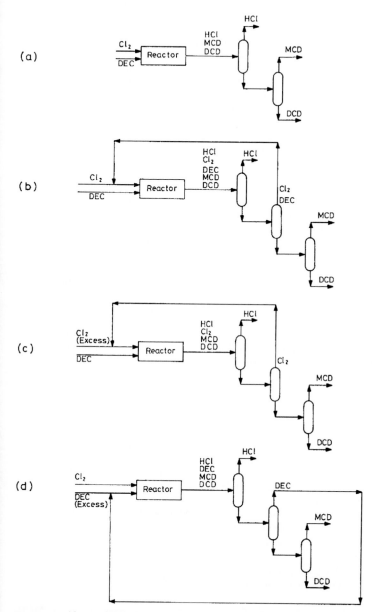

Figure 4.7 Alternative separation and recycle structures for the production of monochlorodecane.

flowsheet in Fig. 4.7a, there are no separators to deal with their presence and no means of recycling unconverted raw materials.

Also, although there are no selectivity data for the reaction, the selectivity losses would be expected to increase with increasing conversion. Complete conversion would tend to produce unacceptable selectivity losses. Finally, the reactor volume required to give a complete conversion would be extremely large.

b. *Incomplete conversion of both feeds.* If complete conversion is not practical, let us consider incomplete conversion. This is shown in Fig. 4.7b. However, in this case, all components are present in the reactor effluent, and one additional separator and a recycle are required. Thus the complexity is somewhat increased compared with complete conversion.

Note that no attempt has been made to separate the chlorine and decane, since they are remixed after recycling to the reactor.

c. *Excess chlorine.* Use of excess chlorine in the reactor can force the decane to effectively complete conversion (see Fig. 4.7c). Now there is effectively no decane in the reactor effluent, and again, three separators and a recycle are required.

In practice, there is likely to be a trace of decane in the reactor effluent. However, this should not be a problem, since it can either be recycled with the unreacted chlorine or leave with the product, monochlorodecane (providing it can still meet product specifications).

At this stage, how great the excess of chlorine should be for Fig. 4.7c to be feasible cannot be specified. Experimental work on the reaction chemistry would be required in order to establish this. However, the size of the excess does not change the basic structure.

d. *Excess decane.* Use of excess decane in the reactor forces the chlorine to effectively complete conversion (see Fig. 4.7d). Now there is effectively no chlorine in the reactor effluent. Again, three separators and a recycle of unconverted raw material are required.

Again, in practice, there is likely to be a trace of chlorine in the reactor effluent. This can be recycled to the reactor with the unreacted decane or allowed to leave with the hydrogen chloride byproduct (providing this meets with the byproduct specification).

It cannot be said at this stage exactly how great an excess of decane would be required in order to make Fig. 4.7d feasible. This would have to be established experimentally, but the size of the excess does not change the basic structure.

2. An arrangement is to be chosen to inhibit the side reaction, i.e., give low selectivity losses. The side reaction is suppressed by starving the reactor of either monochlorodecane or chlorine. Since the reactor is designed to produce monochlorodecane, the former option is not practical. However, it is practical to use an excess of decane.

The last of the four flowsheet options generated above, which features excess decane in the reactor, is therefore preferred (see Fig. 4.7d).

3. The selectivity S is defined by

$$S = \frac{(\text{MCD produced in the reactor})}{(\text{DEC consumed in the reactor})} \times \text{stoichiometric factor}$$

In this case, the stoichiometric factor is 1. This is a measure of the MCD obtained from the DEC consumed. To assess the selectivity losses, the MCD produced in the primary reaction is split into that fraction which will become final product and that which will become the byproduct. Thus the reaction stoichiometry is

$$C_{10}H_{22} + Cl_2 \longrightarrow SC_{10}H_{21}Cl + (1-S)C_{10}H_{21}Cl + HCl$$

and for the byproduct reaction is

$$(1-S)C_{10}H_{21}Cl + (1-S)Cl_2 \longrightarrow (1-S)C_{10}H_{20}Cl_2 + (1-S)HCl$$

Adding the two reactions gives overall

$$C_{10}H_{22} + (2-S)Cl_2 \longrightarrow SC_{10}H_{21}Cl + (1-S)C_{10}H_{20}Cl_2 + (2-S)HCl$$

Considering raw materials costs only, the economic potential (EP) of the process is defined as

$$EP = \text{value of products} - \text{raw materials cost}$$

$$= [176 \times S \times 0.45 + 36 \times (2-S) \times 0.35]$$

$$- [142 \times 1 \times 0.27 + 71 \times (2-S) \times 0.21]$$

$$= 79.2S - 2.31(2-S) - 38.34 \ (\$ \, kmol^{-1} \ \text{decane reacted})$$

The minimum selectivity which can be tolerated is given when the economic potential is just zero:

$$0 = 79.2S - 2.31(2-S) - 38.34$$

$$S = 0.53$$

In other words, the process must convert at least 53 percent of the decane which reacts to monochlorodecane rather than to dichlorodecane for the process to be economic. This figure assumes selling the hydrogen chloride to a neighboring process. If this is not the case, there is no value associated with the hydrogen chloride. Assuming that there are no treatment and disposal costs for the now waste hydrogen chloride, the minimum economic potential is given by

$$0 = (176 \times S \times 0.45) - [142 \times 1 \times 0.27 + 71 \times (2-S) \times 0.21]$$

$$= 79.2S - 14.91(2-S) - 38.34$$

$$S = 0.72$$

Now the process must convert at least 72 percent of the decane to monochlorodecane.

If the hydrogen chloride cannot be sold, it must be disposed of somehow. Alternatively, it could be converted back to chlorine via the reaction

$$2HCl + \tfrac{1}{2}O_2 \rightleftharpoons Cl_2 + H_2O$$

and then recycled to the MCD reactor. Now the overall stoichiometry changes, since the $(2 - S)$ moles of HCl which were being produced as byproduct are now being recycled to substitute fresh chlorine feed:

$$(2 - S)HCl + \tfrac{1}{4}(2 - S)O_2 \longrightarrow \tfrac{1}{2}(2 - S)Cl_2 + \tfrac{1}{2}(2 - S)H_2O$$

Thus the overall reaction now becomes

$$C_{10}H_{22} + \tfrac{1}{2}(2 - S)Cl_2 + \tfrac{1}{4}(2 - S)O_2 \longrightarrow$$

$$SC_{10}H_{21}Cl + (1 - S)C_{10}H_{20}Cl_2 + \tfrac{1}{2}(2 - S)H_2O$$

The economic potential is now given by

$$0 = (176 \times S \times 0.45) - [142 \times 1 \times 0.27 + 71 \times \tfrac{1}{2}(2 - S) \times 0.21]$$

$$= 79.2S - 7.455(2 - S) - 38.34$$

$$S = 0.61$$

The minimum selectivity which can now be tolerated becomes 61 percent.

4.2   Vapor Recycles and Purges

When a mixture contains components with a broad range of volatilities, either a partial condensation from the vapor phase or a partial vaporization from the liquid phase followed by a simple phase split often can produce an effective separation. This is in essence a single-stage distillation process. However, by its very nature, a single-stage separation does not produce pure products; hence further separation of both liquid and vapor streams is often required.

Consider the single-stage phase split shown in Fig. 3.1a. Overall material balances and component material balances can be written as

$$F = V + L \tag{4.5}$$

$$Fz_i = Vy_i + Lx_i \tag{4.6}$$

where F = feed flow rate
V = vapor flow rate from the separator
L = liquid flow rate from the separator
z_i = mole fraction of component i in the feed
y_i = mole fraction of component i in the vapor
x_i = mole fraction of component i in the liquid

The vapor-liquid equilibrium relationship can be defined in terms of K values by

$$y_i = K_i x_i \tag{4.7}$$

Equations (4.5) to (4.7) can now be solved to give expressions for the vapor- and liquid-phase compositions leaving the separator:

$$y_i = \frac{z_i}{\dfrac{V}{F} + \left(1 - \dfrac{V}{F}\right)\dfrac{1}{K_i}} \tag{4.8}$$

$$x_i = \frac{z_i}{(K_i - 1)\dfrac{V}{F} + 1} \tag{4.9}$$

The vapor fraction V/F in Eqs. (4.8) and (4.9) lies in the range $0 \le V/F \le 1$.

If K_i is very large (Douglas[4] has suggested a value greater than 10) in Eq. (4.8), then

$$y_i \approx z_i/(V/F) \tag{4.10}$$

that is, $\qquad\qquad Vy_i \approx Fz_i$

This means that all of component i entering with the feed Fz_i leaves in the vapor phase as Vy_i. Thus, if a component is required to leave in the vapor phase, its K value should be large (typically greater than 10).

On the other hand, if K_i is very small (Douglas[4] has suggested a value less than 0.1) in Eq. (4.9), then

$$x_i \approx Fz_i/L \tag{4.11}$$

that is, $\qquad\qquad Lx_i \approx Fz_i$

This means that all of component i entering with the feed Fz_i leaves with the liquid phase as Lx_i. Thus, if a component is required to leave in the liquid phase, its K value should be small (typically less than 0.1).

Ideally, the K value for the light key component in the phase separation should be greater than 10, and at the same time, the K value for the heavy key should be less than 0.1. Having such circumstances leads to a good separation in a single stage. However, use of phase separators might still be effective in the flowsheet if the K values for the key components are not so extreme. Under such circumstances a more crude separation must be accepted.

Many processes, particularly in the petrochemical industries, produce a reactor effluent which consists of a mixture of low-boiling components such as hydrogen and methane together with much less volatile organic components. In such circumstances, partial condensation followed by a simple phase split produces a good separation. Cooling below cooling water temperature is not desirable; otherwise, refrigeration is required. A simple dew-point calculation at the system pressure reveals whether partial condensation above cooling water temperatures is possible. If partial condensation does not occur, even down to cooling water temperature, increasing the reactor pressure or using refrigeration (or both) can be considered to accomplish a phase split. An increase in reactor pressure is sometimes readily achieved if there is a liquid stream upstream of the reactor. This allows the pressure increase to be achieved using a pump.

Phase separation in this way is most effective if the light key component is significantly above its critical temperature. If a component is above its critical temperature, it does not truly condense. Some, however, "dissolves" in the liquid phase. This means that it is bound to have an extremely high K value.

If a vapor from the phase split is either predominantly product or predominantly byproduct, then it is removed from the process. If the vapor contains predominantly unconverted feed material, it is normally recycled to the reactor. In these cases, there is no need to carry out any separation on the vapor.

If the vapor stream consists of a mixture of unconverted feed material, products, and byproducts, then some separation of the vapor may be needed. The vapor from the phase split is difficult to condense if the feed has been cooled to cooling water temperature. If separation of the vapor is needed, one of the following methods can be used:

1. *Refrigerated condensation.* Separation by condensation relies on differences in volatility between the condensing components. Refrigeration or a combination of high pressure and refrigeration is needed.

2. *Absorption.* If possible, a component which already exists in the flowsheet should be used as a solvent. Introducing an extraneous component into the flowsheet introduces additional complexity and the possibility of increased environmental and safety problems later in the design.

3. *Adsorption.* Adsorption involves the transfer of a component onto a solid surface. An example is the adsorption of organic vapors by activated carbon. Activated carbon is a highly porous form of carbon manufactured from a variety of carbonaceous raw materials such as coal or wood. The adsorbent may need to be

regenerated by a vapor, such as steam. The regeneration stream then needs further separation. Regeneration does not always require a vapor or gas; sometimes it can be performed by a change in pressure or temperature. Such methods, if they are possible, are normally preferred to regeneration by a vapor, since this reduces the overall separation load.

4. *Membrane separation.* Membranes separate gases by means of a pressure gradient across a membrane, typically 40 bar or greater. Some gases permeate through the membrane faster than others and concentrate on the low-pressure side. Low-molecular-weight gases and strongly polar gases have high permeabilities and are known as *fast gases. Slow gases* have higher molecular weight and symmetric molecules. Thus membrane gas separators are effective when the gas to be separated is already at a high pressure and only a partial separation is required.

Rather than send the vapor to one of the separation units described above, a purge can be used. This removes the need for a separator but incurs raw material losses. Not only can these material losses be expensive, but they also can create environmental problems. However, another option is to use a combination of a purge with a separator.

As an example, consider ammonia synthesis. In an ammonia synthesis loop, hydrogen and nitrogen are reacted to ammonia. The reactor effluent is partially condensed to separate ammonia as a liquid. Unreacted gaseous hydrogen and nitrogen are recycled to the reactor. A purge on the recycle prevents the buildup of argon and methane, which enter the system as feed impurities. The purge can be burnt as fuel. Considerable quantities of hydrogen are lost in the purge, and recovery of this hydrogen is often economic. For such hydrogen recovery applications, only two processes are usually viable, cryogenic condensation and membranes. Separation by condensation relies on differences in volatility between the components. Separation of gases in a membrane unit is achieved due to the differences in the rates at which different gases permeate through the membrane. The membrane allows fast gases, such as hydrogen, to be separated from slow gases, such as methane. A fractional recovery of around 90 percent is possible with a single membrane, giving better than 90 percent hydrogen purity.

Example 4.2   Benzene is to be produced from toluene according to the reaction[6]

$$C_6H_5CH_3 + \quad H_2 \quad \longrightarrow \quad C_6H_6 \quad + \quad CH_4$$
$$\text{toluene} \quad \text{hydrogen} \qquad \text{benzene} \quad \text{methane}$$

The reaction is carried out in the gas phase and normally operates at around 700°C and 40 bar. Some of the benzene formed undergoes a secondary reversible reaction to an unwanted byproduct, diphenyl, according to the reaction

$$2C_6H_6 \rightleftharpoons C_{12}H_{10} + H_2$$

benzene   diphenyl   hydrogen

Laboratory studies indicate that a hydrogen-toluene ratio of 5 at the reactor inlet is required to prevent excessive coke formation in the reactor. Even with a large excess of hydrogen, the toluene cannot be forced to complete conversion. The laboratory studies indicate that the selectivity (i.e., fraction of toluene reacted which is converted to benzene) is related to the conversion (i.e., fraction of toluene fed which is reacted) according to[6]

$$S = 1 - \frac{0.0036}{(1-X)^{1.544}}$$

where S = selectivity
X = conversion

The reactor effluent is thus likely to contain hydrogen, methane, benzene, toluene, and diphenyl. Because of the large differences in volatility of these components, it seems likely that partial condensation will allow the effluent to be split into a vapor stream containing predominantly hydrogen and methane and a liquid stream containing predominantly benzene, toluene, and diphenyl.

The hydrogen in the vapor stream is a reactant and hence should be recycled to the reactor inlet (Fig. 4.8). The methane enters the process as a feed impurity and is also a byproduct from the primary reaction and must be removed from the process. The hydrogen-methane separation is likely to be expensive, but the methane can be removed from the process by means of a purge (see Fig. 4.8).

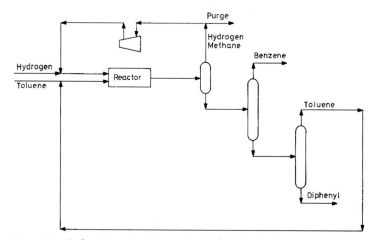

Figure 4.8 A flowsheet for the production of benzene uses a purge to remove the methane, which enters as a feed impurity and also is formed as a byproduct.

The liquid stream can be separated readily into pure components by distillation, the benzene taken off as product, the diphenyl as an unwanted byproduct, and the toluene recycled. It is possible to recycle the diphenyl to improve selectivity, but we will assume that is not done here.

The hydrogen feed contains methane as an impurity at a mole fraction of 0.05. The production rate of benzene required is $265 \, \text{kmol h}^{-1}$.

Assume initially that a phase split can separate the reactor effluent into a vapor stream containing only hydrogen and methane and a liquid stream containing only benzene, toluene, and diphenyl and that the liquid separation system can produce essentially pure products.

For a conversion in the reactor of 0.75,

1. Determine the relation between the fraction of vapor from the phase split sent to purge (α) and the fraction of methane in the recycle and purge (y).
2. Given the assumptions, estimate the composition of the reactor effluent for fraction of methane in the recycle and purge of 0.4.

Solution

1. Following Douglas,[6] let P_B be the production rate of benzene:

$$C_6H_5CH_3 + H_2 \longrightarrow C_6H_6 + C_6H_6 \quad + CH_4$$

$$\frac{P_B}{S} \qquad \frac{P_B}{S} \qquad P_B \qquad P_B\left(\frac{1}{S} - 1\right) \quad \frac{P_B}{S}$$

$$2C_6H_6 \; \rightleftharpoons \; C_{12}H_{10} \; + H_2$$

$$P_B\left(\frac{1}{S} - 1\right) \qquad \frac{P_B}{2}\left(\frac{1}{S} - 1\right) + \frac{P_B}{2}\left(\frac{1}{S} - 1\right)$$

For $X = 0.75$,

$$S = 1 - \frac{0.0036}{(1 - 0.75)^{1.544}}$$

$$= 0.9694$$

Toluene balance

$$\text{Fresh toluene feed} = \frac{P_B}{S}$$

$$\text{Toluene recycle} = R_T$$

$$\text{Toluene entering the reactor} = \frac{P_B}{S} + R_T$$

$$\text{Toluene in reactor effluent} = \left(\frac{P_B}{S} + R_T\right)(1 - X) = R_T$$

for $P_B = 265 \, \text{kmol h}^{-1}$, $X = 0.75$ and $S = 0.9694$.

$$R_T = 91.12 \, \text{kmol h}^{-1}$$

$$\text{Toluene entering the reactor} = \frac{265}{0.9694} + 91.12$$

$$= 364.5 \, \text{kmol h}^{-1}$$

Hydrogen balance

$$\text{Hydrogen entering the reactor} = 5 \times 364.5$$

$$= 1823 \text{ kmol h}^{-1}$$

$$\text{Net hydrogen consumed in reaction} = \frac{P_B}{S} - \frac{P_B}{2}\left(\frac{1}{S} - 1\right)$$

$$= \frac{P_B}{S}\left(1 - \frac{1-S}{2}\right)$$

$$= 269.2 \text{ kmol h}^{-1}$$

$$\text{Hydrogen in reactor effluent} = 1823 - 269.2$$

$$= 1554 \text{ kmol h}^{-1}$$

$$\text{Hydrogen lost in purge} = 1554\alpha$$

$$\text{Hydrogen feed to the process} = 1554\alpha + 269.2$$

Methane balance

$$\text{Methane feed to process as impurity} = (1554\alpha + 269.2)\frac{0.05}{0.95}$$

$$\text{Methane produced by reactor} = \frac{P_B}{S}$$

$$\text{Methane in purge} = \frac{P_B}{S} + (1554\alpha + 269.2)\frac{0.05}{0.95}$$

$$= 81.79\alpha + 287.5$$

$$\text{Total flow rate of purge} = 1554\alpha + 81.79\alpha + 287.5$$

$$= 1636\alpha + 287.5$$

Fraction of methane in the purge (and recycle)

$$y = \frac{81.79\alpha + 287.5}{1636\alpha + 287.5} \tag{4.12}$$

Figure 4.9 shows a plot of Eq. (4.12). As the purge fraction α is increased, the flow rate of purge increases, but the concentration of methane in the purge and recycle decreases. This variation (along with reactor conversion) is an important degree of freedom in the optimization of reaction and separation systems, as we shall see later.

2. **Methane balance**

Mole fraction of methane in vapor from phase separator $= 0.4$

$$\text{Methane in reactor effluent} = \frac{0.4}{0.6} \times 1554$$

$$= 1036 \text{ kmol h}^{-1}$$

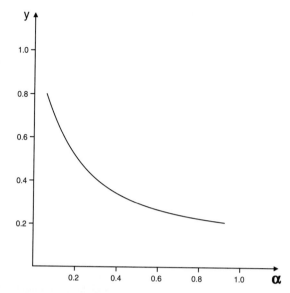

Figure 4.9 Variation of vapor mole fraction of methane with purge fraction.

Diphenyl balance

$$\text{Diphenyl in reactor effluent} = \frac{P_B}{2}\left(\frac{1}{S} - 1\right)$$

$$= 4 \, \text{kmol h}^{-1}$$

The estimated composition of the reactor effluent is given in Table 4.2. This calculation assumes that all separations in the phase split are sharp.

Example 4.3

1. Given the estimate of the reactor effluent in Example 4.2 for fraction of methane in the purge of 0.4, calculate the actual separation in the phase split assuming a temperature in the phase separator of 40°C. Phase equilibrium for this mixture can be represented by the Soave-Redlich-Kwong equation of state. Many computer programs are available commercially to carry out such calculations.

TABLE 4.2 Composition of Reactor Effluent

Component	Flow rate (kmol h^{-1})
Hydrogen	1554
Methane	1036
Benzene	265
Toluene	91
Diphenyl	4

TABLE 4.3 Vapor-Liquid Phase Split Using the
Soave-Redlich-Kwong Equation of State

Component	Vapor flow rate (kmol h^{-1})	Liquid flow rate (kmol h^{-1})
Hydrogen	1550	4
Methane	1021	15
Benzene	16.9	248.1
Toluene	2.1	88.9
Diphenyl	0.0	4.0

2. Repeat the calculation from Example 4.2 with actual phase equilibrium data in the phase split instead of assuming a sharp split.

Solution

1. If such a phase-split calculation is carried out assuming the feed given in Table 4.2, the result is given in Table 4.3. The phase split at 40°C gives a good separation of the hydrogen and methane into the vapor phase and benzene, toluene, and diphenyl in the liquid phase. It is clear that under these conditions the hydrogen and methane are both above their critical temperatures and are effectively noncondensables. However, some hydrogen and methane dissolve in the liquid phase. Also, some aromatics are carried with the vapor. An extremely important consequence of this is that the flowsheet in Fig. 4.8 would need to be modified to separate the hydrogen and methane carried forward with the liquid from the phase split. This would require an extra distillation column to separate the hydrogen and methane from the aromatics before separating the aromatics.

   The reader might wish to check that if the temperature of the phase split is increased or its pressure decreased, the separation between hydrogen, methane, and the other components becomes worse.

2. Assuming that the phase split operates at 40 bar and 40°C, a rigorous solution of the phase equilibrium using the Soave-Redlich-Kwong equation of state and the recycle equations using flowsheet simulation software gives a composition of the reactor effluent given in Table 4.4.

   Comparing this solution with that based on a sharp phase separation in Example 4.2, the errors are surprisingly small. However, looking at the K

TABLE 4.4 Composition of Reactor Effluent
Using a Rigorous Phase Split and Solution
of the Recycle

Component	Flow rate (kmol h^{-1})
Hydrogen	1585
Methane	1083
Benzene	283
Toluene	93
Diphenyl	4

values in the phase separator, it is not so surprising. The K values are given in Table 4.5.

The temperature of the phase split is well above the critical temperatures of both hydrogen and methane, leading to large K values. On the other hand, the K values of the benzene, toluene, and diphenyl are very low, and hence the assumption of a sharp split in Example 4.2 was a good one.

4.3   Vapor versus Liquid Recycles

When recycling material to the reactor for whatever reason, the pressure drop through the reactor, phase separator (if there is one), and the heat exchangers upstream and downstream of the reactor must be overcome. This means increasing the pressure of any material to be recycled.

In the case of a liquid recycle, the cost of this pressure increase is usually small. Pumps usually have low capital and operating costs relative to other plant items. On the other hand, to increase the pressure of material in the vapor phase for recycle requires a compressor. Compressors tend to have a high capital cost and large power requirements giving higher operating costs.

Sometimes it is extremely difficult to avoid vapor recycles without using very high pressures or very low levels of refrigeration, in which case we must accept the expense of a recycle compressor. However, when synthesizing the separation and recycle configuration, vapor recycles should be avoided, if possible, and liquid recycles used instead.

4.4   Batch Processes

In a *batch process,* the main steps operate discontinuously. In contrast with a continuous process, a batch process does not deliver its product continuously but in discrete amounts. This means that

TABLE 4.5   *K* Values for the Phase Split

Component	K_i
Hydrogen	54
Methane	9.5
Benzene	0.0095
Toluene	0.0033
Diphenyl	7.9×10^{-6}

heat, mass, temperature, concentration, and other properties vary with time. In practice, most batch processes are made up of a series of batch and semicontinuous steps.[7] A *semicontinuous step* runs continuously with periodic start-ups and shutdowns.

Batch processes

- Are economical for small volumes.
- Are flexible in accommodating changes in product formulation.
- Are flexible in changing production rate.
- Allow the use of standardized multipurpose equipment for the production of a variety of products from the same plant.
- Are best if equipment needs regular cleaning because of fouling or needs regular sterilization.
- Are amenable to direct scale-up from the laboratory.
- Allow product integrity. Each batch of product can be clearly identified in terms of the feeds involved and conditions of processing. This is particularly important in industries such as pharmaceuticals and foodstuffs.

Consider the simple process shown in Fig. 4.10. Feed material is withdrawn from storage using a pump. The feed material is preheated in a heat exchanger before being fed to a batch reactor. Once the reactor is full, further heating takes place inside the reactor using steam to the reactor jacket before the reaction proceeds. During the later stages of the reaction, cooling water is applied to the reactor jacket. Once the reaction is complete, the reactor product is withdrawn using a pump. The reactor product is cooled in a heat exchanger before going to storage.

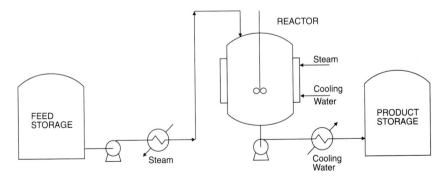

Figure 4.10 A simple batch process.

The process is shown in Fig. 4.11 as a *Gantt* or *time-event chart*.[7] The first two steps, pumping for reactor filling and feed preheat, are both semicontinuous. The heating inside the reactor, the reaction itself, and the cooling using the reactor jacket are all batch. The pumping to empty the reactor and the product cooling steps are again semicontinuous.

High utilization of equipment is one of the goals of batch process design. This can be achieved by *overlapping* batches. Overlapping means that more than one batch, at different processing stages, resides in the plant at any given time. This allows the batch cycle time, i.e., the time interval between producing successive batches of product, to be at its shortest.

The step with the longest time limits the cycle time. Alternatively, if more than one step is carried out in the same equipment, the cycle time is limited by the longest series of steps in the same equipment. The batch cycle time must be at least as long as the longest step. The rest of the equipment other than the limiting step is then idle for some fraction of the batch cycle.

Given the choice of a batch rather than continuous process, does this need a different approach to the synthesis of the reaction and separation and recycle system? In fact, a different approach is not needed. We start by assuming the process to be continuous and then, if choosing to use batch operation, replace continuous steps by batch steps.[8] It is simpler to start with continuous process operation

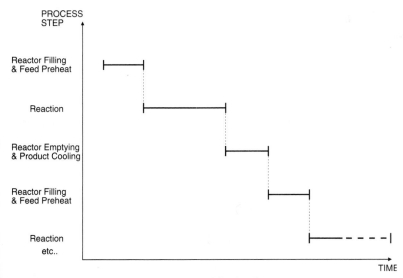

Figure 4.11   A Gantt or time-event chart of the batch process.

because, as will be seen, the time dependency of batch operation simply adds additional constraints over and above continuous operation. Also, batch operations often can be merged into a single piece of equipment. In the example shown in Figs. 4.10 and 4.11, the reactor to some extent operates as both a feed preheater and a product cooler as well as a reactor. However, recycling materials for any reason is probably not possible unless some intermediate storage is provided for the recycle.

The approach is illustrated by the following example.

Example 4.4   Butadiene sulfone (or 3-sulfolene) is an intermediate used for the production of solvents. It can be produced from butadiene and sulfur dioxide according to the reaction[8,9]

$$CH_2{=}CHCH{=}CH_2 \ + \ SO_2 \ \rightleftharpoons \ \begin{matrix} CH{=}CH \\ | \quad | \\ CH_2 \ \ CH_2 \\ \diagdown \diagup \\ SO_2 \end{matrix}$$

butadiene   sulfur   butadiene
                       dioxide   sulfone

This is an exothermic, reversible, homogeneous reaction taking place in a single liquid phase. The liquid butadiene feed contains 0.5 percent normal butane as an impurity. The sulfur dioxide is essentially pure. The mole ratio of sulfur dioxide to butadiene must be kept above 1 to prevent unwanted polymerization reactions. A value of 1.2 is assumed. The temperature in the process must be kept above 65°C to prevent crystallization of the butadiene sulfone but below 100°C to prevent its decomposition. The product must contain less than 0.5 wt% butadiene and less than 0.3 wt% sulfur dioxide.

The normal boiling points of the materials are given in Table 4.6. Synthesize a continuous reaction, separation, and recycle system for the process, bearing in mind that the process will later become batch.

Solution   The reversible nature of the reaction means that neither of the feed materials can be forced to complete conversion. The reactor design in Fig.

TABLE 4.6 Normal Boiling Points of the Components in Example 4.4

Material	Normal boiling point (°C)
Sulfur dioxide	−10
Butadiene	−4
n-Butane	−1
Butadiene sulfone	151

4.12a shows that the reactor product contains a mixture of both feed and product materials together with the n-butane impurity. These must be separated, but how?

If the relative boiling points of the components in the reactor product are considered, there is a wide range of volatilities. The sulfur dioxide, butadiene, and n-butane are all low boilers, and the butadiene sulfone is a much higher boiling material by comparison. Given that the reaction takes place in the liquid phase, a partial vaporization might well give a good separation between the butadiene sulfone and the other components (see Fig. 4.12b).

A vapor-liquid equilibrium calculation shows that a good separation is obtained, but the required product purity of butadiene <0.5 wt% and sulphur

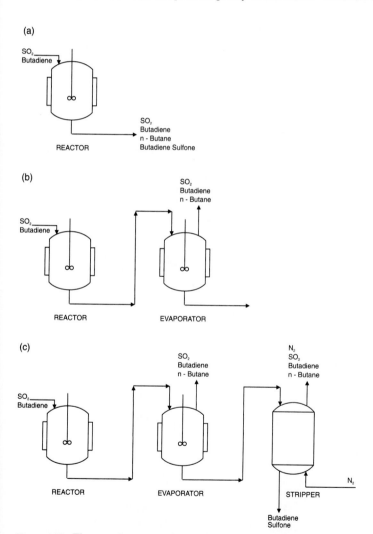

Figure 4.12 The reaction-separation system for the production of buta-diene sulfone.

dioxide <0.3 wt% is not obtained. Further separation of the liquid is needed. Distillation of the liquid is difficult because of the narrow temperature limits between which the distillation must operate. However, the liquid can be stripped using nitrogen (see Fig. 4.12c).

The type of equipment illustrated in Fig. 4.12 is more typical of batch operation than continuous operation even though continuous is being contemplated at the moment. For example, the evaporator is a stirred tank with a heating jacket. In continuous plant, a more elaborate design with tubular heating of some type probably would have been used.

Now consider recycling unconverted feed material to the reactor. Figure 4.13a shows the recycles of unconverted feed material. The recycle from the

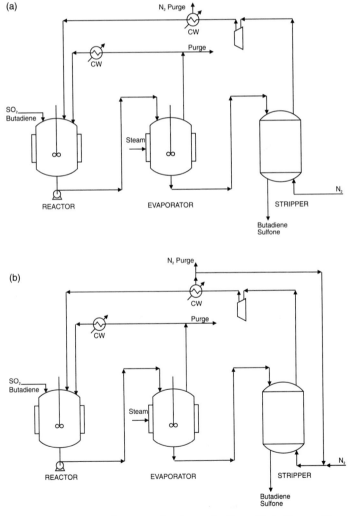

Figure 4.13 The recycle system for the production of butadiene sulfone.

evaporator to the reactor has been made possible by pressurizing the evaporator with the evaporator feed pump. Had this not been done, the vapor recycle would have required a compressor. The stripper works at a lower pressure to allow the unconverted material to be stripped. Thus the recycle requires a compressor. It is then condensed and passed back to the reactor.

One further problem remains. Most of the n-butane impurity which enters with the feed enters the vapor phase in the first separator. Thus the n-butane builds up in the recycle unless a purge is provided (see Fig. 4.13a). Finally, the possibility of a nitrogen recycle should be considered to minimize the use of fresh nitrogen (see Fig. 4.13b).

Example 4.5 Given that a low rate of production is required, convert the continuous process from Example 4.4 into a batch process. Preliminary sizing of the equipment indicates that the duration of the processing steps are given in Table 4.7.[8]

Solution Having synthesized the continuous flowsheet shown in Fig. 4.13b, let us now convert this into batch operation.

The reactor now becomes batch, requiring the reaction to be completed before the separation can take place. Figure 4.14 shows the time-event chart for a repeated batch cycle. Note in Fig. 4.14 that there is a small overlap between the process steps. This is to allow for the fact that emptying of one step and filling of the following step occur at the same time.

Clearly, the time chart shown in Fig. 4.14 indicates that individual items of equipment have a poor utilization; i.e., they are in use for only a small fraction of the batch cycle time. To improve the equipment utilization, overlap batches as shown in the time-event chart in Fig. 4.15. Here, more than one batch, at different processing stages, resides in the process at any given time. Clearly, it is not possible to recycle directly from the separators to the reactor, since the reactor is fed at a time different from that at which the separation is carried out. A storage tank is needed to hold the recycle material. This material is then used to provide part of the feed for the next batch. The final flowsheet for batch operation is shown in Fig. 4.16. Equipment utilization might be improved further by various methods which are considered in Chap. 8 when economic tradeoffs are discussed.

4.5 The Process Yield

Having considered the separation and recycling of material, the streams entering and leaving the process can now be established. Figure 4.17 illustrates typical input and output streams. Feed

TABLE 4.7 **Duration of Processing Steps**

Processing step	Duration (hours)
Reaction	2.1
Evaporation	0.45
Stripping	0.65
Vessel filling	0.25
Vessel emptying	0.25

streams enter the process, and product, byproduct, and purge streams leave after the separation and recycle systems are set up.

Raw materials costs dominate the operating costs of most processes (see App. A). Also, if raw materials are not used efficiently, this creates waste, which then becomes an environmental problem. It is therefore important to have a measure of the efficiency of raw materials use. The process yield is defined as

$$\text{Process yield} = \frac{\text{(desired product produced)}}{\text{(reactant fed to the process)}}$$

$$\times \text{stoichiometric factor} \tag{4.13}$$

where stoichiometric factor is the stoichiometric moles of reactant required per mole of product. When more than one reactant is used (or more than one desired product produced) Eq. (4.13) can be applied to each reactant (or product).

In general terms, there are two sources of yield loss in the process:

- Losses in the reactor due to byproduct formation or unconverted feed material if recycling is not possible.

- Losses from the separation and recycle system.

Addressing in particular the streams entering and leaving the

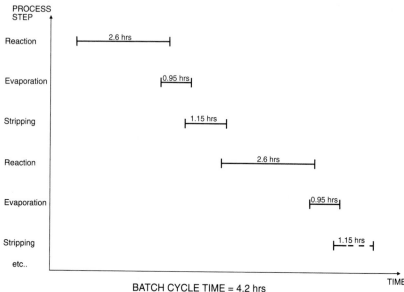

Figure 4.14   Time-event chart for a repeated batch cycle for Example 4.5.

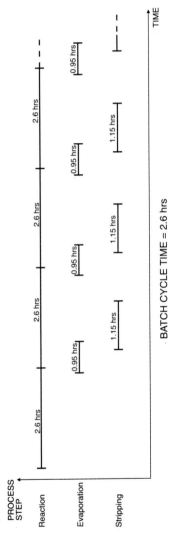

Figure 4.15 Overlapping batches in Example 4.5 reduces the batch cycle time.

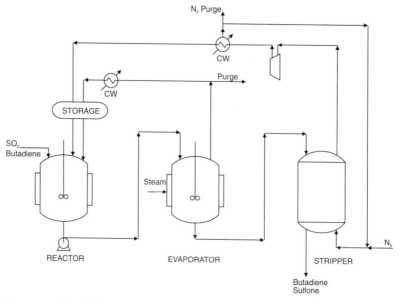

Figure 4.16 Final flowsheet for the production of butadiene sulfone in a batch process.

process in Fig. 4.17, there are material losses in the byproducts and purges to reduce if possible. Thus, before we proceed further, we should consider these questions:

1. Can byproduct formation be avoided or reduced by recycling? This is often possible when the byproduct is formed by secondary reversible reactions.

2. If a byproduct is formed by reaction of feed impurities, can this be avoided or reduced by purification of the feed?

3. Can the byproduct be subjected to further reaction and its value upgraded? For example, most organic chlorination reactions produce hydrogen chloride as a byproduct. If this cannot be sold, it

Figure 4.17 Streams entering and leaving the process.

must be disposed of. Another alternative, as discussed in Example 4.1, is to convert the hydrogen chloride back to chlorine via the reaction.

$$2HCl + \tfrac{1}{2}O_2 \rightleftharpoons Cl_2 + H_2O$$

The chlorine can then be recycled.

4. Can the loss of useful material in the purge streams be avoided or reduced by feed purification? If the purge is required to remove byproducts formed in the reactor, then this is clearly not possible.

5. Can the loss of useful material in the purge be avoided or reduced by additional separation on the purge? The roles of refrigerated condensation, membranes, etc. in this respect have already been discussed.

6. Can the useful material lost in the purge streams be reduced by additional reaction? If the purge stream contains significant quantities of reactants, then placing a reactor and additional separation on the purge can sometimes be justified. This technique is used in some designs of ethylene oxide processes.

Example 4.6 Calculate the process yield of benzene from toluene and benzene from hydrogen for the approximate phase split in Example 4.2.

Solution

$$\text{Benzene yield} = \frac{\text{(benzene produced)}}{\text{(toluene fed to the process)}}$$

$$\times \text{ stoichiometric factor}$$

Stoichiometric factor = stoichiometric moles of toluene required per mole of benzene produced

$$= 1$$

$$\text{Benzene yield from toluene} = \frac{(P_B)}{(P_B/S)} \times 1$$

$$= S$$

$$= 0.97$$

In this case, because there are no raw materials losses in the separation and recycle system, the only yield loss is in the reactor, and the process yield equals the reactor selectivity.

$$\text{Benzene yield from hydrogen} = \frac{\text{(benzene produced)}}{\text{(hydrogen fed to the process)}}$$

$$\times \text{ stoichiometric factor}$$

Stoichiometric factor = stoichiometric moles of hydrogen required per

mole of benzene produced

$$= 1$$

For $y = 0.4$, $\alpha = 0.3013$

$$\text{Benzene yield from hydrogen} = \frac{(P_B)}{(1554\alpha + 269.2)} \times 1$$

$$= \frac{265}{1554 \times 0.3013 + 269.2}$$

$$= 0.36$$

4.6 Synthesis of Reaction-Separation Systems—Summary

The use of excess reactants, diluents, or heat carriers in the reactor design has a significant effect on the flowsheet recycle structure. Sometimes the recycling of unwanted byproduct to the reactor can inhibit its formation at the source. If this can be achieved, it improves the overall use of raw materials and eliminates effluent disposal problems. Of course, the recycling does in itself raise some of the other costs. The general tradeoffs are discussed in Chap. 8.

When a mixture in a reactor effluent contains components with a wide range of volatilities, then a partial condensation from the vapor phase or a partial vaporization from the liquid phase followed by a simple phase split often can produce a good separation. If the vapor from such a phase split is difficult to condense, then further separation needs to be carried out in a vapor separation unit such as a membrane.

Batch processes can be synthesized by first synthesizing a continuous process and then converting it to batch operation. The process yield is an important measure of both raw materials efficiency and environmental impact.

4.7 References

1. Smith, R., and Linnhoff, B., "The Design of Separators in the Context of Overall Processes," *Trans. IChemE, ChERD*, 66: 195, 1988.
2. Powers, G. J., "Heuristic Synthesis in Process Development," *Chem. Eng. Progr.*, 68: 88, 1972.
3. Rudd, D. F., Powers, G. J., and Siirola, J. J., *Process Synthesis*, Prentice-Hall, Englewood Cliffs, N.J., 1973.
4. Douglas, J. M., *Conceptual Design of Chemical Processes*, McGraw-Hill, New York, 1988.
5. Tomlinson, T. R., and Finn, A. J., "Hydrogen from Off-Gases," The Membrane Alternative Energy Implications for Industry, The Watt Committee on Energy, University of Bath, U.K., March 29–30, 1989.

6. Douglas, J. M., "A Hierarchical Design Procedure for Process Synthesis," *AIChEJ,* 31: 353, 1985.
7. Mah, R. S. H., *Chemical Process Structures and Information Flows,* Butterworth, Reading, Mass., 1990.
8. Myriantheos, C. M., "Flexibility Targets for Batch Process Design," M.S. thesis, University of Massachusetts, Amherst, 1986.
9. McKetta, J. J., *Encyclopedia of Chemical Processing and Design,* vol. 5, Marcel Dekker, New York, 1977, p. 2.

5

Distillation Sequencing

In the preceding chapter, the interaction between the design of a reactor and the design of the separation system was discussed. Consider now the particular case in which the reactor design is "fixed" and the reactor effluent is a homogeneous multicomponent mixture that must be separated into essentially pure products. If this is the case, as we have already seen in Chap. 4, generally there is a choice of the order in which components are separated, i.e., the choice of *separation sequence.* Since distillation is the most common method of separating homogeneous mixtures, consideration is restricted to distillation.

5.1 Distillation Sequencing Using Simple Columns

Consider first the design of distillation systems comprising only simple columns. These simple columns employ:

- One feed split into two products
- Key components which are adjacent in volatility
- A reboiler and a condenser

If there is a three-component mixture and simple columns are employed, then the decision is between two sequences, as illustrated in Fig. 5.1. The sequence shown in Fig. 5.1a is called the *direct sequence,* in which the lightest component is taken overhead in each column. The *indirect sequence,* shown in Fig. 5.1b, takes the heaviest component as the bottom product in each column. There may be significant differences in the capital and operating costs of these two sequences.

For a three-component mixture, there are only two alternative sequences. The complexity increases dramatically as the number of components increases. Figure 5.2 shows the alternative sequences for a five-component mixture. Table 5.1 shows the relationship between the number of products and the number of possible sequences for simple columns.[1]

Much work has been carried out to find methods for the synthesis of distillation sequences of simple columns that do not involve heat integration.[2-4] However, heat integration may have a significant

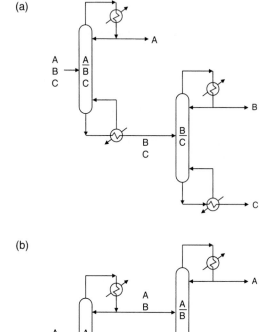

Figure 5.1 The direct and indirect sequences of simple distillation columns for a three-component separation. *(From Smith and Linnhoff, Trans. IChemE, ChERD, 66: 195, 1988; reproduced by permission of the Institution of Chemical Engineers.)*

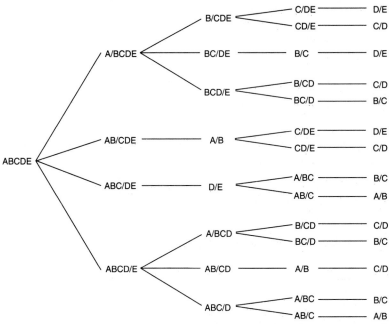

Figure 5.2 Alternative sequences for the separation of a five-component mixture.

effect on operating costs. Heat integration may be within the separation system. For example, in Fig. 5.1*a* the pressure of the second column could be increased relative to the first column to allow the condenser heat from the second column to provide the reboiler heat for the first column. In Fig. 5.1*b* the pressure of the first column could be increased relative to the second column to allow the condenser heat from the first column to provide the reboiler heat for the second column. Alternatively, rather than heat integration

**TABLE 5.1 The Number of Possible
Distillation Sequences Using
Simple Columns**

Number of products	Number of possible sequences
2	1
3	2
4	5
5	14
6	42
7	132
8	429

within the distillation sequence, there can be heat integration with other parts of the process.

This appears to be a complex problem requiring simultaneous solution of the sequence together with heat integration.

5.2   Practical Constraints that Restrict Options

Process constraints often reduce the number of options that can be considered. Examples of constraints of this type are as follows:

1. Safety considerations might dictate that a particularly hazardous component be removed from the process as early as possible to minimize the inventory of that material.

2. Reactive and heat-sensitive components must be removed early.

3. Corrosion problems often dictate that a particularly corrosive component be removed early to minimize the use of expensive materials of construction.

4. If decomposition in the reboilers contaminates the product, then this dictates that no finished products be taken from the bottoms of columns.

5. Some compounds tend to polymerize when distilled unless chemicals are added to inhibit polymerization. These polymerization inhibitors tend to be involatile, ending up in the column bottoms. If this is the case, finished products normally cannot be taken from column bottoms.

6. Often, in the feed to a separation train, some components are difficult to condense. Total condensation of these components might require low-temperature refrigeration and/or very high operating pressures. Under these circumstances, the light components are normally removed from the top of the first column to minimize the use of refrigeration and high pressures in the column sequence.

5.3   Selection of the Sequence for Simple, Nonintegrated Distillation Columns

Probably the most common method used for sequence selection for simple distillation columns is heuristic.[1] Many heuristics have been proposed, but they can be summarized by the following four:[5]

Heuristic 1: Separations in which the relative volatility of the key components is close to unity or which exhibit azeotropic behavior should be performed in the absence of nonkey components. In other words, do the most difficult separation last.

Heuristic 2: Sequences that remove the lightest components alone one by one in column overheads should be favored. In other words, favor the direct sequence.

Heuristic 3: A component composing a large fraction of the feed should be removed first.

Heuristic 4: Favor near-equimolar splits between top and bottom products in individual columns.

These heuristics are based on observations made in many practical applications. In addition to being restricted to simple columns, the observations are based on no heat integration (i.e., all reboilers and condensers are serviced by utilities). Difficulties can arise when the heuristics are in conflict with each other, as the following example illustrates.

Example 5.1   Each component for the mixture in Table 5.2 is to be separated into relatively pure products. Use the heuristics to determine sequences which are candidates for further evaluation.

Solution

Heuristic 1: Do D/E split last, since this separation has the smallest relative volatility.

Heuristic 2: Favor the direct sequence:

$$\frac{A}{\begin{array}{c} B \\ C \\ D \\ E \end{array}}$$

TABLE 5.2   Data for Mixture of Alkanes to Be Separated by Distillation

Component	Flow rate (kmol h^{-1})	Normal boiling point (K)	Relative volatility	Relative volatility between adjacent components
A. Propane	45.4	231	7.98	
B. *i*-Butane	136.1	261	3.99	2.00
C. *n*-Butane	226.8	273	3.00	1.33
D. *i*-Pentane	181.4	301	1.25	2.40
E. *n*-Pentane	317.5	309	1.00	1.25

Heuristic 3: Remove the most plentiful component first:

$$\begin{array}{c} A \\ B \\ C \\ \underline{D} \\ E \end{array}$$

Heuristic 4: Favor near-equimolar splits between top and bottom products:

$$\begin{array}{cl} A & \\ B & 408.3\,\text{kmol}\,\text{h}^{-1} \\ \underline{C} & \\ \overline{D} & 498.9\,\text{kmol}\,\text{h}^{-1} \\ E & \end{array}$$

All four heuristics are in conflict here. Heuristic 1 suggests doing the D/E split last, whereas heuristic 3 suggests doing it first. Heuristic 2 suggests doing the A/B split first, and heuristic 4 suggests doing the C/D split first.

Take one of the candidates and accept, say, the A/B split first.

Heuristic 1: Do D/E split last.

Heuristic 2:

$$\begin{array}{c} \underline{B} \\ \overline{C} \\ D \\ E \end{array}$$

Heuristic 3:

$$\begin{array}{c} B \\ C \\ \underline{D} \\ \overline{E} \end{array}$$

Heuristic 4:

$$\begin{array}{cl} B & 362.9\,\text{kmol}\,\text{h}^{-1} \\ \underline{C} & \\ \overline{D} & 498.9\,\text{kmol}\,\text{h}^{-1} \\ E & \end{array}$$

Again, the heuristics are in conflict. Heuristic 1 again suggests doing the D/E split last, whereas heuristic 3 again suggests doing it first. Heuristic 2 suggests doing the B/C split first, and heuristic 4 suggests doing the C/D first.

This process could be continued and possible sequences identified for further consideration. Some possible sequences would be eliminated, narrowing down the number suggested by Table 5.1.

Clearly, the conflicts that have arisen in this problem have not been too helpful in identifying sequences which are candidates for further evaluation. A little more intelligence could be used in application of the heuristics, and they could be ranked in order of importance. However, the rank order might well change from process to process. Although in the preceding example the heuristics do not give a clear indication of the correct sequence, in some problems they do. It does seem, though, that a more general method than the heuristics is needed.

Rather than relying on heuristics which can be ambiguous or in conflict, a parameter would be preferred that can measure quantitatively the relative performance of different sequences. The vapor flow rate up the column is a good measure of both capital and operating costs. There is clearly a relationship between the heat duty required to run the distillation and the vapor rate, since the latent heat relates these two parameters. However, there is also a link between vapor rate and capital cost, since a high vapor rate leads to a large-diameter column. The high vapor rate also requires large reboilers and condensers. Thus vapor rate is a good measure of both capital and operating costs on individual columns. Consequently, sequences with a lower total vapor load would be preferred to those with a high total vapor load. But how is the total vapor load predicted?

The multicomponent form of the Underwood equation[1] can be used to calculate the vapor flow at minimum reflux in each column of the sequence. The minimum vapor rate in a single column is obtained by alternate use of two equations:

$$\sum_{i=1}^{NC} \frac{\alpha_i x_{F,i}}{\alpha_i - \Theta} = 1 - q \tag{5.1}$$

$$V_{min} = D \sum_{i=1}^{NC} \frac{\alpha_i x_{D,i}}{\alpha_i - \Theta} \tag{5.2}$$

where V_{min} = vapor rate at minimum reflux
$\quad\quad D$ = distillate rate
$\quad\quad q$ = thermal condition of the feed, i.e., heat to vaporize 1 mol feed divided by the molar latent heat of feed
$\quad\quad x_{F,i}$ = mole fraction of component i in the feed
$\quad\quad x_{D,i}$ = mole fraction of component i in the distillate
$\quad\quad \alpha_i$ = relative volatility of component i
$\quad\quad \Theta$ = variable defined by Eq. (5.1)
$\quad\quad NC$ = number of components

To calculate the vapor load for a single column of a sequence, start by assuming a feed condition such that q can be fixed. Initially assume saturated liquid feed (i.e., $q = 1$). Equation (5.1) can be written for all NC components of the feed and solved for the necessary values of Θ. There are $(NC - 1)$ real positive values of Θ which satisfy Eq. (5.1), and each lies between the α values of the components. One more value of Θ is required than there are components between the keys, which lie between the α values for the light and heavy keys. Equation (5.2) is then written for each value of Θ obtained. These are then solved simultaneously for V_{min} and the unknown $x_{D,i}$ values.

Having computed V_{min}, the actual vapor load can be calculated from a simple mass balance around the top of the column, which gives

$$V = D(1 + R) \qquad (5.3)$$

where V = vapor load
R = reflux ratio
D = distillate flow rate

Equation (5.3) can be written at minimum reflux and then at finite reflux, say, 1.1 times minimum reflux. The calculation is then repeated for all columns in the sequence.

The problem with this approach is obvious. It involves a considerable amount of work to generate a measure of the quality of the sequence, the total vapor load, which is only a guideline. There are many other factors to be considered. Indeed, as we shall see later, when variables such as reactor conversion are optimized, the sequence might well need readdressing.

Porter and Momoh[6] have suggested an approximate but simple method of calculating the total vapor rate for a sequence of simple columns. Start by rewriting Eq. (5.3) with the reflux ratio R defined as a proportion relative to the minimum reflux ratio R_{min} (typically $R/R_{min} = 1.1$). Defining R_F to be the ratio R/R_{min} Eq. (5.3) becomes

$$V = D(1 + R_F R_{min}) \qquad (5.4)$$

R_{min} is calculated from the binary form of the Underwood equation applied to the key components:[1]

$$R_{min} = \frac{1}{\alpha - 1} \left(\frac{x_{DLK}}{x_{FLK}} - \alpha \frac{x_{DHK}}{x_{FHK}} \right) \qquad (5.5)$$

where α = relative volatility between the key components
x_{DLK} = mole fraction of light key in the distillate
x_{FLK} = mole fraction of light key in the feed
x_{DHK} = mole fraction of heavy key in the distillate
x_{FHK} = mole fraction of heavy key in the feed

Assuming a sharp separation with only the light key and lighter-than-light key components in the overheads and only the heavy key and heavier-than-heavy key components in the bottoms, then Eq. (5.5) simplifies to

$$R_{\min} = \frac{1}{\alpha - 1}\left(\frac{x_{DLK}}{x_{FLK}}\right) = \frac{1}{\alpha - 1}\left(\frac{F}{D}\right) \qquad (5.6)$$

where F = feed flow rate
D = distillate flow rate

Combining Eqs. (5.4) and (5.6) gives

$$V = D\left[1 + \frac{R_F}{(\alpha - 1)}\frac{F}{D}\right] \qquad (5.7)$$

Because only light key and lighter components go to the distillate and heavy key and heavier components go to the bottoms, Eq. (5.7) can be written in terms of the molar flow rate of each component in the feed:

$$V = (F_A + F_B + \cdots + F_{LK})$$
$$+ (F_A + F_B + \cdots + F_{LK} + F_{HK} + \cdots + F_{NC})\frac{R_F}{\alpha - 1} \qquad (5.8)$$

where F_A, F_B, etc. are the molar flow rates of component A, component B, and so on.

There is now a simple explicit expression for the vapor rate in a single column in terms of the feed to the column. In order to use this expression to screen column sequences, the vapor rate in each column must be calculated according to Eq. (5.8), assuming a sharp separation in each column, and the individual vapor rates summed.

Equation (5.8) tends to predict vapor loads slightly higher than those predicted by the full multicomponent form of the Underwood equation.[6] The important thing, however, is not the absolute value but the relative values of the alternative sequences. Porter and Momoh[6] have demonstrated that the rank order of total vapor load follows the rank order of total cost.

Example 5.2 Table 5.3 gives the data for a ternary separation of benzene,

TABLE 5.3 Data for Ternary Mixture of Aromatics to Be Separated by Distillation

Component	Flow rate (kmol h^{-1})	Relative volatility	Relative volatility between adjacent components
Benzene	269	3.53	1.96
Toluene	282	1.80	1.80
Ethyl benzene	57	1.0	

toluene, and ethyl benzene. Using Eq. (5.8), determine whether the direct or indirect sequence should be used. Assume a ratio of actual to minimum reflux of 1.1.

Solution For the direct sequence (see Fig. 5.3a):

$$\sum V = 269 + (269 + 282 + 57)\frac{1.1}{(1.96 - 1)} + 282 + (282 + 57)\frac{1.1}{(1.8 - 1)}$$

$$= 965.7 + 748.1$$

$$= 1713.8 \text{ kmol h}^{-1}$$

For the indirect sequence (see Fig. 5.3b):

$$\sum V = (269 + 282) + (269 + 282 + 57)\frac{1.1}{(1.8 - 1)} + 269 + (269 + 282)\frac{1.1}{(1.96 - 1)}$$

$$= 1387 + 900.4$$

$$= 2287.4 \text{ kmol h}^{-1}$$

Hence we should use the direct sequence.

Example 5.3 Using Eq. (5.8), determine the best sequence for the mixture of alkanes in Table 5.2. Assume the ratio of actual to minimum reflux to be 1.1.

(a)

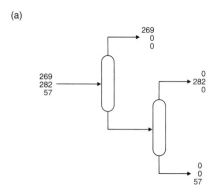

(b)

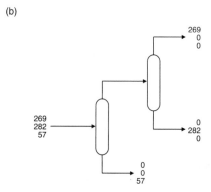

Figure 5.3 The direct and indirect sequence for Example 5.2.

TABLE 5.4 Total Vapor Flows for the 14 Possible Sequences for the Separation of a Mixture of Alkanes

Rank order	Total vapor flow $(kmol\,h^{-1})$
1	5285.2
2	5338.0
3	5805.7
4	6624.1
5	7225.6
6	7240.6
7	7293.3
8	7411.8
9	7515.9
10	7521.1
11	7599.4
12	7773.4
13	7877.5
14	8200.9

Solution   Table 5.4 gives the total vapor flow for different sequences in rank order. The sequence with the lowest total vapor flow is shown in Fig. 5.4.

It is interesting to note that this is the sequence that would have been obtained had only heuristic 4 been used (favor near-equimolar splits between top and bottom products) throughout.

Had only one of the four heuristics been used throughout, the rank order for the sequence from each heuristic would have been

Heuristic	Rank order of sequence from Table 5.4
1	3
2	5
3	10
4	1

However, it would be extremely dangerous from this one calculation to assume that heuristic 4 is the most important, as the next example shows.

Example 5.4   The data for an aromatics separation are shown in Table 5.5. Assuming the ratio of actual to minimum reflux to be 1.1, determine the best sequence using Eq. (5.8).

Solution   Table 5.6 gives the total vapor flow for different sequences in rank order. The sequence with the lowest overall vapor flow is shown in Fig. 5.5.

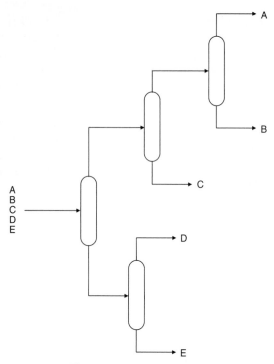

Figure 5.4 The sequence with the lowest total vapor flow for Example 5.3.

TABLE 5.5 Data for Five-Component Mixture of Aromatics to Be Separated by Distillation

Component	Flow rate (kmol h^{-1})	Relative volatility	Relative volatility between adjacent components
A. Benzene	269	6.24	
B. Toluene	282	3.28	1.90
C. Ethyl benzene	57	1.86	1.76
D. Xylenes	215	1.76	1.06
E. C9s	42	1.00	1.76

In Example 5.3, heuristic 4 seemed to be the most important. Had heuristic 4 been used exclusively for this example, the sequence ranked fifth in Table 5.6 would have been obtained.

Examples 5.3 and 5.4 show that the relative importance of the various heuristics changes between problems. Equation (5.8)

**TABLE 5.6   Total Vapor Flows for the 14
Possible Sequences for the Separation
of a Mixture of Aromatics**

Rank order	Total vapor flow (kmol h^{-1})
1	8,241.0
2	8,515.6
3	8,870.4
4	8,871.5
5	9,146.1
6	9,477.4
7	9,803.3
8	13,951.5
9	14,011.2
10	14,618.2
11	18,838.1
12	19,426.8
13	19,556.1
14	20,144.8

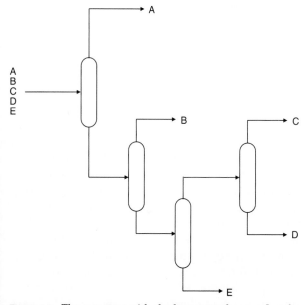

Figure 5.5 The sequence with the lowest total vapor flow for
Example 5.4.

provides a more reliable method of indicating the best sequence. The multicomponent version of the Underwood equation [Eqs. (5.1) and (5.2)] gives a better measure of the total vapor load but is more complex to calculate.

However, use of total vapor rate is still only a guide and might not give the correct rank order in some cases. In fact, given some computational aids, it is a practical proposition to size and cost all the alternative sequences using a shortcut sizing calculation, such as the Fenske-Gilliland-Underwood approach,[1] together with cost correlations. Even though practical problems might involve a large number of components, it is rare for them to have more than six products, which means 42 possible sequences from Table 5.1. In addition, process constraints often reduce this number.

Whatever the method used to screen possible sequences, it is important not to give exclusive attention to the one that appears to have the lowest vapor load or lowest total cost. There is often little to choose in this respect between the best few sequences, particularly when the number of possible sequences is large. Other considerations such as heat integration, safety, and so on also might have an important bearing on the final decision. Thus the screening of sequences should focus on the best few sequences rather than exclusively on the single best sequence.

5.4   Heat Integration of Sequences of Simple Distillation Columns

Having found the best nonintegrated sequence, most designers would then heat integrate. In other words, the total problem is not solved simultaneously but in two steps. Moving outward from the center of the onion (see Fig. 1.6), the separation layer is addressed first, followed by the heat exchanger network layer.

Whether this approach works in practice is easily tested. We can take a problem and design all possible nonintegrated sequences and then heat integrate those sequences and compare. Freshwater and Ziogou[7] and Stephanopoulos, Linnhoff, and Sophos[5] have carried out extensive numerical studies on sequences of simple distillation columns both with and without heat integration. One interesting result from the study of Freshwater and Ziogou[7] was that the configuration that achieved the greatest energy saving by integration often already had the lowest energy requirement prior to integration. When this was not so, the difference in energy consumption between the integrated configuration with the lowest energy import and the one based on the nonintegrated configuration that required the least energy was usually minimal.

In the study carried out by Stephanopoulos, Linnhoff, and Sophos,[5] in each of three examples these workers found that the optimal nonintegrated sequence turned out to be the optimal integrated sequence in terms of minimum total cost (capital and operating). In addition, it was observed that higher heat loads tended to occur in sequences that also had wider spans in temperature between reboiler and condenser. These studies suggest that the two problems of separation sequencing and heat integration can, for all practical purposes, be decoupled. The result is illustrated in Fig. 5.6. The best nonintegrated sequence turns out to be among the best few integrated sequences in terms of total cost (capital and operating).

This result is important, since in practice we should not focus exclusively on the single sequence with the lowest overall cost. Rather, because of the uncertainties in the calculations and the fact that other factors need to be considered in a more detailed evaluation, the best few sequences should be evaluated in more detail. Thus there is no need to solve the separation sequence and heat integration problems simultaneously. Rather, decouple the two problems and tackle them separately, simplifying considerably the overall task. It must be emphasized strongly that the decoupling depends on the absence of significant constraints limiting the heat integration potential within the distillation sequence. For example, there may be limitations on the pressure of some of the columns due to product decomposition, etc. that limit the heat integration potential. In these

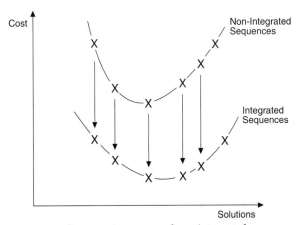

Figure 5.6 Comparative costs of nonintegrated versus heat-integrated sequences of simple columns. *(From Smith and Linnhoff, Trans. IChemE, ChERD, 66: 195, 1988; reproduced by permission of the Institution of Chemical Engineers.)*

circumstances, a method needs to be adopted that solves the sequencing and heat integration simultaneously.[8,9]

If the problem in the absence of significant constraints can be decoupled in this way, there must be some mechanism which allows this, and that mechanism should be explored.

5.5 Internal Mass Flows in Sequences of Simple Distillation Columns

Take two different sequences for the separation of a four-component mixture[4] (Fig. 5.7). Summing the feed flow rates of the key components to each column in the sequence, the total flow rate is the same in both cases:

$$\sum_{\text{Keys}} (m) = m_A + 2(m_B + m_C) + m_D \tag{5.9}$$

However, the flow of nonkeys is different. The arrangement shown in

(a)

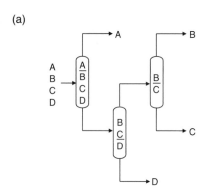

(b)

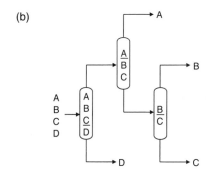

Figure 5.7 Two possible configurations of simple columns for the separation of a four-component mixture. *(From Smith and Linnhoff, Trans. IChemE, ChERD, 66: 195, 1988; reproduced by permission of the Institution of Chemical Engineers.)*

Fig. 5.7*a* has a flow rate of nonkeys given by

$$\sum_{\text{Nonkeys}} (m) = m_B + m_C + m_D \tag{5.10}$$

However, the case shown in Fig. 5.7*b* has a flow rate of nonkeys given by

$$\sum_{\text{Nonkeys}} (m) = m_A + m_B + m_C \tag{5.11}$$

In general, the flow of key components is constant and independent of the sequence, while the flow of nonkey components varies according to the choice of sequence,[4] as illustrated in Fig. 5.8.

It thus appears that the flow rate of the nonkey components may account for the differences between sequences. Essentially, nonkey components have two effects on a separation. They cause

1. An unnecessary load on the separation, leading to higher heat loads and vapor rates.

2. A widening of the temperature differences across columns, since light nonkey components cause a decrease in condenser temperature and heavy nonkey components cause an increase in the reboiler temperature.

These effects can be expressed quantitatively in temperature-heat

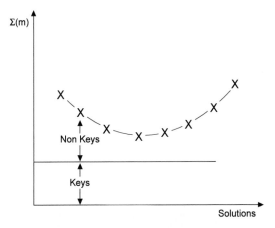

Figure 5.8 The overall flow rate of key components is constant for any sequence of simple columns. The overall flow rate of nonkey components varies. (*From Smith and Linnhoff, Trans. IChemE, ChERD, 66: 195, 1988: reproduced by permission of the Institution of Chemical Engineers.*)

profiles illustrated in Fig. 5.9. A high flow rate of nonkey components leads simultaneously to higher loads and more extreme levels (see Fig. 5.9).

Whether heat integration is restricted to the separation system or allowed with the rest of the process, integration always benefits from colder reboiler streams and hotter condenser streams. This point is dealt with in more general terms in Chap. 12. In addition, when column pressures are allowed to vary, columns with smaller temperature differences are easier to integrate, since smaller changes in pressure are required to achieve suitable integration. This second point is explained in more detail in Chap. 14.

Having established that there is apparently a mechanism whereby the problems of sequencing and heat integration can be decoupled for simple columns on the basis of energy costs, it is interesting to consider whether there is any conflict with capital cost. A column sequence that handles a large amount of heat must have a high capital cost for two reasons:

1. Large heat loads to be transferred result in large reboilers and condensers.

2. Large heat loads will cause high vapor rates, and these require large column diameters.

It is thus unlikely that a distillation sequence has a small capital cost and a large operating cost, or vice versa. A nonintegrated sequence with a large heat load has high vapor rates, large heat transfer areas, and large column diameters. Even if the heat requirements of this sequence are substantially reduced by integration, the large heat transfer areas and large diameters must still be

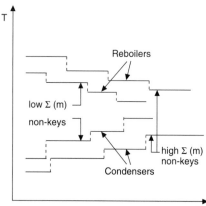

Figure 5.9 The temperature-heat profiles of a sequence with low overall flow rate of nonkey components is favorable. *(From Smith and Linnhoff, Trans. IChemE, ChERD, 66: 195, 1988; reproduced by permission of the Institution of Chemical Engineers.)*

provided. Any savings in utility costs achieved by heat integration must compensate for these high capital costs before this sequence can become competitive with a nonintegrated sequence starting with a smaller heat load.

Thus capital cost considerations reinforce the argument that the nonintegrated sequence with the lowest heat load is that with the lowest total cost.

The mechanism by which nonkey components affect a given separation is more complex in practice than the broad arguments presented here. There are complex interrelationships between the volatility of the key and nonkey components, etc. Although the argument presented is thus not rigorous, it is broadly correct.

It is interesting to note that heuristics 2, 3, and 4 from Sec. 5.2 tend to minimize the flow rate of nonkey components. Heuristic 1 relates to special circumstances when there is a particularly difficult separation.[5]

5.6 Distillation Sequencing Using Columns with More than Two Products

When separating a three-component mixture using simple columns, there are only two possible sequences (see Fig. 5.1). Consider the first characteristic of simple columns. A single feed is split into two products. As a first alternative to two simple columns, the possibilities shown in Fig. 5.10 can be considered. Here, three products are taken from one column. The designs are in fact both feasible and cost-effective when compared with simple arrangements on a stand-alone basis (i.e., reboilers and condensers operating on utilities) for certain ranges of conditions. If the feed is dominated by the middle product (typically more than 50 percent of the feed) and the heaviest product is present in small quantities (typically less than 5 percent), then the arrangement shown in Fig. 5.10a can be an attractive option.[10] The heavy product must find its way down the column past the sidestream. Unless the heavy product has a small flow and the middle product a high flow, a reasonably pure middle product cannot be achieved. In these circumstances, the sidestream is usually taken as a vapor product to obtain a reasonably pure sidestream.

If the feed is dominated by the middle product (typically more than 50 percent) and the lightest product is present in small quantities (typically less than 5 percent), then the arrangement shown in Fig. 5.10b can be an attractive option.[10] This time the light product must find its way up the column past the sidestream. Again, unless the light product is a small flow and the middle product a high flow, a reasonably pure middle product cannot be achieved. This time the sidestream is taken as a liquid product to obtain a reasonably pure sidestream. In summary, single-column sidestream arrangements can be attractive when the middle product is in excess and one of the other components is present in only minor quantities. Thus the sidestream column only applies to special circumstances for the feed composition. More generally applicable arrangements are possible by relaxing the restriction that separations must be between adjacent key components.

Consider a three-product separation as in Fig. 5.11a in which the lightest and heaviest components are chosen to be the key separation in the first column. Two further columns are required to produce pure products (see Fig. 5.11a). However, note from Fig. 5.11a that the bottoms and overheads of the second and third columns are both pure B. Hence the second and third columns could simply be connected and product B taken as a sidestream (see Fig. 5.11b). The arrangement in Fig. 5.11b is known as a *prefractionator* arrangement. Note that the first column in Fig. 5.11b, the prefractionator, has a partial condenser to reduce the overall energy consumption. Comparing the prefractionator arrangement in Fig. 5.11b with the conventional

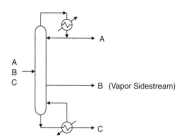

(a) More than 50% middle component and less than 5% heaviest component.

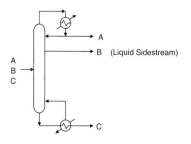

(b) More than 50% middle component and less than 5% lightest component.

Figure 5.10 Distillation columns with three products. *(From Smith and Linnhoff, Trans. IChemE, ChERD, 66: 195, 1988; reproduced by permission of the Institution of Chemical Engineers.)*

arrangements in Fig. 5.1, turns out that the prefractionator arrangement typically requires 30 percent less energy than conventional arrangements for the same separation duty. The reason for this difference is rooted in the fact that the prefractionator arrangement is fundamentally thermodynamically more efficient than a simple arrangement. Let us consider why.

Consider the sequence of simple columns shown in Fig. 5.12. In the direct sequence shown in Fig. 5.12, the composition of component B in the first column increases below the feed as the more volatile component A decreases. However, moving further down the column, the composition of component B decreases again as the composition of the less volatile component C increases. Thus the composition of component B reaches a peak only to be remixed.[11]

Similarly, with the first column in the indirect sequence, the composition of component B first increases above the feed as the less volatile component C decreases. It reaches a maximum only to decrease as the more volatile component A increases. Again, the composition of component B reaches a peak only to be remixed.

This remixing which occurs in both sequences of simple distillation columns is a source of inefficiency in the separation. By contrast,

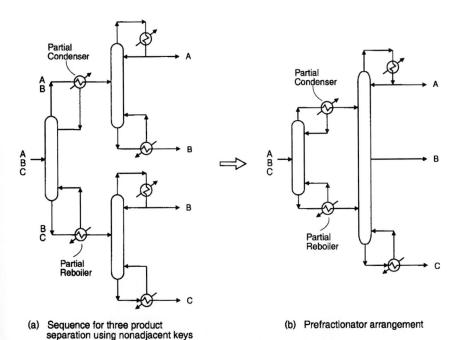

(a) Sequence for three product   (b) Prefractionator arrangement
     separation using nonadjacent keys

Figure 5.11 Choosing nonadjacent keys leads to the prefractionator arrangement.

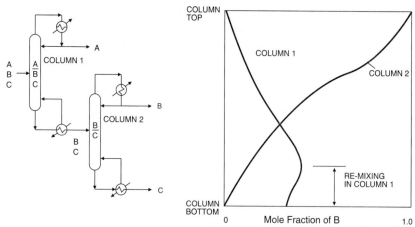

Figure 5.12 Composition profiles for the middle product in the columns of the direct sequence show remixing effects. *(From Triantafyllou and Smith, Trans. IChemE, part A, 70: 118, 1992; reproduced by permission of the Institution of Chemical Engineers.)*

consider the prefractionator arrangement shown in Fig. 5.13. In the prefractionator, a crude split is performed so that component B is distributed between the top and bottom of the column. The upper section of the prefractionator separates AB from C, while the lower section separates BC from A. Thus both sections remove only one component from the product of that column section, and this is also true for all four sections of the main column. In this way, the

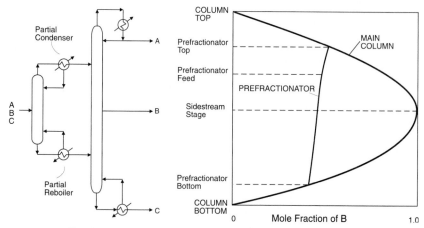

Figure 5.13 Composition profiles for the middle product in the prefractionator arrangement show that there are no remixing effects. *(From Triantafyllou and Smith, Trans. IChemE, part A, 70: 118, 1992; reproduced by permission of the Institution of Chemical Engineers.)*

remixing effects which are a feature of both simple column sequences are avoided.[11]

In addition, one other feature of the prefractionator arrangement is important in reducing mixing effects. Losses occur in distillation operations due to mismatches between the composition of the column feed and the composition on the feed tray. Because the prefractionator distributes component B top and bottom, this allows greater freedom to match the feed composition with one of the trays in the column to reduce mixing losses at the feed tray.

The elimination of mixing losses in a prefractionator arrangement means that it is inherently more efficient than an arrangement using simple columns.

5.7 Distillation Sequencing Using Thermal Coupling

The final restriction of simple columns stated earlier was that they should have a reboiler and a total condenser. It is possible to use materials flow to provide some of the necessary heat transfer by direct contact. This transfer of heat via direct contact is known as *thermal coupling.*

First consider thermal coupling of the simple sequences from Fig. 5.1. Figure 5.14a shows a thermally coupled direct sequence. The reboiler of the first column is replaced by a thermal coupling. Liquid from the bottom of the first column is transferred to the second as before, but now the vapor required by the first column is supplied by the second column instead of by a reboiler on the first column. The four column sections are marked as 1, 2, 3, and 4 in Fig. 5.14a. In

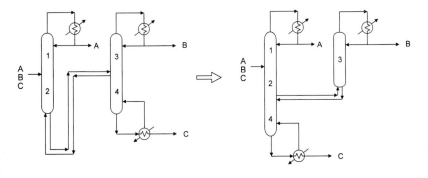

(a) Thermally-coupled direct
      sequence

(b) Side - rectifier arrangement

Figure 5.14 Thermal coupling of the direct sequence.

Fig. 5.14*b*, the four column sections are rearranged to form a *side-rectifier* arrangement.[12]

Similarly, Fig. 5.15*a* shows a thermally coupled indirect sequence. The condenser of the first column is replaced by a thermal coupling. The four column sections are again marked as 1, 2, 3, and 4 in Fig. 5.15*a*. In Fig. 5.15*b*, the four column sections are arranged to form a *side-stripper* arrangement.[12]

Both the side-rectifier and side-stripper arrangements have been shown to reduce the energy consumption compared with simple two-column arrangements.[10,13] This results from reduced mixing losses in the first (main) column. As with the first column of the simple sequence, a peak in composition occurs with the middle product. Now, however, advantage of the peak is taken by transferring material to the side-rectifier or side-stripper.

Consider now thermal coupling of the prefractionator arrangement from Fig. 5.11*b*. Figure 5.16*a* shows a prefractionator arrangement with partial condenser and reboiler on the prefractionator. Figure 5.16*b* shows the equivalent thermally coupled prefractionator arrangement sometimes known as a *Petlyuk column*. To make the two arrangements in Fig. 5.16 equivalent, the thermally coupled prefractionator requires extra plates to substitute for the prefractionator condenser and reboiler.[14]

Various studies[11,13,15–18] have compared the thermally coupled arrangement in Fig. 5.16*b* with a conventional arrangement using simple columns on a stand-alone basis. These studies show that the thermally coupled arrangement in Fig. 5.16*b* typically requires 30 percent less energy than a conventional arrangement using simple columns. The fully thermally coupled column in Fig. 5.16*b* also

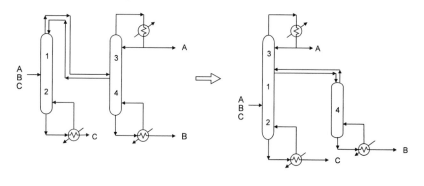

(a) Thermally-coupled indirect
     Sequence

(b) Side - Stripper arrangement

Figure 5.15 Thermal coupling of the indirect sequence.

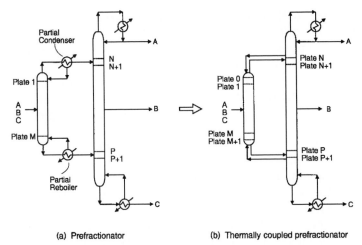

(a) Prefractionator   (b) Thermally coupled prefractionator

Figure 5.16 Thermal coupling of the prefractionator arrangement.

requires less energy than the side-rectifier and side-stripper arrangements for the same separation.[13] The energy saving for the thermally coupled prefractionator arrangement is the same as that for the prefractionator, with reduced mixing losses, as illustrated in Fig. 5.13.

The prefractionator arrangement in Fig. 5.17a and the thermally coupled prefractionator (Petlyuk column) in Fig. 5.17b are equivalent in terms of total heating and cooling duties.[14] Note that although the total heating and cooling duties are the same, there are differences in the temperatures at which the heat is supplied and rejected. Figure 5.17c shows an alternative configuration for the thermally coupled prefractionator which uses a single shell with a vertical baffle

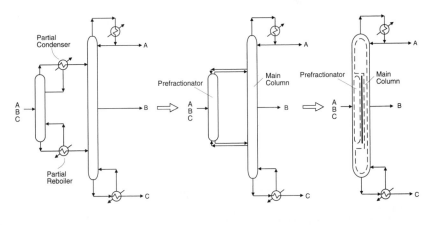

(a) Prefractionator arrangement   (b) Thermally coupled prefractionator   (c) Dividing wall column
                                              (Petlyuk Column)

Figure 5.17 The thermally coupled prefractionator can be arranged in a single shell.

dividing the central section of the shell into two parts, known as the *dividing-wall column.* The arrangements in Fig. 5.17 all require the same energy consumption, which is typically 30 percent less than a conventional arrangement. In the case of the prefractionator in Fig. 5.17a, the heat load is supplied at two points and rejected from two points. In addition, the dividing-wall column in Fig. 5.17c typically requires 30 percent less capital cost than a two-column arrangement of simple columns.

Although side-stripper arrangements are common in the petroleum industry, designers have been reluctant to use the fully thermally coupled arrangements in practical applications until recently.[17,18]

When the integration of sequences of simple columns was considered, it was observed that sequences with higher heat loads occurred simultaneously with more extreme levels. Heat integration always benefits from low heat loads and less extreme levels, as we shall see later in Chap. 12. Now consider the effect of thermal coupling arrangements on loads and levels. Figure 5.18 compares a

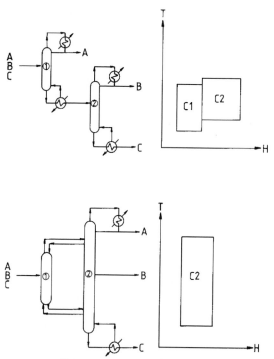

Figure 5.18 Relationship between heat load and level in simple and prefractionator sequences. *(From Smith and Linnhoff, Trans. IChemE, ChERD, 66: 195, 1988; reproduced by permission of the Institution of Chemical Engineers.)*

conventional and a thermally coupled arrangement in terms of temperature and enthalpy. In the conventional arrangement there is freedom to choose the pressures of the two columns independently, and thus the temperatures of the two condensers or the two reboilers can be varied independently. In the case of the thermally coupled arrangement, no such freedom exists. It can be seen in Fig. 5.18 that although the thermally coupled arrangement requires a smaller heat load than the conventional arrangement, more of the duties are at extreme levels. The smaller duties work to the benefit of heat integration, but the more extreme levels work to its detriment. Thus, if a thermally coupled arrangement is to be integrated, then the smaller loads and the more extreme levels may work to advantage or disadvantage depending on the circumstances.[4]

It is thus recommended that in a first pass through a design, thermal coupling should not be considered. Rather, simple columns should be used until a first overall design has been established. Only when the full heat-integration context has been understood should thermal coupling be considered.

5.8 Optimization of a Reducible Structure

Given the possibilities for changing the sequence of simple columns with the introduction of prefractionators, side-strippers, side-rectifiers, and fully thermally coupled arrangements, it is apparent that the problem is extremely complex with many structural alternatives. The problem can be tackled using the approach based on optimization of a reducible structure. As discussed in Chap. 1, this approach starts by setting up a "grand" flowsheet in which all candidates for an optimal solution are embedded. A reducible structure for a four-component mixture is shown in Fig. 5.19a. This structure is then subjected to optimization, and during the optimization, some features of the design are discarded, as shown in Fig. 5.19b. Methods also have been developed which consider sequencing, thermal coupling, and heat integration.[20]

As discussed earlier, the application of such techniques should be restricted until later in the design when the full heat-integration context both within and outside the distillation system has been established.

5.9 Distillation Sequencing—Summary

Unless there are constraints severely restricting heat integration, sequencing of simple distillation columns can be carried out in two steps: (1) identify the best few nonintegrated sequences and (2) study the heat integration. In most cases there is no need to solve the problem simultaneously.

The best few nonintegrated sequences can be identified most simply using the total vapor load as a criterion. If this is not satisfactory, then the alternative sequences can be sized and costed using shortcut techniques.

Complex column arrangements, such as the Petlyuk column, offer large potential savings in energy compared with sequences of simple columns. The dividing-wall also offers large potential savings in capital cost. However, it is recommended that complex column arrangements only be considered on a second pass through the

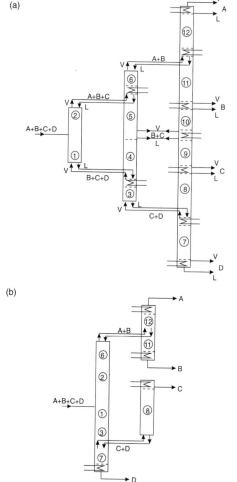

Figure 5.19 The approach based on optimization of a reducible structure starts with the most general configuration and simplifies. *(From Eliceche and Sargent, IChemE Symp. Series No. 61: 1, 1981; reproduced by permission of the Institution of Chemical Engineers*

design after first establishing a complete design with simple columns. Once this first complete design is established, then thermally coupled arrangements can be evaluated in the context of the overall design.

5.10   References

1. King, C. J., *Separation Processes,* 2d ed., McGraw-Hill, New York, 1980.
2. Nishida, G., Stephanopoulos, G., and Westerberg, A. W., "A Review of Process Synthesis," *AIChEJ,* 27: 321, 1981.
3. Westerberg, A. W., "The Synthesis of Distillation-Based Separation Systems," *Comp. Chem. Eng.,* 9: 421, 1985.
4. Smith, R., and Linnhoff, B., "The Design of Separators in the Context of Overall Processes," *Trans. IChemE. ChERD,* 66: 195, 1988.
5. Stephanopoulos, G., Linnhoff, B., and Sophos, A., "Synthesis of Heat Integrated Distillation Sequences," *IChemE Symp. Ser.,* 74: 111, 1982.
6. Porter, K. E., and Momoh, S. O., "Finding the Optimum Sequence of Distillation Columns—An Equation to Replace the Rules of Thumb (Heuristics)," *Chem. Engg. J.,* 46: 97, 1991.
7. Freshwater, D. C., and Ziogou, E., "Reducing Energy Requirements in Unit Operations," *Chem. Engg., J.,* 11: 215, 1976.
8. Morari, M., and Faith, D. C., "The Synthesis of Distillation Trains with Heat Integration," *AIChEJ,* 26: 916, 1980.
9. Andrecovich, M. J., and Westerberg, A. W., "An MILP Formulation of Heat Integrated Distillation Sequences," *AIChEJ,* 31: 1461, 1985.
10. Tedder, D. W., and Rudd, D. F., "Parametric Studies in Industrial Distillation," *AIChEJ,* 24: 303, 1978.
11. Triantafyllou, C., and Smith, R., "The Design and Optimization of Fully Thermally Coupled Distillation Columns," *Trans. IChemE,* Part A, 70: 118, 1992.
12. Calberg, N. A., and Westerberg, A. W., "Temperature-Heat Diagrams for Complex Columns: 2. Underwood's Method for Side Strippers and Enrichers," *Ind. Eng. Chem. Res.,* 28(9): 1379, 1989.
13. Glinos, K., and Malone, M. F., "Optimality Regions for Complex Column Alternatives in Distillation Columns," *Trans. IChemE ChERD,* 66: 229, 1988.
14. Aichele, P., "Sequencing Distillation Operations Being Serviced by Multiple Utilities," M.Sc. dissertation, UMIST, U.K., 1992.
15. Petlyuk, F. B., Platonov, V. M., and Slavinskii, D. M., "Thermodynamically Optimal Method for Separating Multicomponent Mixtures," *Int. Chem. Eng.,* 5: 555, 1965.
16. Stupin, W. J., and Lockhart, F. J., "Thermally Coupled Distillation—A Case History," *Chem. Eng. Progr.,* 68: 71, 1972.
17. Kaibel, G., "Distillation Columns with Vertical Partitions," *Chem. Eng. Technol.,* 10: 92, 1987.
18. Kaibel, G., "Distillation Column Arrangements with Low Energy Consumption," *IChemE Symp. Ser.,* 109: 43, 1988.
19. Eliceche, A. M., and Sargent, R. W. H., "Synthesis and Design of Distillation Systems," *IChemE Symp. Ser.,* 61(1): 1, 1981.
20. Kakhu, A. I., and Flower, J. R., "Synthesising Heat-Integrated Distillation Systems Using Mixed Integer Programming," *Trans. IChemE ChERD,* 66: 241, 1988.

6

Heat Exchanger
Network and Utilities:
Energy Targets

The design philosophy started at the heart of the onion with the reactor and moved out to the next layer, which is the separation and recycle system. Acceptance of the major processing steps (reactors, separators, and recycles) in the inner two layers fixes the material and energy balance. Thus the heating and cooling duties for the outer two layers of the onion (i.e., the heat exchanger network and utilities) are now known. To complete a first pass through the total problem, a design for these two outer layers must now be provided. However, completing the design of the heat exchanger network is not necessary in order to assess the completed design. *Targets* can be set for the heat exchanger network and utilities to assess the performance of the complete process design without actually having to carry out the design. These targets allow both energy and capital costs of the outer two layers to be assessed. Moreover, the targets allow the designer to suggest process changes for the inner layers of the onion (reactor, separation and recycle systems) which improve the targets for energy and capital costs of the two outer layers (heat exchanger network and utilities).

Using targets rather than design for the outer layers allows many design alternatives to be screened quickly and conveniently. Screening many design alternatives by complete designs is usually simply not practical in terms of the time and effort required. Using targets to suggest design changes works inward to the center of the onion and *evolves* the design for the inner layers. First, consider the details of how to set energy targets. Capital cost targets are considered in the next chapter. In later chapters we shall consider how the targets are used to suggest design improvements to the reaction, separation, and recycle systems.

6.1 Composite Curves

The analysis of the heat exchanger network first identifies sources of heat (termed *hot streams*) and sinks (termed *cold streams*) from the material and energy balance. Consider first a very simple problem with just one hot stream (heat source) and one cold stream (heat sink). The initial temperature (termed *supply temperature*), final temperature (termed *target temperature*), and enthalpy change of both streams are given in Table 6.1.

Steam is available at 180°C and cooling water at 20°C. Clearly, it is possible to heat the cold stream using steam and cool the hot stream using cooling water. However, this would incur excessive energy costs. It is also incompatible with the goals of sustainable industrial activity, which call for use of the minimum energy practicable. Instead, it is preferable to try to recover heat, if this is possible. The scope for heat recovery can be identified by plotting both streams on temperature-enthalpy axes. For feasible heat exchange between the two streams, the hot stream must be hotter than the cold stream at all points. Figure 6.1a shows the temperature-enthalpy plot for this problem with a minimum temperature difference $\Delta T_{\min}$ of 10°C. The region of overlap between the two streams in Fig. 6.1a identifies the amount of heat recovery possible (for $\Delta T_{\min} = 10$°C). For this problem, the heat recovery Q_{REC} is 11 MW. The part of the cold stream which extends beyond the start of the hot stream in Fig. 6.1a cannot be heated by recovery and requires steam. This is the minimum hot utility or *energy target* $Q_{H\min}$, which for this problem is 3 MW. The part of the hot stream which extends beyond the start of the cold stream in Fig. 6.1a cannot be cooled by heat recovery and requires cooling water. This is the minimum cold utility $Q_{C\min}$, which for this problem is 1 MW. Also shown at the bottom of Fig. 6.1a is the arrangement of heat exchangers that corresponds to the temperature-enthalpy plot.

TABLE 6.1 Two-Stream Heat Recovery Problem

Stream	Type	Supply temp. T_S (°C)	Target temp. T_T (°C)	ΔH (MW)
1	Cold	30	100	14
2	Hot	150	30	−12

The temperatures or enthalpy change for the streams (and hence their slope) cannot be changed, but the relative position of the two streams can be changed by moving them horizontally relative to each other. This is possible because the reference enthalpy for the hot stream can be changed independently from the reference enthalpy for the cold stream. Figure 6.1*b* shows the same two streams moved to a different relative position such that $\Delta T_{\min}$ is now 20°C. The amount of overlap between the streams is reduced (and hence heat recovery is reduced) to 10 MW. More of the cold stream extends beyond the start of the hot stream, and hence the amount of steam is increased to 4 MW. Also, more of the hot stream extends beyond the start of the cold stream, increasing the cooling water demand to 2 MW. Thus this approach of plotting a hot and a cold stream on the same temperature-enthalpy axis can determine hot and cold utility for a given value of $\Delta T_{\min}$. Let us now extend this approach to many hot and cold streams.

Consider the simple flowsheet shown in Fig. 6.2. Flow rates, temperatures, and heat duties for each stream are shown. Two of the streams in Fig. 6.2 are sources of heat (hot streams) and two are sinks for heat (cold streams). Assuming that heat capacities are constant, the hot and cold streams can be extracted as given in Table 6.2. Note that the heat capacities CP are total heat capacities and

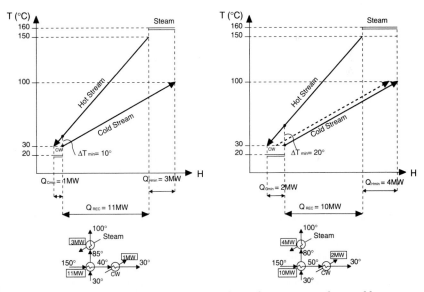

Figure 6.1 A simple heat recovery problem with one hot stream and one cold stream.

are the product of mass flow rate and specific heat capacity ($CP = mC_P$). Had the heat capacities varied significantly, we would have to represent the nonlinear temperature-enthalpy behavior by a series of linear *segments*.[1]

Instead of dealing with individual streams as given in Table 6.1, an overview of the process is needed. Figure 6.3a shows the two hot streams presented individually on temperature-enthalpy axes. How these hot streams behave overall can be quantified by combining them in given temperature ranges.[1] The temperature ranges in question are defined where an alteration occurs in the overall rate of change of enthalpy with temperature. If heat capacity is constant, then alterations will occur only when streams start or finish. Thus in Fig. 6.3 the temperature axis is divided into ranges defined by the supply and target temperatures of the streams.

Within each temperature range the streams are combined to produce a composite hot stream. This composite hot stream has a CP

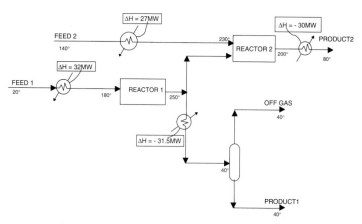

Figure 6.2 A simple flowsheet with two hot streams and two cold streams.

TABLE 6.2 Heat Exchange Stream Data for the Flowsheet in Fig. 6.2

Stream	Type	Supply temp. T_S (°C)	Target temp. T_T (°C)	ΔH (MW)	Heat capacity flow rate CP (MW °C^{-1})
1. Reactor 1 feed	Cold	20	180	32.0	0.2
2. Reactor 1 product	Hot	250	40	−31.5	0.15
3. Reactor 2 feed	Cold	140	230	27.0	0.3
4. Reactor 2 product	Hot	200	80	−30.0	0.25

in any temperature range that is the sum of the individual streams. In any temperature range, the enthalpy change of the composite is the sum of the enthalpy changes of the individual streams. Figure 6.3*b* shows the *composite curve* of the hot streams.[1] The composite stream represents how the individual streams would behave if they were a single stream. Similarly, the composite curve of the cold streams for the problem can be produced (Fig. 6.4).

The composite hot and cold curves can now be plotted on the same

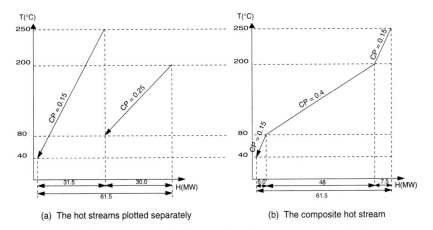

(a) The hot streams plotted separately   (b) The composite hot stream

Figure 6.3 The hot streams can be combined to obtain a composite hot stream.

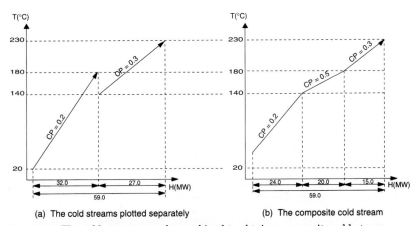

(a) The cold streams plotted separately   (b) The composite cold stream

Figure 6.4 The cold streams can be combined to obtain a composite cold stream.

axes (Fig. 6.5a). The curves in Fig. 6.5a are set to have a minimum temperature difference ΔT_{min} of 10°C. Where the curves overlap in Fig. 6.5a, heat can be rejected vertically from the hot streams which comprise the hot composite curve into the cold streams which comprise the cold composite curve. The way in which the composite curves are constructed (i.e., monotonically decreasing hot composite curve and monotonically increasing cold composite curve) allows maximum overlap between the curves and hence maximum heat recovery. In this problem for $\Delta T_{min} = 10°C$, the maximum heat recovery Q_{REC} is 51.5 MW.

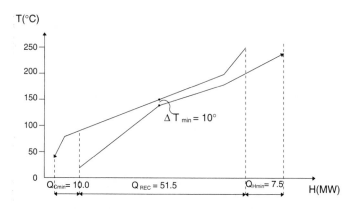

(a) The hot and cold composite curves plotted together at $\Delta T_{min} = 10°C$.

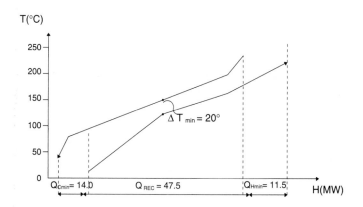

(b) Increasing ΔT_{min} from 10° to 20°C increases the hot and cold utility targets.

Figure 6.5 Plotting the hot and cold composite curves together allows the targets for hot and cold utility to be obtained.

Where the cold composite curve extends beyond the start of the hot composite curve in Fig. 6.5a, heat recovery is not possible, and the cold composite curve must be supplied with an external hot utility such as steam. This represents the target for hot utility (Q_{Hmin}).[1] For this problem, with $\Delta T_{min} = 10°C$, $Q_{Hmin} = 7.5$ MW. Where the hot composite curve extends beyond the start of the cold composite curve in Fig. 6.5a, heat recovery is again not possible, and the hot composite curve must be supplied with an external cold utility such as cooling water. This represents the target for cold utility (Q_{Cmin}).[1] For this problem, with $\Delta T_{min} = 10°C$, $Q_{Cmin} = 10.0$ MW.

Specifying the hot utility or cold utility or ΔT_{min} fixes the relative position of the two curves. As with the simple problem in Fig. 6.2, the relative position of the two curves is a degree of freedom at our disposal. Again, the relative position of the two curves can be changed by moving them horizontally relative to each other. Clearly, to consider heat recovery from hot streams into cold, the hot composite must be in a position such that everywhere it is above the cold composite for feasible heat transfer. Thereafter, the relative position of the curves can be chosen. Figure 6.5b shows the curves set to $\Delta T_{min} = 20°C$. The hot and cold utility targets are now increased to 11.5 and 14 MW, respectively.

Figure 6.6 illustrates what happens to the cost of the system as the relative position of the composite curves is changed over a range of values of ΔT_{min}. When the curves just touch, there is no driving force for heat transfer at one point in the process, which would require an

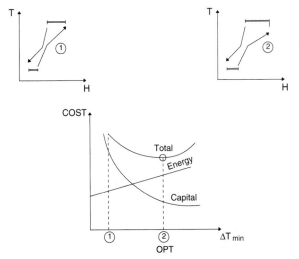

Figure 6.6 The correct setting for ΔT_{min} is fixed by economic tradeoffs.

infinite heat transfer area and hence infinite capital cost. As the energy target (and hence ΔT_{min} between the curves) is increased, the capital cost decreases. This results from increased temperature differences throughout the process, decreasing the heat transfer area. On the other hand, the energy cost increases as ΔT_{min} increases. There is a tradeoff between energy and capital cost; i.e., there is an economic degree of energy recovery. Chapter 7 explains how this tradeoff can be carried out using energy and capital cost targets.

However, care should be taken not to ignore practical constraints when setting ΔT_{min}. To achieve a small ΔT_{min} in a design requires heat exchangers that exhibit pure countercurrent flow. With shell-and-tube heat exchangers this is not possible, even if single-shell pass and single-tube pass designs are used, because the shell-side stream takes periodic cross-flow. Consequently, operating with a ΔT_{min} less than 10°C is illadvised unless special circumstances prevail.[2] A smaller value of 5°C can be achieved with plate heat exchangers, and the value can go as low as 1 to 2°C with plate-fin designs.[2] It should be noted, however, that such constraints apply only to those exchangers which occur around the point of closest approach between the composite curves. Additional constraints apply if the vaporization or condensation is occurring at the point of closest approach.[2]

6.2   The Heat Recovery Pinch

As discussed earlier, the correct setting for the composite curves is determined by an economic tradeoff between energy and capital corresponding to an economic minimum temperature difference ΔT_{min} between the curves. Accepting for the moment that the correct economic ΔT_{min} is known, this fixes the relative position of the composite curves and hence the energy target. The ΔT_{min} for the composite curves and its location have important implications for design if the energy target is to be achieved in the design of a heat exchanger network. The ΔT_{min} is normally observed at only one point between the hot and the cold composite curves, called the *heat recovery pinch*.[1] The pinch point has a special significance.

The tradeoff between energy and capital in the composite curves suggests that, on average, individual exchangers should have a temperature difference no smaller than ΔT_{min}. A good initialization in heat exchanger network design is to assume that no individual heat exchanger has a temperature difference smaller than the ΔT_{min} between the composite curves.

With this rule in mind, divide the process at the pinch as shown in

Fig. 6.7*a*. Above the pinch (in temperature terms), the process is in heat balance with the minimum hot utility $Q_{H\min}$. Heat is received from hot utility, and no heat is rejected. The process acts as a heat sink. Below the pinch (in temperature terms), the process is in heat balance with the minimum cold utility $Q_{C\min}$. No heat is received, but heat is rejected to cold utility. The process acts as a heat source.

Consider now the possibility of transferring heat between these two systems (see Fig. 6.7*b*). Figure 6.7*b* shows that it is possible to transfer heat from hot streams above the pinch to cold streams below. The pinch temperature for hot streams for the problem is 150°C, and that for cold streams is 140°C. Transfer of heat from above the pinch to below as shown in Fig. 6.7*b* transfers heat from hot streams with a temperature of 150°C or greater into cold streams with a temperature of 140°C or less. This is clearly possible. By contrast, Fig. 6.7*c* shows that transfer from hot streams below the pinch to cold streams above is not possible. Such transfer requires heat being transferred from hot streams with a temperature of 150°C or less into cold streams with a temperature of 140°C or greater. This is clearly not possible (without violating the $\Delta T_{\min}$ constraint).

In choosing to transfer heat, say *XP*, from the system above the pinch to the system below the pinch, as shown in Fig. 6.8*a*, then above the pinch there is a heat deficit of *XP*. The only way this can

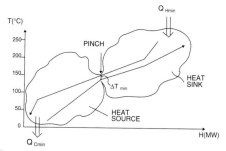

(a) The pinch divides the process into a heat source and a heat sink.

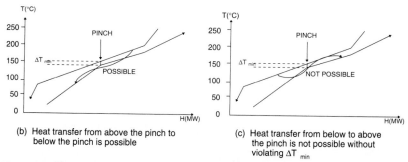

(b) Heat transfer from above the pinch to below the pinch is possible

(c) Heat transfer from below to above the pinch is not possible without violating $\Delta T_{\min}$

Figure 6.7 The composite curves set the energy target and the location of the pinch.

be corrected is by import of an extra *XP* from a hot utility. Likewise, an excess of *XP* below the pinch leads to export of an extra *XP* to a cold utility.[1]

Analogous effects are caused by the inappropriate use of utilities. Utilities are appropriate if they are necessary to satisfy the enthalpy imbalance in that part of the process. Above the pinch in Fig. 6.7*a*, steam is needed to satisfy the enthalpy imbalance. Figure 6.8*b* illustrates what happens if inappropriate use of utilities is made and some cooling water is used to cool hot streams above the pinch, say, *XP*. To satisfy the enthalpy imbalance above the pinch, an import of $(Q_{H\text{min}} + XP)$ is needed from steam. Overall, $(Q_{C\text{min}} + XP)$ of cooling water is used.[1]

An alternative inappropriate use of utilities involves heating of some of the cold streams below the pinch by steam. Below the pinch, cooling water is needed to satisfy the enthalpy imbalance. Figure

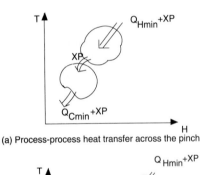

(a) Process-process heat transfer across the pinch

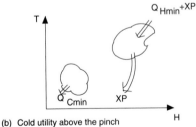

(b) Cold utility above the pinch

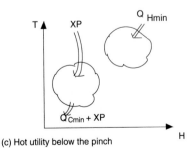

(c) Hot utility below the pinch

Figure 6.8 Three forms of cross-pinch heat transfer.

6.8c illustrates what happens if *XP* of steam is used below the pinch. Q_{Hmin} must still be supplied above the pinch to satisfy the enthalpy imbalance above the pinch. Overall, $(Q_{Hmin} + XP)$ of steam is used as well as $(Q_{Cmin} + XP)$ of cooling water.[1]

In other words, to achieve the energy target set by the composite curves, the designer must *not* transfer heat across the pinch by

1. Process-to-process heat transfer

2. Inappropriate use of utilities

These rules are both necessary and sufficient to ensure that the target is achieved, providing the initialization rule is adhered to that no individual heat exchanger should have a temperature difference smaller than ΔT_{min}.

Figure 6.9a shows a design corresponding to the flowsheet in Fig. 6.2 which achieves the target of $Q_{Hmin} = 7.5$ MW and $Q_{Cmin} = 10$ MW for $\Delta T_{min} = 10°C$. Figure 6.9b shows an alternative representation of the flowsheet, known as the *grid diagram*.[3] The grid diagram shows only heat transfer operations. Hot streams are at the top running left to right. Cold streams are at the bottom running right to left. A heat exchange match is represented by a vertical line joining two circles on the two streams being matched. An exchanger using a hot utility is represented by a circle with a *H*. An exchanger using cold utility is represented by a circle with a *C*. The importance of the grid diagram is clear in Fig. 6.9b, since the pinch, and how it divides the process into two parts, is easily accommodated. Dividing the process into two parts on a conventional diagram such as that shown in Fig. 6.9a is both difficult and extremely cumbersome.

Details of how this design was developed in Fig. 6.9 are included in Chap. 16. For now, simply take note that the targets set by the composite curves are achievable in design, providing that the pinch is recognized, there is no transfer of heat across it, and no inappropriate use of utilities occurs. However, insight into the pinch is needed to analyze some of the important decisions still to be made before network design is addressed.

6.3   Threshold Problems

Not all problems have a pinch to divide the process into two parts.[1] Consider the composite curves in Fig. 6.10a. At this setting, both steam and cooling water are required. As the composite curves are moved closer together, both the steam and cooling water requirements decrease until the setting shown in Fig. 6.10b results. At this setting, the composite curves are in alignment at the hot end, indicating that there is no longer a demand for a hot utility. Moving the curves closer together, as shown in Fig. 6.10c, decreases the cold utility demand at the cold end but opens up a demand for a cold utility at the hot end corresponding with the decrease at the cold end. In other words, as the curves are moved closer together, beyond the setting in Fig. 6.10b, the utility demand is constant. The setting shown in Fig. 6.10b marks a threshold, and problems that exhibit this feature are known as *threshold problems.*[1]

In some threshold problems, the hot utility requirement disappears, as in Fig. 6.10. In others, the cold utility disappears, as shown in Fig. 6.11.

Considering the capital/energy tradeoff for threshold problems, two possible outcomes are shown in Fig. 6.12. Below the threshold

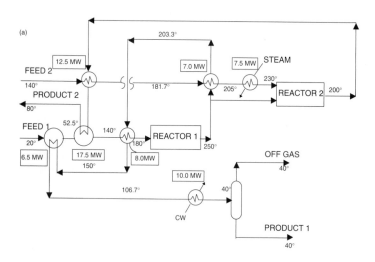

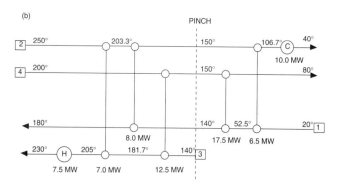

Figure 6.9 A design that achieves the energy target.

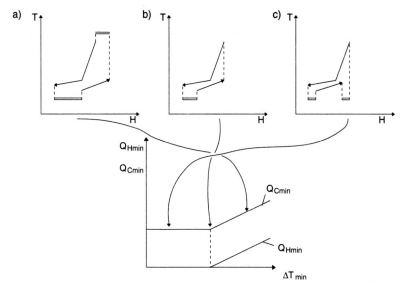

Figure 6.10 As ΔT_{min} is varied, some problems require only a cold utility below a threshold value.

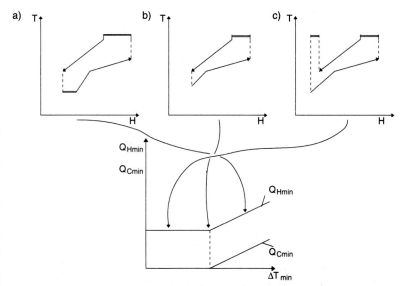

Figure 6.11 In some threshold problems, only a hot utility is required below the threshold value of ΔT_{min}.

$\Delta T_{\min}$, energy costs are constant, since utility demand is constant. Figure 6.12a shows a situation where the optimum occurs at the threshold $\Delta T_{\min}$. Figure 6.12b shows a situation where the optimum occurs above the threshold $\Delta T_{\min}$. The flat profile of energy costs below the threshold $\Delta T_{\min}$ means that the optimum can never occur below the threshold value. It can only be at or above the threshold value.

In a situation as shown in Fig. 6.12a, with the optimal $\Delta T_{\min}$ at the threshold, then there is no pinch. On the other hand, in a situation as shown in Fig. 6.12b, with the optimum above the threshold value, there is a demand for both utilities, and there is a pinch.

It is interesting to note that threshold problems are quite common in practice, and although they do not have a process pinch, pinches

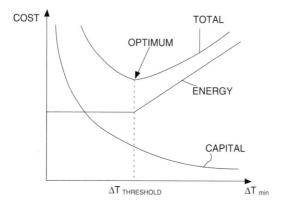

(a) The capital - energy trade-off can lead to an optimum at Threshold $\Delta T_{\min}$

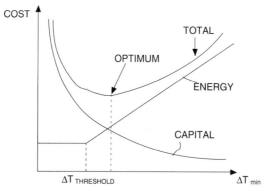

(b) The capital - energy trade-off can lead to an optimum above $\Delta T_{\text{THRESHOLD}}$

Figure 6.12 The optimal setting of the capital/energy tradeoff for threshold problems.

are introduced into the design when multiple utilities are added. Figure 6.13*a* shows composite curves similar to the composite curves in Fig. 6.10 but with two levels of cold utility instead of one. In this case, the second cold utility is steam generation. The introduction of this second utility causes a pinch. This is known as a *utility pinch* because it is caused by the introduction of an additional utility.[1]

Figure 6.13*b* shows composite curves similar to those in Fig. 6.11 but with two levels of steam used. Again, the introduction of a second steam level causes a utility pinch.

In design, the same rules must be obeyed around a utility pinch as around a process pinch. Heat should not be transferred across it by process-to-process transfer, and there should be no inappropriate use of utilities. In Fig. 6.13*a* this means that the only utility to be used above the utility pinch is steam generation and only cooling water below. In Fig. 6.13*b* this means that the only utility to be used above

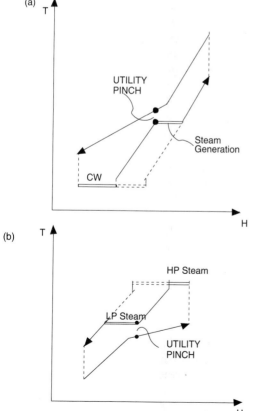

Figure 6.13 Threshold problems are turned into a pinch problem when additional utilities are added.

the utility pinch is high-pressure steam and only low-pressure steam below.

6.4   The Problem Table Algorithm

Although composite curves can be used to set energy targets, they are inconvenient because they are based on a graphic construction. A method of calculating energy targets directly without the need for graphic construction can be developed.[3] The process is first divided into temperature intervals in the same way as was done for construction of the composite curves. Figure 6.14a shows that it is not possible to recover all the heat in each temperature interval, since temperature driving forces are not feasible throughout the interval. This problem can be overcome if, purely for the purposes of construction, the hot composite is pretended to be $\Delta T_{min}/2$ colder than it is in practice and the cold composite is pretended to be $\Delta T_{min}/2$ hotter than it is in practice (see Fig. 6.14b). The *shifted composite curves* now touch at the pinch. Carrying out a heat balance between the shifted composite curves within a *shifted temperature interval* shows that heat transfer is feasible throughout each shifted temperature interval, since hot streams in practice are actually $\Delta T_{min}/2$ hotter and cold streams $\Delta T_{min}/2$ colder. Within each shifted interval the hot streams are in reality hotter than the cold streams by just ΔT_{min}.

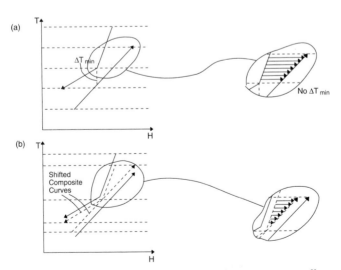

Figure 6.14 Shifting the composite curves in temperature allows complete heat recovery within temperature intervals.

It is important to note that shifting the curves vertically does not alter the horizontal overlap between the curves. It therefore does not alter the amount by which the cold composite curve extends beyond the start of the hot composite curve at the hot end of the problem and the amount by which the hot composite curve extends beyond the start of the cold composite curve at the cold end. The shift simply removes the problem of ensuring temperature feasibility within temperature intervals.

This shifting technique can be used to develop a strategy to calculate the energy targets without having to construct composite curves:[3]

1. Set up shifted temperature intervals from the stream supply and target temperatures by subtracting $\Delta T_{min}/2$ from the hot streams and adding $\Delta T_{min}/2$ to the cold streams (as in Fig. 6.14b).

2. In each shifted temperature interval, calculate a simple energy balance from

$$\Delta H_i = \left(\sum_{\substack{\text{All cold streams}}} CP_C - \sum_{\substack{\text{All hot streams}}} CP_H \right)_i \Delta T_i \qquad (6.1)$$

where ΔH_i is the heat balance for shifted interval i and ΔT_i is the temperature difference across it. If cold streams dominate hot streams, the interval has a net deficit of heat, and ΔH is positive. If hot streams dominate cold streams, the interval has a net surplus of heat, and ΔH is negative. This is consistent with standard thermodynamic convention (e.g., for an exothermic reaction ΔH is negative). If no hot utility is used, this is equivalent to constructing the shifted composite curves shown in Fig. 6.15a.

3. The overlap in the shifted curves as shown in Fig. 6.15a means that heat transfer is infeasible. At some point this overlap is a maximum. This maximum overlap is added as a hot utility to correct the overlap. The shifted curves now touch at the pinch, as shown in Fig. 6.15b. Since the shifted curves just touch, the actual curves are separated by ΔT_{min} at this point (see Fig. 6.15b).

This basic approach can be developed into a formal algorithm known as the *problem table algorithm*.[3] To illustrate the algorithm, it can be developed using the data from Fig. 6.2 given in Table 6.2 for $\Delta T_{min} = 10°C$.

First, determine the shifted temperature intervals T^* from actual supply and target temperatures. Hot streams are shifted down in temperature by $\Delta T_{min}/2$ and cold streams up by $\Delta T_{min}/2$, as detailed in Table 6.3. The stream population is shown in Fig. 6.16 with a vertical temperature scale. The interval temperatures are shown in

Fig. 6.16 set to $\Delta T_{min}/2$ below hot stream temperatures and $\Delta T_{min}/2$ above cold stream temperatures.

Next, we carry out a heat balance within each shifted temperature interval according to Eq. (6.1). The result is given in Fig. 6.17. Some of the shifted intervals in Fig. 6.17 are seen to have a surplus of heat

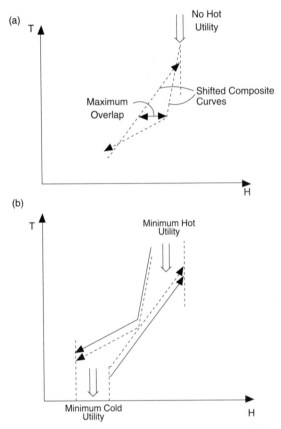

Figure 6.15 The utility target can be determined from the maximum overlap between the shifted composite curves.

TABLE 6.3 Shifted Temperatures for the Data from Table 6.2

Stream	Type	T_S	T_T	T_S^*	T_T^*
1	Cold	20	180	25	185
2	Hot	250	40	245	35
3	Cold	140	230	145	235
4	Hot	200	80	195	75

Interval
Temperature
(°C)

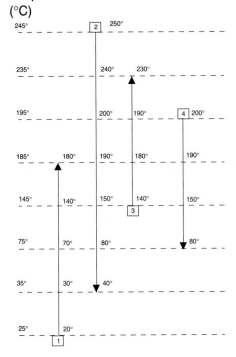

Figure 6.16 The stream population for the data in Table 6.2.

Interval Temperature	Stream Population	$\Delta T_{INTERVAL}$	ΣCP_C $-\Sigma CP_H$	$\Delta H_{INTERVAL}$	Surplus/ Deficit
245°C					
		10	- 0.15	- 1.5	Surplus
235°C					
		40	0.15	6.0	Deficit
195°C					
		10	- 0.1	- 1.0	Surplus
185°C					
		40	0.1	4.0	Deficit
145°C					
		70	- 0.2	- 14.0	Surplus
75°C					
		40	0.05	2.0	Deficit
35°C					
		10	0.2	2.0	Deficit
25°C					

Figure 6.17 The temperature-interval heat balances.

and some a deficit. The heat balance within each shifted interval allows maximum heat recovery within each interval. However, recovery also must be allowed between intervals.

Now *cascade* any surplus heat down the temperature scale from interval to interval. This is possible because any excess heat available from the hot streams in an interval is hot enough to supply a deficit in the cold streams in the next interval down. Figure 6.18 shows the cascade for the problem. First, assume that no heat is supplied to the first interval from a hot utility (Fig. 6.18a). The first interval has a surplus of 1.5 MW, which is cascaded to the next interval. This second interval has a deficit of 6 MW, which reduces the heat cascaded from this interval to −4.5 MW. In the third interval the process has a surplus of 1 MW, which leaves −3.5 MW to be cascaded to the next interval, and so on.

Looking at the heat flows in Fig. 6.18a, some are negative, which is infeasible. Heat cannot be transferred up the temperature scale. To

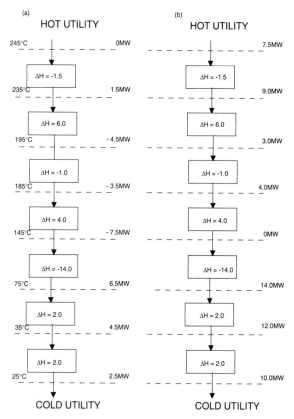

Figure 6.18 The problem table cascade.

make the cascade feasible, sufficient heat must be added from a hot utility to make the heat flows at least zero. The smallest amount of heat needed from a hot utility is the largest negative heat flow from Fig. 6.18a, i.e., 7.5 MW. In Fig. 6.18b, 7.5 MW is added to the first interval from a hot utility. This does not change the heat balance within each interval but increases all the heat flows between intervals by 7.5 MW, giving one heat flow of just zero at an interval temperature of 145°C.

More than 7.5 MW could be added from a hot utility to the first interval, but the objective is to find the minimum hot and cold utility. Thus from Fig. 6.18b, $Q_{Hmin} = 7.5$ MW and $Q_{Cmin} = 10$ MW. This corresponds with the values obtained from the composite curves in Fig. 6.5a. One further important piece of information can be deduced from the cascade in Fig. 6.18b. The point where the heat flow goes to zero at $T^* = 145°C$ corresponds to the pinch. Thus the actual hot and cold stream pinch temperatures are 150 and 140°C. Again, this agrees with the result from the composite curves in Fig. 6.5a.

The initial setting for the heat cascade in Fig. 6.18a corresponds to the shifted composite curve setting in Fig. 6.15a where there is an overlap. The setting of the heat cascade for zero or positive heat flows in Fig. 6.18b corresponds to the shifted composite curve setting in Fig. 6.15b.

The composite curves are useful in providing conceptual understanding of the process, but the problem table algorithm is a more convenient calculation tool.

Example 6.1 The flowsheet for a low-temperature distillation process is shown in Fig. 6.19. Calculate the minimum hot and cold utility requirements and the location of the pinch assuming $\Delta T_{min} = 5°C$.

Solution First, extract the stream data from the flowsheet. This is given in Table 6.4.

Next, calculate the shifted interval temperatures. Hot streams are shifted down by 2.5°C, and cold streams are shifted up by 2.5°C (Table 6.5).

Now, carry out a heat balance within each shifted temperature interval, as shown in Fig. 6.20.

Finally, the heat cascade is shown in Fig. 6.21. Figure 6.21a shows the cascade with zero hot utility. This leads to negative heat flows, the largest of which is -1.84 MW. Adding 1.84 MW from a hot utility as shown in Fig. 6.21b gives $Q_{Hmin} = 1.84$ MW, $Q_{Cmin} = 1.84$ MW, hot stream pinch temperature = $-19°C$, and cold stream pinch temperature = $-24°C$.

6.5 Process Constraints

So far it has been assumed that any hot stream could, in principle, be matched with any cold stream, providing there is feasible temperature difference between the two. Often, however, practical constraints prevent this. For example, it might be the case that if two

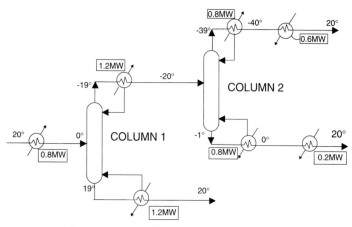

Figure 6.19 A low-temperature distillation process.

TABLE 6.4   Stream Data for Low-Temperature Distillation Process

Stream	Type	Supply temp. T_S (°C)	Target temp. T_T (°C)	ΔH (MW)	Heat capacity flow rate CP (MW °C^{-1})
1. Feed to column 1	Hot	20	0	0.8	0.04
2. Column 1 condenser	Hot	-19	-20	1.2	1.2
3. Column 2 condenser	Hot	-39	-40	0.8	0.8
4. Column 1 reboiler	Cold	19	20	1.2	1.2
5. Column 2 reboiler	Cold	-1	0	0.8	0.8
6. Column 2 bottoms	Cold	0	20	0.2	0.01
7. Column 2 overheads	Cold	-40	20	0.6	0.01

TABLE 6.5   Shifted Temperatures for the Data in Table 6.4

Stream	Type	T_S	T_T	T_S^*	T_T^*
1	Hot	20	0	17.5	-2.5
2	Hot	-19	-20	-21.5	-22.5
3	Hot	-39	-40	-41.5	-42.5
4	Cold	19	20	21.5	22.5
5	Cold	-1	0	1.5	2.5
6	Cold	0	20	2.5	22.5
7	Cold	-40	20	-37.5	22.5

streams are matched in a heat exchanger and a leak develops such that the two streams come into contact, this might produce an unacceptably hazardous situation. If this was the case, then no doubt a constraint would be imposed to prevent the two streams being matched. Another reason for a constraint might be that two streams are expected to be geographically very distant from each other, leading to unacceptably long pipe runs. Potential control and start-up problems also might call for constraints. There are many reasons why constraints might be imposed.

One common reason for imposing constraints results from *areas of integrity*.[4] A process is often normally designed to have logically identifiable sections or areas. An example might be "reaction area" and "separation area" of the process. These areas are kept separate for reasons such as start-up, shutdown, operational flexibility, safety, etc. The areas are often made operationally independent through the use of intermediate storage of process materials between the areas. Such independent areas are generally described as areas of integrity and impose constraints on the ability to transfer heat. Clearly, to maintain operational indepedence, two areas cannot be dependent on each other for heating and cooling by recovery.

The question now is, given that there are often constraints to deal with, how do we evaluate the effect of these constraints on the system performance? The problem table algorithm cannot be used directly if constraints are imposed. However, often the effect of constraints on

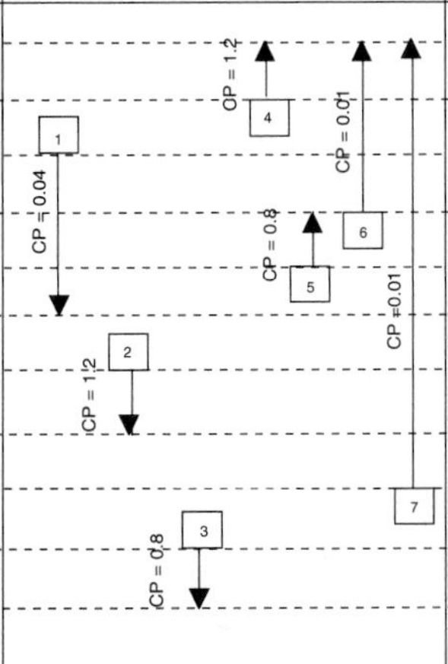

Interval Temperature	Stream Population	$\Delta T_{INTERVAL}$	$\Sigma CP_C - \Sigma CP_H$	$\Delta H_{INTERVAL}$	Surplus/ Deficit
22.5°C					
		1	1.22	1.22	Deficit
21.5°C					
		4	0.02	0.08	Deficit
17.5°C					
		15	-0.02	-0.30	Surplus
2.5°C					
		1	0.77	0.77	Deficit
1.5°C					
		4	-0.03	-0.12	Surplus
-2.5°C					
		19	0.01	0.19	Deficit
-21.5°C					
		1	-1.19	-1.19	Surplus
-22.5°C					
		15	0.01	0.15	Deficit
-37.5°C					
		4	0	0	----------
-41.5°C					
		1	-0.8	-0.8	Surplus
-42.5°C					

Figure 6.20 Temperature-interval heat balances for Example 6.1.

the energy performance can be evaluated using the problem table algorithm together with a little common sense. The following example illustrates how.[4]

Example 6.2 A process is to be divided into two operationally independent areas of integrity, area A and area B. The stream data for the two areas are given in Table 6.6.[4] Calculate the penalty in utility consumption to maintain the two areas of integrity for $\Delta T_{\min} = 20°C$.

Solution To identify the penalty, first calculate the utility consumption of the two areas separate from each other, as shown in Fig. 6.22a. Next, combine all the streams from both areas and again calculate the utility consumption (see Fig. 6.22b). Figure 6.23a shows the problem table cascade for area A, the cascade for area B is shown in Fig. 6.23b, and that for areas A and B combined is shown in Fig. 6.23c.

With areas A and B separate, the total hot utility consumption is (1400 +

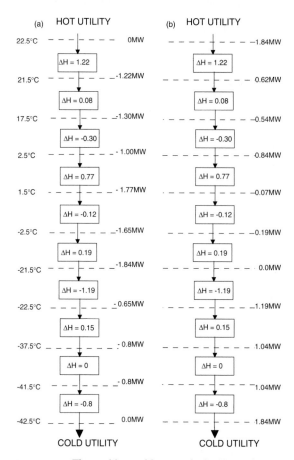

Figure 6.21 The problem table cascade for Example 6.1.

0) = 1400 kW and the total cold utility consumption is (0 + 1350) = 1350 kW. With areas A and B combined, the total utility consumption is 950 kW of hot utility and 900 kW of cold utility. The penalty for maintaining the areas of integrity is thus (1400 − 950) = 450 kW of hot utility and (1350 − 900) = 450 kW of cold utility.

The penalty as a result of the constraint having been identified enables judgment as to whether it is acceptable or too expensive. If it is too expensive, there is a choice between two options:

1. Reject the areas of integrity, and operate the process as a single system.

TABLE 6.6 Stream Data for Heat Recovery Between Two Areas of Integrity

	Area A				Area B		
Stream	T_S (°C)	T_T (°C)	CP (kW °C^{-1})	Stream	T_S (°C)	T_T (°C)	CP (kW °C^{-1})
1	190	110	2.5	3	140	50	20.0
2	90	170	20.0	4	30	120	5.0

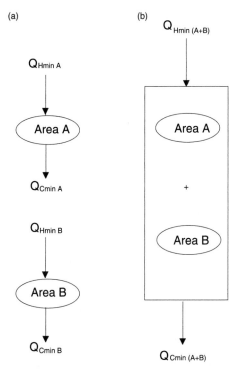

(a)

$Q_{Hmin\ A}$

Area A

$Q_{Cmin\ A}$

$Q_{Hmin\ B}$

Area B

$Q_{Cmin\ B}$

(b)

$Q_{Hmin\ (A+B)}$

Area A

+

Area B

$Q_{Cmin\ (A+B)}$

Figure 6.22 The areas of integrity can be targeted separately and then the combination targeted.

2. Find a way to overcome the constraint while still maintaining the areas. This is often possible by using indirect heat transfer between the two areas. The simplest option is via the existing utility system. For example, rather than have a direct match between two streams, one can perhaps generate steam to be fed into the steam mains and the other use steam from the same mains. The utility system then acts as a buffer between the two areas. Another possibility might be to use a heat transfer medium such as a hot oil which circulates between the two streams being matched. To maintain operational independence, a standby heater and cooler supplied by utilities is needed in the hot oil circuit such that if either area is not operational, utilities could substitute heat recovery for short periods.

Most constraints can be evaluated by scoping the problem with different boundaries, as illustrated in Example 6.2. If this approach cannot be applied, then mathematical programming must be used to obtain the energy target.[5,6]

6.6   Utility Selection

After maximizing heat recovery in the heat exchanger network, those heating duties and cooling duties not serviced by heat recovery must be provided by external utilities. The outer-most layer of the onion model is now being addressed, but still dealing with targets.

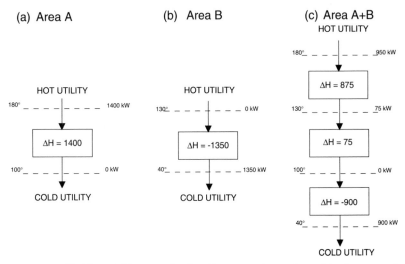

Figure 6.23 Problem table cascade for the separate and combined areas of integrity.

Utilities are varied. The most common hot utility is steam. It is usually available at several levels. High-temperature heating duties require furnace flue gas or a hot oil circuit. Cold utilities might be refrigeration, cooling water, air cooling, furnace air preheating, boiler feedwater preheating, or even steam generation at higher temperatures.

Although the composite curves can be used to set energy targets, they are not a suitable tool for the selection of utilities. The *grand composite curve*[1] is a more appropriate tool for understanding the interface between the process and the utility system. It is also, as is shown in later chapters, a useful tool for study of the interaction between heat-integrated reactors and separators and the rest of the process.

The grand composite curve is obtained by plotting the problem table cascade. A typical grand composite curve is shown in Fig. 6.24. It shows the heat flow through the process against temperature. It should be noted that the temperature plotted here is *shifted temperature* T^* and not actual temperature. Hot streams are represented $\Delta T_{min}/2$ colder and cold streams $\Delta T_{min}/2$ hotter than they are in practice. Thus an allowance for ΔT_{min} is built into the construction.

The point of zero heat flow in the grand composite curve in Fig. 6.24 is the pinch. The open "jaws" at the top and bottom represent Q_{Hmin} and Q_{Cmin}, respectively. Thus the heat sink above the pinch and heat source below the pinch can be identified as shown in Fig.

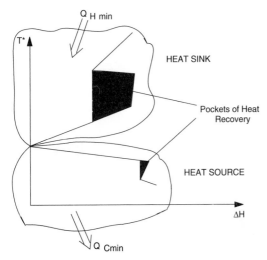

Figure 6.24 The grand composite curve shows the utility requirements in both enthalpy and temperature terms.

6.24. The shaded areas in Fig. 6.24, known as *pockets,* represent areas of additional process-to-process heat transfer. Remember that the profile of the grand composite curve represents residual heating and cooling demands after recovering heat within the shifted temperature intervals in the problem table algorithm. In these pockets in Fig. 6.24, a local surplus of heat in the process is used at temperature differences in excess of $\Delta T_{\min}$ to satisfy a local deficit.[1]

Figure 6.25a shows the same grand composite curve with two levels of saturated steam used as a hot utility. The steam system in Fig. 6.25a shows the low-pressure steam being desuperheated by injection of boiler feedwater after pressure reduction to maintain saturated conditions. Figure 6.25b shows again the same grand composite curve but with hot oil used as a hot utility.

Example 6.3 The problem table cascade for the process in Fig. 6.2 is given in Fig. 6.18. Using the grand composite curve:

a. For two levels of steam at saturation conditions and temperatures of 240 and 180°C, determine the loads on the two levels which maximizes use of the lower-pressure steam.

b. Instead of using steam, a hot oil circuit is to be used with a supply temperature of 280°C and $C_P = 2.1 \text{ kJ kg}^{-1}\text{K}^{-1}$. Calculate the minimum flow rate of hot oil.

Solution

a. For $\Delta T_{\min} = 10°C$, the two steam levels are plotted on the grand composite

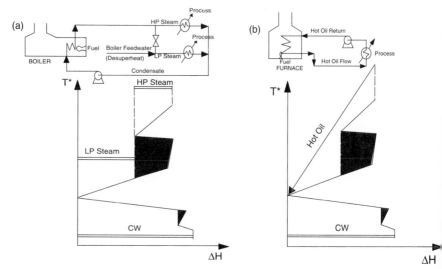

Figure 6.25. The grand composite curve allows alternative utility mixes to be evaluated.

curve at temperatures of 235 and 175°C. Figure 6.26a shows the loads which maximize use of the lower-pressure steam. Calculate the load on the low-pressure steam by interpolation of the cascade heat flows. At $T^* = 175$°C,

$$\text{Load of } 180°\text{C steam} = \frac{175 - 145}{185 - 145} \times 4$$

$$= 3 \text{ MW}$$

$$\text{Load of } 240°\text{C steam} = 7.5 - 3$$

$$= 4.5 \text{ MW}$$

b. Figure 6.26b shows the grand composite curve with hot oil providing the hot utility requirements. If the minimum flow rate is required, then this corresponds to the steepest slope and minimum return temperature. For this problem the minimum return temperature for the hot oil is the pinch temperature ($T^* = 145$°C, $T = 150$°C for hot streams). In other problems, the

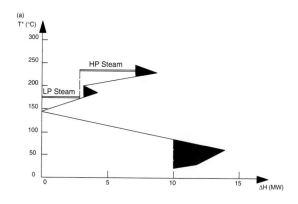

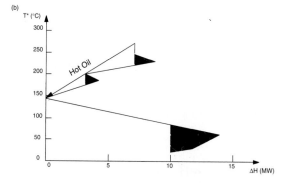

Figure 6.26 Alternative utility mixes for the process in Fig. 6.2.

process away from the pinch could have limited the flow rate. Thus

$$\text{Minimum flow rate} = 7.5 \times 10^3 \times \frac{1}{2.1} \times \frac{1}{(280 - 150)}$$

$$= 27.5 \text{ kg s}^{-1}$$

6.7 Furnaces

When a hot utility needs to be at a high temperature and/or provide high heat fluxes, radiant heat transfer is used from combustion of fuel in a furnace. Furnace designs vary according to the function of the furnace, heating duty, type of fuel, and method of introducing combustion air. Sometimes the function is to purely provide heat; sometimes the furnace is also a reactor and provides heat of reaction. However, process furnaces have a number of features in common. In the chamber where combustion takes place, the heat is transferred mainly by radiation to tubes around the walls of the chamber, through which passes the fluid to be heated. After the flue gas leaves the combustion chamber, most furnace designs extract further heat from the flue gas in a convection section before the flue gas is vented to the atmosphere.

Figure 6.27 shows a grand composite curve with a flue gas matched against it to provide a hot utility.[7] The flue gas starts at its *theoretical flame temperature* (shifted for ΔT_{min} on the grand composite curve) and presents a sloping profile because it is giving up sensible heat. Theoretical flame temperature is the temperature attained when a fuel is burnt in air or oxygen without loss or gain of heat. Methods are presented elsewhere for its calculation.[8] It should be emphasized that the theoretical and real flame temperatures will be quite different. The real flame temperature will be lower than the theoretical flame temperature because, in practice, heat is lost from the flame and because part of the heat released provides heat for a variety of endothermic dissociation reactions that occur at high temperatures, such as

$$CO_2 \rightleftharpoons CO + 1/2O_2$$

$$H_2O \rightleftharpoons H_2 + 1/2O_2$$

$$H_2O \rightleftharpoons 1/2H_2 + OH$$

etc.

However, as the temperature of the flue gas decreases as heat is extracted from it, the dissociation reactions reverse, and the heat is released. Thus, although theoretical flame temperature does not reflect the true flame temperature, it does provide a convenient reference to indicate how much heat is actually released by the combustion as the flue gas is cooled.[7] In Fig. 6.27 the temperature differences between the flue gas and process are indicated to be larger than they are in practice at the high-temperature end of the flue gas. As the flue gas is cooled and passes through the convection section of the furnace, the temperature differences indicated are more representative of what they would be in practice. The temperature differences in the radiant zone of the furnace cannot be represented accurately by simple models, and accurate representation requires a detailed simulation model. However, this does not usually present a problem in using the simple model represented in Fig. 6.27 for scoping and screening alternative designs, since temperature differences in the radiant zone are very large anyway. Let us emphasize again, however, that the simple model in Fig. 6.27 does allow the correct heat duty to be targeted.

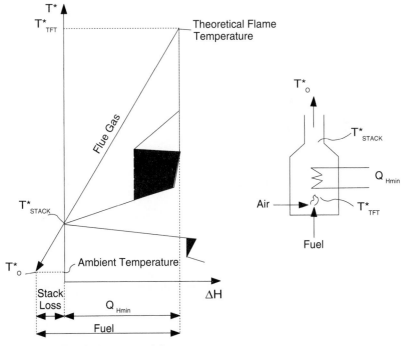

Figure 6.27 Simple furnace model.

In Fig. 6.27, the flue gas is cooled to pinch temperature before being released to the atmosphere. The heat released from the flue gas between pinch and ambient temperature is the stack loss. Thus, in Fig. 6.27, for a given grand composite curve and theoretical flame temperature, the heat from fuel and stack loss can be determined.

All combustion processes work with an excess of air or oxygen to ensure complete combustion of the fuel. Excess air typically ranges between 5 and 20 percent depending on the fuel, burner design, and furnace design. As excess air is reduced, theoretical flame temperature increases as shown in Fig. 6.28. This has the effect of reducing the stack loss and increasing the thermal efficiency of the furnace for a given process heating duty. Alternatively, if the combustion air is preheated (say, by heat recovery), then again the theoretical flame temperature increases as shown in Fig. 6.28, reducing the stack loss.

Although the higher flame temperatures in Fig. 6.28 reduce the fuel consumption for a given process heating duty, there is one significant disadvantage. The higher flame temperatures increase the formation of oxides of nitrogen, which are environmentally harmful. We shall return to this point in Chaps. 10 and 11.

In Figs. 6.27 and 6.28, the flue gas is capable of being cooled to pinch temperature before being released to the atmosphere. This is

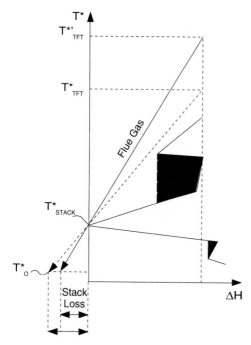

Figure 6.28 Increasing the theoretical flame temperature by reducing excess air or combustion air preheat reduces the stack loss.

not always the case. Figure 6.29*a* shows a situation where the flue gas is released to the atmosphere above pinch temperature for practical reasons. There is a practical minimum to which a flue gas can be cooled without condensation causing corrosion in the stack; this is known as the *acid dew point*. The minimum stack temperature in Fig. 6.29*a* is fixed by the acid dew point.[7] Another case is shown in Fig. 6.29*b*, where the process away from the pinch limits the slope of the flue gas line and hence the stack loss.[7]

Example 6.4 The process in Fig. 6.2 is to have its hot utility supplied by a furnace. The theoretical flame temperature for combustion is 1800°C, and the acid dew point for the flue gas is 160°C. Ambient temperature is 10°C. Assume $\Delta T_{min} = 10°C$ for process-to-process heat transfer but $\Delta T_{min} = 30°C$ for flue-gas-to-process heat transfer. A high value for ΔT_{min} for flue-gas-to-process heat transfer has been chosen because of poor heat transfer coefficients in the convection bank of the furnace. Calculate the fuel required, stack loss, and furnace efficiency.

Solution The first problem is that a different value of ΔT_{min} is required for different matches. The problem table algorithm is easily adapted to accommodate this.[1] This is achieved by assigning ΔT_{min} *contributions* to streams. If the process streams are assigned a contribution of 5°C and flue gas a contribution of 25°C, then a process-process match has a ΔT_{min} of $5 + 5 = 10°C$ and a

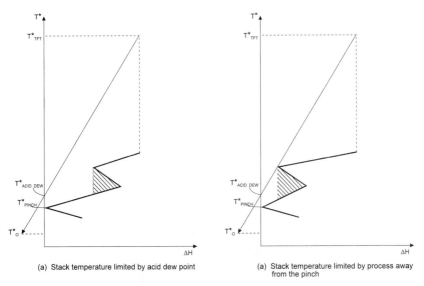

(a) Stack temperature limited by acid dew point

(a) Stack temperature limited by process away from the pinch

Figure 6.29 Furnace stack temperature can be limited by other factors than pinch temperature.

process–flue gas match has a $\Delta T_{\min}$ of $5 + 25 = 30°C$. When setting up the interval temperatures in the problem table algorithm, the interval boundaries are now set at hot stream temperatures minus their $\Delta T_{\min}$ contribution rather than half the global $\Delta T_{\min}$. Similarly, boundaries are now set based on cold stream temperatures plus their $\Delta T_{\min}$ contribution

Figure 6.30 shows the grand composite curve plotted from the problem table cascade in Fig. 6.18*b*. The starting point for the flue gas is an actual temperature of 1800°C, which corresponds to a shifted temperature of $(1800 - 25) = 1775°C$ on the grand composite curve. The flue gas profile is not restricted above the pinch and can be cooled to pinch temperature corresponding to a shifted temperature of 145°C before venting to the atmosphere. The actual stack temperature is thus $145 + 25 = 170°C$. This is just above the acid dew point of 160°C. Now calculate the fuel consumption:

$$Q_{H\min} = 7.5 \text{ MW}$$

$$CP_{\text{FLUE GAS}} = \frac{7.5}{1775 - 145}$$

$$= 0.0046 \text{ MW/°C}$$

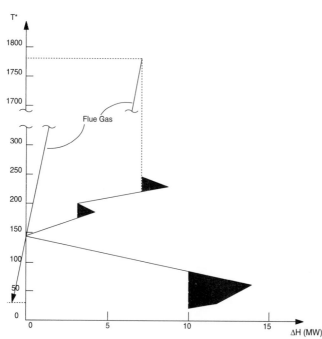

Figure 6.30 Flue gas matched against the grand composite curve of the process in Fig. 6.2.

The fuel consumption is now calculated by taking the flue gas from theoretical flame temperature to ambient temperature:

$$\text{Fuel required} = 0.0046\,(1800 - 10)$$

$$= 8.23\text{ MW}$$

$$\text{Stack loss} = 0.0046\,(170 - 10)$$

$$= 0.74\text{ MW}$$

$$\text{Furnace efficiency} = \frac{Q_{Hmin}}{\text{fuel required}} \times 100$$

$$= \frac{7.49}{8.23} \times 100$$

$$= 91\%$$

6.8 Combined Heat and Power (Cogeneration)

A more complex utility is combined heat and power (or cogeneration). Here, the heat rejected by a heat engine such as a steam turbine, gas turbine, or diesel engine is used as the hot utility.

Fundamentally, there are two possible ways to integrate a heat engine exhaust.[9] In Fig. 6.31 the process is represented as a heat sink and heat source separated by the pinch. Integration of the heat engine across the pinch as shown in Fig. 6.31a is counterproductive. The process still requires Q_{Hmin}, and the heat engine performs no

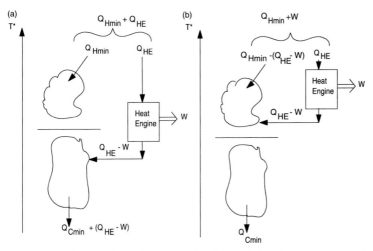

Figure 6.31 Heat engine exhaust can be integrated either across or not across the pinch.

better than if operated stand-alone. There is no savings by integrating a heat engine across the pinch.[9]

Figure 6.31b shows the heat engine integrated entirely above the pinch. In rejecting heat above the pinch it is rejecting heat into the part of the process which is overall a heat sink. In so doing, it is exploiting the temperature difference that exists between the utility source and the process sink, producing work at high efficiency. The net effect in Fig. 6.31b is the import of extra energy W from heat sources to produce W shaftwork. Because the process and heat engine are acting as one system, apparently conversion of heat to work at 100 percent efficiency is achieved.[9]

Now let us take a closer look at the two most commonly used heat engines (steam and gas turbines) to see whether they achieve this efficiency in practice. To make a quantitative assessment of any combined heat and power scheme, the grand composite curve should be used and the heat engine exhaust treated like any other utility.

1. *Steam turbine integration.* Figure 6.32 shows a steam turbine expansion on an enthalpy-entropy plot. In an ideal turbine, steam

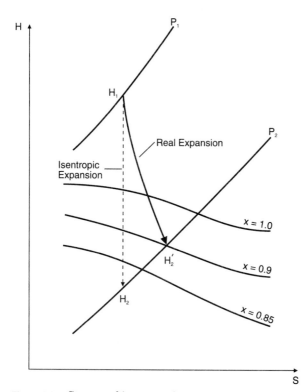

Figure 6.32 Steam turbine expansion.

with an initial pressure P_1 and enthalpy H_1 expands isentropically to pressure P_2 and enthalpy H_2. In such circumstances, the ideal work output is $(H_1 - H_2)$. Because of the frictional effects in the turbine nozzle and blade passages, the exit enthalpy is greater than it would be in an ideal turbine, and the work output is consequently less, given by H_2' in Fig. 6.32. The actual work output is given by $(H_1 - H_2')$.

The turbine isentropic efficiency η_T measures the ratio of the actual to ideal work obtained:

$$\eta_T = \frac{H_1 - H_2'}{H_1 - H_2} \tag{6.2}$$

The output from the turbine might be superheated or partially condensed, as is the case in Fig. 6.32. If the exhaust steam is to be used for process heating, ideally it should be close to saturated conditions. If the exhaust steam is significantly superheated, it can be desuperheated by direct injection of boiler feedwater, which vaporizes and cools the steam. However, if saturated steam is fed to a steam main, with significant potential for heat losses from the main, then it is desirable to retain some superheat rather than desuperheat the steam to saturated conditions. If saturated steam is fed to the main, then heat losses will cause excessive condensation in the main, which is not desirable. On the other hand, if the exhaust steam from the turbine is partially condensed, the condensate is separated and the steam used for heating.

Figure 6.33 shows a steam turbine integrated with the process above the pinch. Heat Q_{HP} is taken into the process from high-pressure steam. The balance of the hot utility demand Q_{LP} is taken

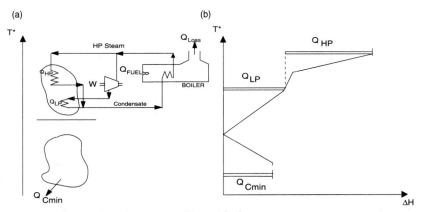

Figure 6.33 Integration of a steam turbine with the process.

from the steam turbine exhaust. In Fig. 6.33a, flash steam is recovered after pressure reduction of the high-pressure steam condensate.

In Fig. 6.33a. heat Q_{FUEL} is taken into the boiler from fuel. An overall energy balance gives

$$Q_{FUEL} = Q_{HP} + Q_{LP} + W + Q_{LOSS} \qquad (6.3)$$

The process requires $(Q_{HP} + Q_{LP})$ to satisfy its enthalpy imbalance above the pinch. If there were no losses from the boiler, then fuel W would be converted to shaftwork W at 100 percent efficiency. However, the boiler losses Q_{LOSS} reduce this to below 100 percent conversion. In practice, in addition to the boiler losses, there also can be significant losses from the steam distribution system. Figure 6.33b shows how the grand composite curve can be used to size steam turbine cycles.[1]

2. *Gas turbine integration.* Figure 6.34 shows a simple gas turbine matched against a process.[1] The machine is essentially a rotary compressor mounted on the same shaft as a turbine. Air enters the compressor, where it is compressed before entering a combustion chamber. Here the combustion of fuel increases its temperature. The mixture of air and combustion gases is expanded in the turbine. The input of energy to the combustion chamber allows enough shaftwork to be developed in the turbine to both drive the compressor and provide useful work. The expanded gas may be discharged to the atmosphere directly or may first be used to preheat the air to the

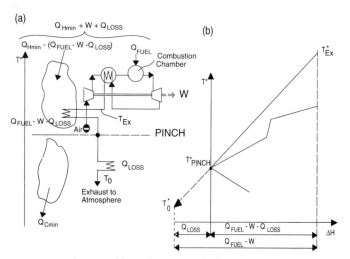

Figure 6.34 A gas turbine exhaust matched against the process.

combustion chamber before discharge, as shown in Fig. 6.34. The performance of the machine is specified in terms of the power output, air flow rate through the machine, efficiency of conversion of heat to work, and temperature of the exhaust. Gas turbines are normally only used for relatively large-scale applications.

As with the steam turbine, if there was no stack loss to the atmosphere (i.e., if Q_{LOSS} was zero), then W heat would be turned into W shaftwork. The stack losses in Fig. 6.34 reduce the efficiency of conversion of heat to work. The overall efficiency of conversion of heat to power depends on the turbine exhaust profile, the pinch temperature, and the shape of the process grand composite.

Example 6.5 The stream data for a heat recovery problem are given in Table 6.7. A problem table analysis for $\Delta T_{\text{min}} = 20°C$ results in the heat cascade given in Table 6.8. The process also has a requirement for 7 MW of power. Two alternative combined heat and power schemes are to be compared economically.

a. A steam turbine with its exhaust saturated at 150°C used for process heating is one of the options to be considered. Superheated steam is generated in the central boilerhouse at 41 bar with a temperature of 300°C. This superheated steam can be expanded in a single-stage turbine with an isentropic efficiency of 85 percent. Calculate the maximum generation of shaftwork possible by matching the exhaust steam against the process.

TABLE 6.7 **Stream Data for Example 6.5**

Stream		T_S (°C)	T_T (°C)	Heat capacity flow rate (MW °C^{-1})
No.	Type			
1	Hot	450	50	0.25
2	Hot	50	40	1.5
3	Cold	30	400	0.22
4	Cold	30	400	0.05
5	Cold	120	121	22.0

TABLE 6.8 **Problem Table Cascade for Example 6.5**

T^* (°C)	Cascade heat flow (MW)
440	21.9
410	29.4
131	23.82
130	1.8
40	0
30	15

b. A second possible scheme uses a gas turbine with a flow rate of air of $97 \, kg \, s^{-1}$ which has an exhaust temperature of 400°C. Calculate the shaftwork generation if the turbine has an efficiency of 30 percent. Ambient temperature is 10°C.

c. The cost of heat from fuel for the gas turbine is $4.5 \, GW^{-1}$. The cost of imported electricity is $19.2 \, GW^{-1}$. Electricity can be exported with a value of $14.4 \, GW^{-1}$. The cost of fuel for steam generation is $3.2 \, GW^{-1}$. The overall efficiency of steam generation and distribution is 60 percent. Which scheme is most cost-effective, the steam turbine or the gas turbine?

Solution

a. This is shown in Fig. 6.35a. Steam condensing interval temperature is 140°C.

$$\text{Heat flow required from the turbine exhaust} = 21.9 \, \text{MW}$$

From steam tables, inlet conditions at $T_1 = 300°C$ and $P_1 = 41$ bar are

$$h_1 = 2959 \, \text{kJ} \, \text{kg}^{-1}$$

$$s_1 = 6.349 \, \text{kJ} \, \text{kg}^{-1} \, \text{K}^{-1}$$

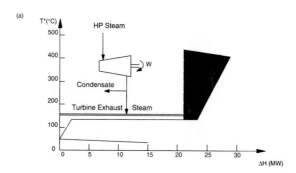

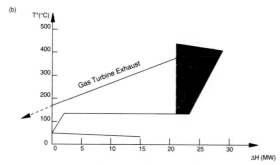

Figure 6.35 Alternative combined heat and power schemes for Example 6.4.

Turbine outlet conditions for isentropic expansion to 150°C from steam tables are

$$P_2 = 4.77 \, \text{bar}$$

$$s_2 = 6.349 \, \text{kJ kg}^{-1} \, \text{K}^{-1}$$

The wetness fraction x can be calculated from

$$s_2 = xs_l + (1-x)s_v$$

where s_l and s_v are the saturated liquid and vapor entropies. Taking saturated liquid and vapor entropies from steam tables at 150°C and 4.77 bar,

$$6.349 = 1.842x + 6.838(1-x)$$

$$x = 0.098$$

The turbine outlet enthalpy can now be calculated from

$$h_2 = xh_l + (1-x)h_v$$

where h_l and h_v are the saturated liquid and vapor enthalpies. Taking saturated liquid and vapor enthalpies from steam tables,

$$h_2 = 0.098 \times 632 + (1 - 0.098)2747$$

$$= 2540 \, \text{kJ kg}^{-1}$$

For a single-stage expansion with isentropic efficiency of 85 percent, from Eq. (6.2),

$$h_2' = h_1 - \eta_T(h_1 - h_2)$$

$$= 2959 - 0.85(2959 - 2540)$$

$$= 2603 \, \text{kJ kg}^{-1}$$

The wetness fraction x can be calculated from

$$h_2' = xh_l + (1-x)h_v$$

where h_l and h_v are the saturated liquid and vapor enthalpies. Thus

$$h_2' = 2603 = 632x + 2747(1-x)$$

$$x = 0.07$$

$$\text{Steam flow to process} = \frac{21.9 \times 10^3}{2747 - 632}$$

$$= 10.35 \, \text{kg s}^{-1}$$

$$\text{Steam flow through turbine} = \frac{10.35}{(1 - 0.07)}$$

$$= 11.13 \, \text{kg s}^{-1}$$

$$\text{Shaftwork generated } W = 11.13(2959 - 2603) \times 10^{-3}$$

$$= 3.96 \, \text{MW}$$

b. The exhaust from the turbine is primarily air with a small amount of combustion gases. Hence the CP of the exhaust can be approximated to be that of the airflow. Assuming C_P for air $= 1.03\,\text{kJ}\,\text{kg}^{-1}\,\text{K}^{-1}$,

$$CP_{\text{EXHAUST}} = 97 \times 1.03$$

$$= 100\,\text{kW}\,\text{K}^{-1}$$

The gas turbine option is shown in Fig. 6.35b.

$$Q_{\text{EXHAUST}} = (400 - 10)0.1$$

$$= 39\,\text{MW}$$

$$Q_{\text{FUEL}} = \frac{39}{0.7}$$

$$= 55.71\,\text{MW}$$

$$W = 16.71\,\text{MW}$$

c. *Steam turbine economics:*

$$\text{Cost of fuel} = (21.9 + 3.96) \times \frac{3.2 \times 10^{-3}}{0.6}$$

$$= \$0.14\,\text{s}^{-1}$$

$$\text{Cost of imported electricity} = (7 - 3.96) \times 19.2 \times 10^{-3}$$

$$= \$0.06\,\text{s}^{-1}$$

$$\text{Net cost} = \$0.20\,\text{s}^{-1}$$

Gas turbine economics:

$$\text{Cost of fuel} = 55.71 \times 4.5 \times 10^{-3}$$

$$= \$0.25\,\text{s}^{-1}$$

$$\text{Electricity credit} = (16.71 - 7) \times 14.4 \times 10^{-3}$$

$$= \$0.14\,\text{s}^{-1}$$

$$\text{Net cost} = \$0.11\,\text{s}^{-1}$$

Hence the gas turbine is the most profitable in terms of energy costs. However, this is only part of the story, since the capital cost of a gas turbine installation is likely to be significantly higher than that of a steam turbine installation.

Example 6.6 The problem table cascade for a process is given in Table 6.9 for $\Delta T_{min} = 10°C$. It is proposed to provide process cooling by steam generation from boiler feedwater with a temperature of 100°C.

a. Determine how much steam can be generated at a saturation temperature of 230°C.

b. Determine how much steam can be generated at a saturation temperature of 230°C and superheated to the maximum temperature possible against the process.

c. Calculate how much power can be generated from the superheated steam from part b assuming that a single-stage condensing steam turbine is to be used with an isentropic efficiency of 85 percent. Cooling water is available at 20°C and is to be returned to the cooling tower at 30°C.

Solution

a. Heat available for steam generation at 235°C interval temperature is 12.0 MW. From steam tables, the latent heat of water at a saturated temperature of 230°C is 1812 kJ kg^{-1}.

$$\text{Steam production} = 12.0 \times \frac{10^3}{1812}$$

$$= 6.62 \text{ kg s}^{-1}$$

Taking the heat capacity of water to be 4.3 kJ kg^{-1} K^{-1}, heat duty on boiler feedwater preheating is

$$6.62 \times 4.3 \times 10^{-3}(230 - 100) = 3.70 \text{ MW}$$

The profile of steam generation is shown against the grand composite curve

TABLE 6.9 Problem Table Cascade

Interval temperature (°C)	Heat flow (MW)
495	3.6
455	9.2
415	10.8
305	4.2
285	0
215	16.8
195	17.6
185	16.6
125	16.6
95	21.1
85	18.1

in Fig. 6.36a. The process can support both boiler feedwater preheat and steam generation.

b. Maximum superheat temperature is 285°C interval and 280°C actual. The profile is shown against the grand composite curve in Fig. 6.36b. From steam tables, enthalpy of superheated steam at 280°C and 28 bar is 2947 kJ kg^{-1}, and enthalpy of saturated water at 230°C and 28 bar is 991 kJ kg^{-1}. Thus

$$\text{Steam production} = \frac{12.0 \times 10^3}{(2947 - 991)}$$
$$= 6.13 \text{ kg s}^{-1}$$

c. In a condensing turbine, the exhaust from the turbine is condensed under vacuum against cooling water. The lower the condensing temperature, the greater is the power generation. The lowest condensing temperature for this problem is cooling water temperature plus ΔT_{min}, i.e., $30 + 10 = 40$°C. From steam tables, inlet conditions at $T_1 = 280$°C and $P_1 = 28$ bar are

$$h_1 = 2947 \text{ kJ kg}^{-1}$$
$$s_1 = 6.488 \text{ kJ kg}^{-1} \text{ K}^{-1}$$

Turbine outlet conditions for isentropic expansion to 40°C from steam tables are

$$P_2 = 0.074 \text{ bar}$$

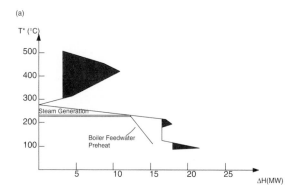

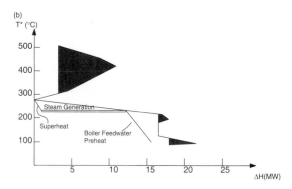

Figure 6.36 Alternative cold utilities for Example 6.5.

For $s_2 = 6.488\,\text{kJ kg}^{-1}\,\text{K}^{-1}$, the wetness fraction x and outlet enthalpy h_2 can be calculated as shown in Example 6.4:

$$x = 0.23$$

$$h_2 = 2020\,\text{kJ kg}^{-1}$$

For a single-stage expansion with isentropic efficiency of 85 percent, from Eq. (6.2),

$$h_2' = 2947 - 0.85(2947 - 2020)$$

$$= 2159\,\text{kJ kg}^{-1}$$

The shaftwork generation W is given by

$$W = 6.13(2947 - 2159) \times 10^{-3}$$

$$= 4.8\,\text{MW}$$

The wetness fraction x for the real expansion is given by

$$h_2' = 2159 = xh_l + (1 - x)h_v$$

$$= 167.5x + 2574(1 - x)$$

$$x = 0.17$$

This wetness fraction is possibly too high, since high levels of wetness can cause damage to the turbine. To allow a lower wetness fraction of, say, $x = 0.15$, then the outlet pressure of the turbine must be raised 0.2 bar, corresponding to a condensing temperature of 60°C. However, in so doing, the shaftwork generation decreases to 4.2 MW.

6.9   Integration of Heat Pump

A schematic of a simple vapor compression heat pump is shown in Fig. 6.37. A heat pump is a device that absorbs heat at a low

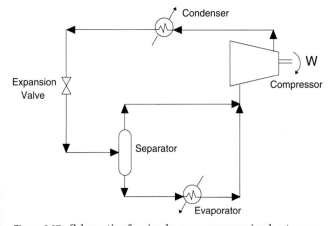

Figure 6.37 Schematic of a simple vapor compression heat pump.

temperature in the evaporator, consumes shaftwork when the working fluid is compressed, and rejects heat at a higher temperature in the condenser. The condensed working fluid is expanded and partially vaporizes. The cycle then repeats. The working fluid is usually a pure component, which means that the evaporation and condensation take place isothermally. Just as with heat engine integration, there are appropriate and inappropriate ways to integrate heat pumps.

Again, there are two fundamental ways in which a heat pump can be integrated with the process: across and not across the pinch.[9] Integration not across (above) the pinch is illustrated in Fig. 6.38a. This arrangement imports W shaftwork and saves W hot utility. In other words, the system converts power into heat, which is normally never economically worthwhile. Another integration not across (below) the pinch is shown in Fig. 6.38b. The result is worse economically. Power is turned into waste heat.[9] Integration across the pinch is illustrated in Fig. 6.38c. This arrangement brings about a genuine saving. It also makes overall sense because heat is pumped from the part of the process which is overall a heat source to the part which is overall a heat sink.

Thus the appropriate placement of heat pumps is that they should be placed across the pinch.[9] Note that the principle needs careful interpretation if there are utility pinches. In such circumstances, heat pump replacement above the process pinch or below it can be economic, providing that the heat pump is placed across a utility pinch. Such considerations are outside the scope of the present text.

Figure 6.39 shows a heat pump appropriately integrated against a process. Figure 6.39a shows the overall balance. Figure 6.39b illustrates how the grand composite curve can be used to size the heat pump. How the heat pump performs determines its coefficient of performance. The coefficient of performance for a heat pump generally can be defined as the useful energy delivered to the process divided by the shaftwork expended to produce this useful energy. From Fig. 6.39a,

$$COP_{HP} = \frac{Q_{HP} + W}{W} \qquad (6.4)$$

where COP_{HP} is the heat pump coefficient of performance, Q_{HP} is the heat absorbed at low temperature, and W is the shaftwork consumed.

For any given type of heat pump, a higher COP_{HP} leads to better economics. Having a better COP_{HP} and hence better economics means working across a small temperature lift with the heat pump. The smaller the temperature lift, the better is the COP_{HP}. For most applications, a temperature lift greater than 25°C is rarely economical. Attractive heat pump application normally requires a lift much less than 25°C.

Using the grand composite curve, the loads and temperatures of

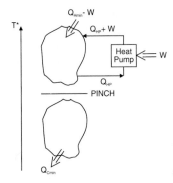

(a) Integration of heat pump above the pinch.

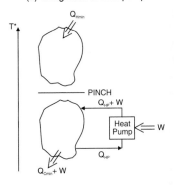

(b) Integration of heat pump below the pinch.

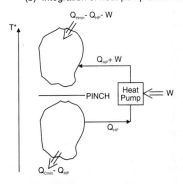

(c) Integration of heat pump across the pinch.

Figure 6.38 Integration of heat pumps with the process.

the cooling and heating duties and hence the COP_{HP} of integrated heat pumps can be readily assessed.

6.10 Integration of Refrigeration Cycles

A refrigeration system is a heat pump in which heat is absorbed below ambient temperature. Thus the appropriate placement principle for heat pumps applies in exactly the same way as for refrigeration cycles. The appropriate placement for refrigeration cycles is that they also should be across the pinch.[9] As with heat pumps, refrigeration cycles also can be appropriately placed across utility pinches. It is common for refrigeration cycles to be placed across a utility pinch caused by maximizing cooling water duty.

Most refrigeration systems are essentially the same as the heat pump cycle shown in Fig. 6.37. Heat is absorbed at low temperature, servicing the process, and rejected at higher temperature either directly to ambient (cooling water or air cooling) or to heat recovery in the process. Heat transfer takes place essentially over latent heat profiles. Such cycles can be much more complex if more than one refrigeration level is involved.

As with heat pumping, the grand composite curve is used to assess how much heat from the process needs to be extracted into the refrigeration system and where, if appropriate, the process can

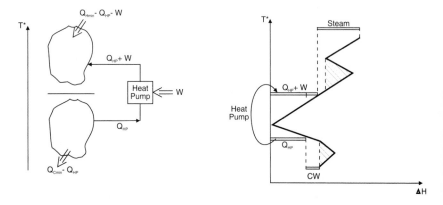

(a) Appropriately placed heat pump

(b) The heat pump against the grand composite curve

Figure 6.39 The grand composite curve can be used to size heat pump cycles. *(From Smith and Linnhoff, Trans. IChemE, ChERD, 66: 195, 1988; reproduced by permission of the Institution of Chemical Engineers.)*

accept rejected heat.[9] Again, the coefficient of performance determines how the refrigeration system performs. In the case of refrigeration systems, the coefficient of performance COP_{REF} is generally defined in terms of the heat extracted from the process divided by the shaftwork consumed:[9]

$$COP_{REF} = \frac{Q_{HP}}{W} \qquad (6.5)$$

The higher the COP_{REF}, the better are the economics.

The cost of shaftwork required to run a refrigeration system can be estimated approximately as a multiple of the shaftwork required for an ideal system.[1] The performance of an ideal system is given by

$$\frac{W_{IDEAL}}{Q_C} = \frac{T_H - T_C}{T_C} \qquad (6.6)$$

where W_{IDEAL} = ideal shaftwork required for the refrigeration cycle
Q_C = the cooling duty
T_C = temperature at which heat is taken into the refrigeration cycle (K)
T_H = temperature at which heat is rejected from the refrigeration cycle (K)

The ratio of ideal to actual shaftwork is usually around 0.6. Thus

$$W = \frac{Q_C}{0.6}\left(\frac{T_H - T_C}{T_C}\right) \qquad (6.7)$$

where W is the actual shaftwork required for the refrigeration cycle.

Example 6.6 Determine the refrigeration requirements for the low-temperature distillation process from Example 6.1 shown in Fig. 6.19 for $\Delta T_{min} = 5°C$.

a. Plot the grand composite curve from the heat cascade given in Fig. 6.21b, and determine the temperature and duties of the refrigeration if two levels of refrigeration are to be used. Assume that both vaporization and condensation of the refrigerant occur isothermally.

b. Calculate the power requirements for heat rejection to cooling water operating between 20 and 25°C. The power required for the refrigeration system can be approximated by Eq. (6.7).

c. Heat rejection from the refrigeration system into the process can be used to reduce the shaftwork requirements. Suggest a scheme, and calculate the reduced power requirement.

Solution

a. Figure 6.40a shows a plot of the grand composite curve from the problem table cascade given in Fig. 6.21b. Also shown in Fig. 6.40a are two refrigeration profiles:

	T^* (°C)	T (°C)	Duty (MW)
Level 1	−22.5	−25	1.04
Level 2	−42.5	−45	(1.84 − 1.04) = 0.8

A schematic of the two-level refrigeration system is shown in Fig. 6.40b. It should be noted that the single exchangers represented in Fig. 6.40b might in practice be several exchangers.

b. For heat rejection to cooling water,

$$T_H = 25 + 5 = 30°C = 303 \text{ K}$$

$$W_1 = \frac{1.04}{0.6}\left(\frac{303 - 248}{248}\right) = 0.38 \text{ MW}$$

$$W_2 = \frac{0.8}{0.6}\left(\frac{303 - 228}{228}\right) = 0.44 \text{ MW}$$

Total power for heat rejection to cooling water = 0.38 + 0.44 = 0.82 MW

c. Now assume that part of level 2 can be rejected to the process above the pinch, as shown in Fig. 6.41a. A schematic of the refrigeration system with heat rejection to the process is shown in Fig. 6.41b. Again, the heat exchangers represented in Fig. 6.41b might be several exchangers in practice. The rejection load is fixed at 0.54 MW.

$$W = \frac{Q_C}{0.6}\left(\frac{T_H - T_C}{T_C}\right)$$

$$W = \frac{Q_H - W}{0.6}\left(\frac{T_H - T_C}{T_C}\right)$$

$$W = \frac{0.54 - W}{0.6}\left[\frac{(5 + 273) - 228}{228}\right]$$

$$W = 0.14 \text{ MW}$$

Process cooling by level 2 by this arrangement across the pinch is 0.54 − 0.14 = 0.40 MW. The balance of the cooling demand on level 2, 0.8 − 0.4 = 0.4 MW, together with the load from level 1, must be either rejected to the process at a higher temperature above the pinch or to cooling water.

The process has a heating demand at 20°C, which means that heat could be rejected at 25°C. However, there seems little advantage in such an arrangement, since the heat can be rejected to cooling water at 30°C and rejection to the process would add to the complexity of both design and operation. Assume that the rest of the rejection heat goes to cooling water.

$$W_1 = \frac{1.04}{0.6}\left(\frac{303 - 248}{248}\right) = 0.38 \text{ MW} \qquad \text{(as before)}$$

$$W_2 = \frac{0.8 - 0.4}{0.6}\left(\frac{303 - 228}{228}\right) = 0.22 \text{ MW}$$

Total shaftwork for part rejection of level 2 to the process is $0.14 + 0.38 + 0.22 = 0.74$ MW. Thus the saving in shaftwork by integration with the process is $0.82 - 0.74 = 0.08$ MW.

It must be emphasized that Eq. (6.7) is only an approximate method for calculating the performance of refrigeration cycles. If greater accuracy is required, the refrigeration cycle must be followed using thermodynamic properties of the refrigerant being used.[10]

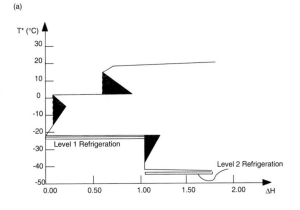

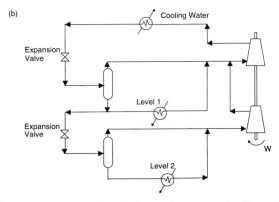

Figure 6.40 A two-level refrigeration system for Example 6.6 with heat rejection to cooling water.

6.11   Heat Exchanger Network and Utilities Energy Targets—Summary

The energy cost of the process can be set without having to design the heat exchanger network and utility system. These energy targets can be calculated directly from the material and energy balance. Thus

(a)

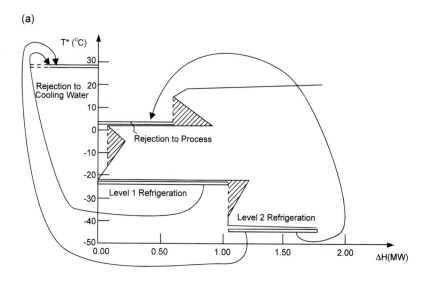

(b)

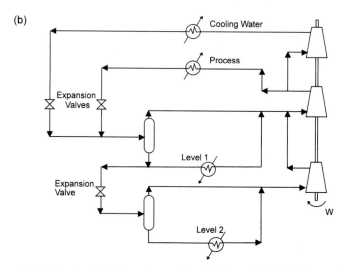

Figure 6.41   A two-level refrigeration system for Example 6.6 with partial heat rejection to the process.

energy costs can be established without design for the outer layers of the onion. Using the grand composite curve, alternative utility scenarios can be screened quickly and conveniently.

6.12 References

1. Linnhoff, B., Townsend, D. W., and Boland, D., et al., "A User Guide on Process Integration for the Efficient Use of Energy," *IChemE UK,* 1982.
2. Polley, G. T., "Heat Exchanger Design and Process Integration," *Chem. Eng.,* 8: 16, 1993.
3. Linnhoff, B., and Flower, J. R., "Synthesis of Heat Exchanger Networks," *AIChEJ,* 24: 633, 1978.
4. Ahmad, S., and Hui, D. C. W., "Heat Recovery Between Areas of Integrity," *Computers Chem. Eng.,* 15: 809, 1991.
5. Cerda, J., Westerberg, A. W., Mason, D., and Linnhoff, B., "Minimum Utility Usage in Heat Exchanger Network Synthesis—A Transportation Problem," *Chem. Eng. Sci.,* 38: 373 1983.
6. Papoulias, S. A., and Grossmann, I. E., "A Structural Optimization Approach in Process Synthesis: II. Heat Recovery Networks," *Computers Chem. Eng.,* 7: 707, 1983.
7. Linnhoff, B., and de Leur, J., "Appropriate Placement of Furnaces in the Integrated Process," *IChemE Symp. Series No 109,* 1988.
8. Hougen, O. A., Watson, K. M., and Ragatz, R. A., *Chemical Process Principles,* part I: *Material and Energy Balances,* 2d ed., Wiley, New York, 1954.
9. Townsend, D. W., and Linnhoff, B., "Heat and Power Networks in Process Design," *AIChEJ,* 29: 742, 1983.
10. Hougen, O. A., Watson, K. M., and Ragatz, R. A., *Chemical Process Principles,* part II: *Thermodynamics,* 2d ed., Wiley, New York, 1959.

Heat Exchanger Network and Utilities: Capital and Total Cost Targets

In addition to being able to predict the energy costs of the heat exchanger network and utilities directly from the material and energy balance, it would be useful to be able to calculate the capital cost, if this is possible. The principal components that contribute to the capital cost of the heat exchanger network are

- Number of units
- Heat exchange area
- Number of shells
- Materials of construction
- Equipment type
- Pressure rating

Let us take each of these components in turn and explore whether they can be accounted for from the material and energy balance without having to perform heat exchanger network design.

7.1 Number of Heat Exchange Units

To understand the minimum number of matches or units in a heat exchanger network, some basic results of *graph theory*[1] can be used. A *graph* is any collection of points in which some pairs of points are connected by lines. Figure 7.1 gives two examples of graphs. Note that the lines *BG, CE,* and *CF* in Fig. 7.1*a* are not supposed to cross; i.e., the diagram should be drawn in three dimensions. This is true for the other lines in Fig. 7.1 which appear to cross.

In this context, the points correspond to process and utility streams and the lines to heat exchange matches between the heat sources and heat sinks.

A *path* is a sequence of distinct lines which are connected to each other. For example, in Fig. 7.1*a*, *AECGD* is a path. A graph forms a single *component* if any two points are joined by a path. Thus Fig. 7.1*b* has two components and Fig. 7.1*a* has only one.

A *loop* is a path which begins and ends at the same point, such as *CGDHC* in Fig. 7.1*a*. If two loops have a line in common, they can be linked to form a third loop by deleting the common line. In Fig. 7.1*a*, for example, *BGCEB* and *CGDHC* can be linked to give *BGDHCEB*. In this case, this last loop is said to be dependent on the other two.

From graph theory the main result needed in the present context is that the number of independent loops for a graph is given by[1]

$$N_{\text{UNITS}} = S + L - C \qquad (7.1)$$

where N_{UNITS} = number of matches or units (lines in graph theory)
S = number of streams including utilities (points in graph theory)

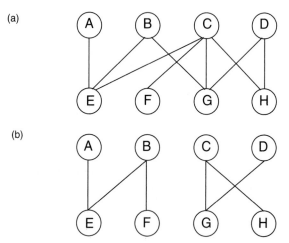

(a)

(b)

Figure 7.1 Two alternative "graphs." *(Reprinted from Linnhoff, Mason, and Wardle, "Understanding Heat Exchanger Networks," Computers Chem. Engg., 3: 295, 1979; with permission from Elsevier Science, Ltd.)*

L = number of independent loops
C = number of components

In general, the final network design should be achieved in the minimum number of units to keep down the capital cost (although this is not the only consideration to keep down the capital cost). To minimize the number of units in Eq. (7.1), L should be zero and C should be a maximum. Assuming L to be zero in the final design is a reasonable assumption. However, what should be assumed about C? Consider the network in Fig. 7.1b, which has two components. For there to be two components, the heat duties for streams A and B must exactly balance the duties for streams E and F. Also, the heat duties for streams C and D must exactly balance the duties for streams G and H. Such balances are likely to be unusual and not easy to predict. The safest assumption for C thus appears to be that there will be one component only, i.e., $C = 1$. This leads to an important special case when the network has a single component and is loop-free. In this case,[1]

$$N_{\text{UNITS}} = S - 1 \qquad (7.2)$$

Equation (7.2) put in words states that the minimum number of units required is one less than the number of streams (including utility streams).

This is a useful result, since if the network is assumed to be loop-free and has a single component, the minimum number of units can be predicted simply by knowing the number of streams. If the problem does not have a pinch, then Eq. (7.2) predicts the minimum number of units. If the problem has a pinch, then Eq. (7.2) is applied on each side of the pinch separately:

$$N_{\text{UNITS}} = (S_{\text{ABOVE PINCH}} - 1) + (S_{\text{BELOW PINCH}} - 1) \qquad (7.3)$$

Example 7.1   For the process in Fig. 6.2, calculate the minimum number of units given that the pinch is at 150°C for the hot streams and 140°C for the cold streams.

Solution   Figure 7.2 shows the stream grid with the pinch in place dividing the process into two parts. Above the pinch there are five streams, including the steam. Below the pinch there are four streams, including the cooling water. Applying Eq. (7.3),

$$N_{\text{UNITS}} = (5 - 1) + (4 - 1)$$

$$= 7$$

Looking back at the design presented for this problem in Fig. 6.9, it does in fact use the minimum number of units of 7. We shall see in Chap. 16 how to design for the minimum number of units.

7.2 Heat Exchange Area Targets

In addition to giving the necessary information to predict energy targets, the composite curves also contain the necessary information to predict network area. To calculate the network area from the composite curves, utility streams must be included with the process streams in the composite curves to obtain the *balanced composite curves,*[2,3] going through the same procedure as illustrated in Figs. 6.3 and 6.4 but including the utility streams. The resulting balanced composite curves should have no residual demand for utilities. The balanced composite curves are divided into vertical *enthalpy intervals,* as shown in Fig. 7.3. Let us assume initially that the overall heat transfer coefficient U is constant throughout the process. Assuming true countercurrent heat transfer, then the area requirement for enthalpy interval k for this *vertical heat transfer* is given by

$$A_{\text{NETWORK}\,k} = \frac{\Delta H_k}{U\,\Delta T_{LMk}} \qquad (7.4)$$

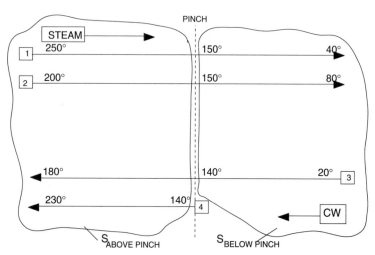

Figure 7.2 To target the number of units for pinched problems, the streams above and below the pinch must be counted separately, with the appropriate utilities included.

where $A_{\text{NETWORK}\,k}$ = heat exchange area for vertical heat transfer required by interval k

ΔH_k = enthalpy change over interval k

ΔT_{LMk} = log mean temperature difference for interval k

U = overall heat transfer coefficient

To obtain the network area, Eq. (7.4) can be applied to all enthalpy intervals:

$$A_{\text{NETWORK}} = \frac{1}{U} \sum_{k}^{\text{INTERVALS}\,K} \frac{\Delta H_k}{\Delta T_{LMk}} \qquad (7.5)$$

where A_{NETWORK} = heat exchange area for vertical heat transfer for the whole network

K = total number of enthalpy intervals

The problem with Eq. (7.5) is that the overall heat transfer coefficient is not constant throughout the process. Is there some way to extend this model to deal with the individual heat transfer coefficients?

The effect of individual stream film transfer coefficients can be included using the following expression, which is derived in App. B[2,3]

$$A_{\text{NETWORK}} = \sum_{k}^{\text{INTERVALS}\,K} \frac{1}{\Delta T_{LMk}} \left(\sum_{i}^{\text{HOT STREAMS}\,I} \frac{q_i}{h_i} + \sum_{j}^{\text{COLD STREAMS}\,J} \frac{q_j}{h_j} \right)$$

$$(7.6)$$

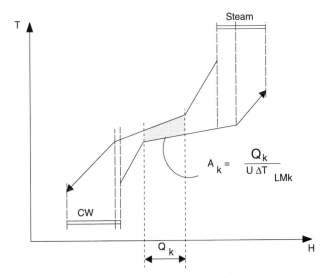

Figure 7.3 To determine the network area, the balanced composite curves are divided into enthalpy intervals.

where q_i = stream duty on hot stream i in enthalpy interval k
  q_j = stream duty on cold stream j in enthalpy interval k
  h_i, h_j = film transfer coefficients for hot stream i and cold stream j
    (including wall and fouling resistances)
  I = total number of hot streams in enthalpy interval k
  J = total number of cold streams in enthalpy interval k
  K = total number of enthalpy intervals

This simple formula allows the network area to be targeted, based on a vertical heat exchange model, if film transfer coefficients vary. However, if there are large variations in film transfer coefficients, Eq. (7.6) does not predict the true minimum network area. If film transfer coefficients vary significantly, then deliberate nonvertical matching is required to achieve minimum area. Consider Fig. 7.4a. Hot stream A with a low coefficient is matched against cold stream C with a high coefficient. Hot stream B with a high coefficient is matched with cold stream D with a low coefficient. In both matches, the temperature difference is taken to be the vertical separation between the curves. This arrangement requires $1616\,\text{m}^2$ of area overall.

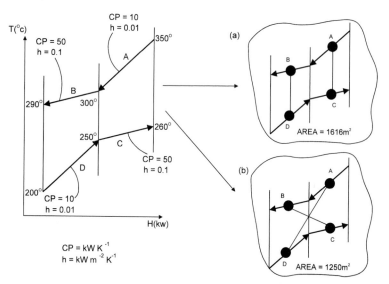

Figure 7.4 If film transfer coefficients differ significantly, then nonvertical heat transfer is necessary to achieve the minimum area. *(Reprinted from Linnhoff and Ahmad, "Cost Optimum Heat Exchanger Networks: I. Minimum Energy and Capital Using Simple Models for Capital Cost," Computers Chem. Engg., 7: 729, 1990; with permission from Elsevier Science, Ltd.)*

By constrast, Fig. 7.4*b* shows a different arrangement. Hot stream *A* with a low coefficient is matched with cold stream *D,* which also has a low coefficient but uses temperature differences greater than vertical separation. Hot stream *B* is matched with cold stream *C,* both with high heat transfer coefficients but with temperature differences less than vertical. This arrangement requires $1250 \, m^2$ of area overall, less than the vertical arrangement.

Thus, if film transfer coefficients vary significantly, then Eq. (7.6) does not predict the true minimum network area. The true minimum area must be predicted using linear programming.[4,5] However, Eq. (7.6) is still a useful basis to calculate the network area for the purposes of capital cost estimation for the following reasons:

1. Providing film coefficients vary by less than one order of magnitude, then Eq. (7.6) has been found to predict network area to within 10 percent of the actual minimum.[5]

2. Network designs tend *not* to approach the true minimum in practice because a minimum area design is usually too complex to be practical. Putting the argument the other way around, starting with the complex design required to achieve minimum area, then a significant reduction in complexity usually only requires a small penalty in area.

3. The area target being predicted here is used for predesign optimization of the capital/energy tradeoff evaluation of alternative flowsheet options such as different reaction/separator configurations. Thus the area prediction is used in conjunction with capital cost data for heat exchangers which often have a considerable degree of uncertainty. The capital cost predictions obtained later from Eq. (7.6) are likely to be more reliable than the capital cost predictions for the major items of equipment such as reactors and distillation columns.

One problem remains: where to get the film transfer coefficients h_i and h_j. There are three possibilities:

1. Tabulated experience values.[6]

2. By assuming a reasonable fluid velocity, together with fluid physical properties, standard heat transfer correlations can be used.

3. If the pressure drop available for the stream is known, the expressions of Polley et al.[7,8] can be used.

Example 7.2 For the process in Fig. 6.2, calculate the target for network heat transfer area. Steam at 240°C and condensing to 239°C is to be used for the hot

TABLE 7.1   Complete Stream and Utility Data for the process in Fig. 6.2

Stream	Supply temp. T_s (°C)	Target temp. T_T (°C)	ΔH (MW)	Heat capacity flow rate CP (MW °C^{-1})	Heat transfer coefficient h (MW m^{-2} °C)
1. Reactor 1 feed	20	180	32.0	0.2	0.0006
2. Reactor 1 product	250	40	−31.5	0.15	0.0010
3. Reactor 2 feed	140	230	27.0	0.3	0.0008
4. Reactor 2 product	200	80	−30.0	0.25	0.0008
5. Steam	240	239	−7.5	7.5	0.0030
6. Cooling water	20	30	10.0	1.0	0.0010

utility. Cooling water at 20°C and returning to the cooling tower at 30°C is to be used for the cold utility. Table 7.1 presents the stream data together with utility data and stream heat transfer coefficients. Calculate the heat exchange area target for the network.

Solution   First, we must construct the balanced composite curves using the complete set of data from Table 7.1. Figure 7.5 shows the balanced composite curves. Note that the steam has been incorporated within the construction of the hot composite curve to maintain the monotonic nature of composite curves. The same is true of the cooling water in the cold composite curve. Figure 7.5 also shows the curves divided into enthalpy intervals where there is either a

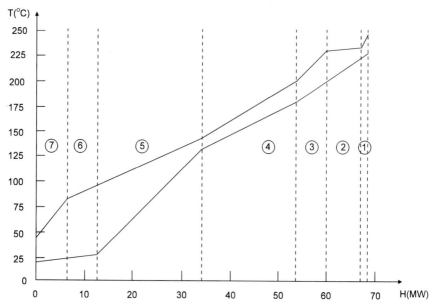

Figure 7.5   The enthalpy intervals for the balanced composite curves of Example 7.2.

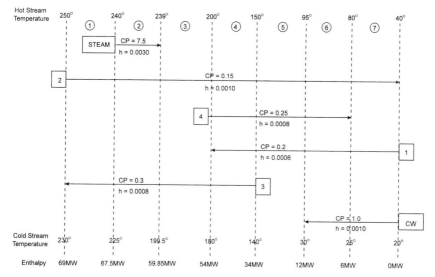

Figure 7.6 The enthalpy interval stream population for Example 7.2.

change of slope on the hot composite curve or a change of slope on the cold composite curve.

Figure 7.6 now shows the stream population for each enthalpy interval together with the hot and cold stream temperatures. Now set up a table to compute Eq. (7.6). This is shown in Table 7.2. Thus the network area target for this problem for $\Delta T_{\min} = 10°C$ is 7410 m².

The network design in Fig. 6.9 already achieves minimum energy consumption, and it is now possible to judge how close the area target is to design if the area for the individual units in Fig. 6.9 is calculated. Using the same heat transfer coefficients as given in Table 7.1, the design in Fig. 6.9 requires some 8341 m², which is 13 percent above target. Remember that no attempt was made to steer the design in Fig. 6.9 toward minimum area. Instead, the design achieved the energy target in the minimum number of units, which tends to lead to simple designs.

TABLE 7.2   Network Area target for the Process in Fig. 6.2

Enthalpy interval	ΔT_{LMk}	Hot streams $\Sigma (q_i/h_i)_k$	Cold streams $\Sigma (q_j/h_j)_k$	A_k
1	17.38	1,500	1,875	194.2
2	25.30	2,650	9,562.5	482.7
3	28.65	5,850	7,312.5	459.4
4	14.43	23,125	28,333.3	3566.1
5	29.38	25,437.5	36,666.7	2113.8
6	59.86	6,937.5	6,666.7	227.3
7	34.60	6,000	6,666.7	366.1
			ΣA_k	7,409.6

7.3 Number of Shells Target

The shell-and-tube heat exchanger is probably the most common type of exchanger used in the chemical and process industries. The simplest type of such device is the 1-1 design (1 shell pass, 1 tube pass), as illustrated in Fig. 7.7a. Of all shell-and-tube types, this comes closest to pure countercurrent flow and is designed using the basic countercurrent equation:

$$Q = UA\,\Delta T_{LM} \qquad (7.7)$$

For a given duty and overall heat transfer coefficient, the 1-1 design offers the lowest requirement for surface area. Many flow arrangements other than the 1-1 design exist, the most common of which is the 1-2 design (1 shell pass, 2 tube passes), as illustrated in Fig. 7.7b. Because the flow arrangement involves part countercurrent and part cocurrent flow, the effective temperature difference for heat exchange is reduced compared with a pure counter-current device. This is accounted for in design by introduction of the F_T factor into the basic heat exchanger design equation[9]

$$Q = UA\,\Delta T_{LM}F_T \qquad \text{where } 0 < F_T < 1 \qquad (7.8)$$

Thus, for a given exchanger duty and overall heat transfer coefficient, the 1-2 design needs a larger area than the 1-1 design. However, the 1-2 design offers many practical advantages. These

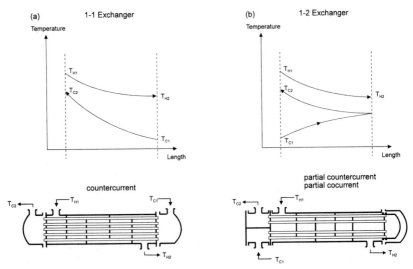

Figure 7.7 1-1 shells approach pure countercurrent flow, whereas 1-2 shells exhibit partial countercurrent and partial cocurrent flow.

include, in particular, allowance for thermal expansion, easy mechanical cleaning, and good heat transfer coefficients on the tube side (due to higher velocity).

The F_T correction factor is usually correlated in terms of two dimensionless ratios, the ratio of the two heat capacity flow rates R and the thermal effectiveness P of the exchanger:

$$F_T = f(R, P) \tag{7.9}$$

where

$$R = CP_C/CP_H = (T_{H1} - T_{H2})/(T_{C2} - T_{C1}) \tag{7.10}$$

and

$$P = (T_{C2} - T_{C1})/(T_{H1} - T_{C1}) \tag{7.11}$$

Take note that F_T can therefore be regarded as depending only on the inlet and outlet temperatures of the streams in a 1-2 exchanger. Three basic situations can be encountered when using 1-2 exchangers (Fig. 7.8):

1. The final temperature of the hot stream is higher than the final temperature of the cold stream, as illustrated in Fig. 7.8a. This is called a *temperature approach*. This situation is straightforward to design for, because it can always be accommodated in a single 1-2 shell.

2. The final temperature of the hot stream is slightly lower than the final temperature of the cold stream, as illustrated in Fig. 7.8b. This is called a *temperature cross*. This situation is also usually straightforward to design for, providing the temperature cross is small, because, again, it can probably be accommodated in a single shell.

3. As the amount of temperature cross increases, however, problems are encountered, as illustrated in Fig. 7.8c. Local reversal of heat flow may be encountered, which is wasteful in heat transfer area. The design may even become infeasible.

The maximum temperature cross which can be tolerated is normally set by rules of thumb, e.g., $F_T \geq 0.75$[10]. It is important to ensure that $F_T > 0.75$, since any violation of the simplifying assumptions used in the approach tends to have a particularly significant effect in areas of the F_T chart where slopes are particularly steep. Any uncertainties or inaccuracies in design data also have a more significant effect when slopes are steep. Consequently, to be confident in a design, those parts of the F_T chart where slopes are steep should be avoided, irrespective of $F_T \geq 0.75$.

A simple method to achieve this is based on the fact that for any value of R there is a maximum asymptotic value for P, say, P_{max}, which is given as F_T tends to $-\infty$ and is given by[11]

$$P_{max} = \frac{2}{R + 1 + \sqrt{R^2 + 1}} \qquad (7.12)$$

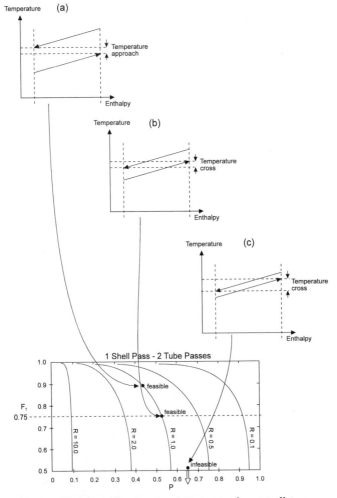

Figure 7.8 Designs with a temperature approach or small temperature cross can be accommodated in a single 1-2 shell, whereas designs with a large temperature cross become infeasible. *(From Ahmad, Linnhoff, and Smith, Trans. ASME, J. Heat Transfer, 110: 304, 1988; reproduced by permission of the American Society of Mechanical Engineers.)*

Equation (7.12) is derived in App. C. Practical designs will be limited to some fraction of P_{max}, that is,

$$P = X_P P_{max} \qquad 0 < X_P < 1 \qquad (7.13)$$

where X_P is a constant defined by the designer.

A line of constant X_P is compared with a line of constant F_T in Fig. 7.9. It can be seen that the line of constant X_P avoids the regions of steep slope.

Situations are often encountered where the F_T is too low or the F_T slope too large. If this happens, either different types of shells or multiple shell arrangements (Fig. 7.10) must be considered. We shall concentrate on multiple shell arrangements of the 1-2 type. By using 1-2 shells in series (Fig. 7.10), the temperature cross in each individual shell is reduced below that for a single 1-2 shell for the same duty. The profiles shown in Fig. 7.10 could in principle be achieved either by two 1-2 shells in series or by a single 2-4 shell. Traditionally, the designer would approach a design for an individual unit by trial and error. Starting by assuming one shell, the F_T can be evaluated. If the F_T is not acceptable, then the number of shells in series is progressively increased until a satisfactory value of F_T is obtained for each shell.

Before suggesting an approach for predicting the minimum number of shells for an entire network, a more convenient method for determining the number of shells in a single unit must first be found. Adopting the design criterion given by Eq. (7.13) as the basis, then any need for trial and error can be eliminated, since an explicit

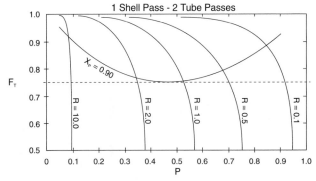

Figure 7.9 The X_P parameter avoids steep slopes on the F_T curves, whereas minimum F_T does not. (*Reprinted from Ahmad, Linnhoff, and Smith, "Cost Optimum Heat Exchanger Networks: II. Targets and Design for Detailed Capital Cost Models," Computers Chem. Engg., 7: 751, 1990; with permission from Elsevier Science, Ltd.)*

expression for the number of shells for a given unit is derived in App. D[11]

$R \neq 1$:

$$N_{\text{SHELLS}} = \frac{\ln\left(\dfrac{1 - RP}{1 - P}\right)}{\ln W} \tag{7.14}$$

where

$$W = \frac{R + 1 + \sqrt{R^2 + 1} - 2RX_P}{R + 1 + \sqrt{R^2 + 1} - 2X_P} \tag{7.15}$$

$R = 1$:

$$N_{\text{SHELLS}} = \frac{\left(\dfrac{P}{1 - P}\right)\left(1 + \dfrac{\sqrt{2}}{2} - X_P\right)}{X_P} \tag{7.16}$$

X_P is chosen to satisfy the minimum allowable F_T (e.g., for $F_{T_{\min}} > 0.75$, $X_P = 0.9$ is used). Once the real (noninteger) number of shells is calculated from Eq. (7.14), this is rounded up to the next largest number to obtain the number of shells.

Example 7.3 A hot stream is to be cooled from 410 to 110°C by exchange with

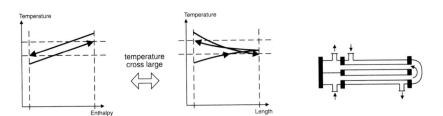

(a) A single 1-2 shell is infeasible.

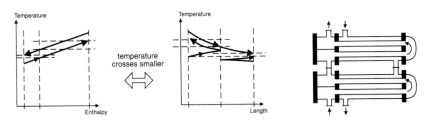

(b) Putting shells in series reduces the temperature cross in individual exchangers.

Figure 7.10 A large overall temperature cross requires shells in series to reduce the cross in individual exchangers. *(From Ahmad, Linnhoff, and Smith, Trans. ASME, J. Heat Transfer, 110: 304, 1988; reproduced by permission of the American Society of Mechanical Engineers.)*

a cold stream being heated from 0 to 360°C in a single unit. 1-2 shell and tube exchangers are to be used subject to $X_P = 0.9$. Calculate the number of shells required.

Solution

$$R = \frac{T_{H1} - T_{H2}}{T_{C2} - T_{C1}}$$

$$= \frac{410 - 110}{360 - 0}$$

$$= 0.8333$$

$$P_{N-2N} = \frac{T_{C2} - T_{C1}}{T_{H1} - T_{C1}}$$

$$= \frac{360 - 0}{410 - 0}$$

$$= 0.8780$$

$$W = \frac{(R + 1 + \sqrt{R^2 + 1} - 2RX_P)}{(R + 1 + \sqrt{R^2 + 1} - 2X_P)}$$

$$= 1.225$$

$$N_{\text{SHELLS}} = \frac{\ln\left[\frac{(1 - RP_{N-2N})}{(1 - P_{N-2N})}\right]}{\ln W}$$

$$= 3.9$$

Thus the unit requires 4 shells.

If exchangers are countercurrent devices, then the number of units equals the number of shells, providing individual shells do not exceed some practical upper size limit. If, however, equipment is used that is not completely countercurrent, as with the 1-2 shell and tube heat exchanger, then

$$N_{\text{SHELLS}} \geq N_{\text{UNITS}} \tag{7.17}$$

Since the number of shells can have a significant influence on the capital cost, it would be useful to be able to predict it as a target ahead of design.

A simple algorithm can be developed (see App. E) to target the minimum total number of shells (as a real, i.e., noninteger number) for a stream set based on the temperature distribution of the composite curves. The algorithm starts by dividing the composite curves into enthalpy intervals in the same way as the area target algorithm. The resulting number of shells is[12]

$$N_{\text{SHELLS}} = \sum_{k}^{\text{INTERVALS } K} N_k(S_k - 1) \tag{7.18}$$

where N_{SHELLS} = total number of shells over K enthalpy intervals
$\quad N_k$ = real (or fractional) number of shells resulting from the temperatures of enthalpy interval k
$\quad S_k$ = number of streams in enthalpy interval k

N_k is given by the application of Eqs. (7.14) to (7.16) to interval k.

In practice, the integer number of shells is evaluated from Eq. (7.18) for each side of the pinch. This maintains consistency between achieving maximum energy recovery and the corresponding minimum number of units target N_{UNITS}. In summary, the number of shells target can be calculated from the basic stream data and an assumed value of X_P (or equivalently, $F_{T_{\min}}$).

The F_T correction factor for each enthalpy interval depends both on the assumed value of X_P and the temperatures of the interval on the composite curves. It is possible to modify the simple area target formula to obtain the resulting increased overall area A_{NETWORK} for a network of 1-2 exchangers[12]

$$A_{\text{NETWORK,1-2}} = \sum_{k}^{\text{INTERVALS } K} \frac{1}{\Delta T_{LMk} \, F_{Tk}}$$
$$\times \left[\sum_{i}^{\text{HOT STREAMS } I} \frac{q_i}{h_i} + \sum_{j}^{\text{COLD STREAMS } J} \frac{q_j}{h_j} \right] \tag{7.19}$$

Furthermore, the average area per shell ($A_{\text{NETWORK,1-2}}/N_{\text{SHELL}}$) also can be considered at the targeting stage. If this is greater than the maximum allowable area per shell A_{SHELLmax}, then the shells target needs to be increased to the next largest integer above $A_{\text{NETWORK,1-2}}/A_{\text{SHELLmax}}$. Again, this can be applied to each side of the pinch.

7.4   Capital Cost Targets

To predict the capital cost of a network, it must first be assumed that a single heat exchanger with surface area A can be costed according to a simple relationship such as

$$\text{Installed capital cost of exchanger} = a + bA^c \tag{7.20}$$

where a, b, and c are cost law constants which vary according to materials of construction, pressure rating, and type of exchanger.

When cost targeting, the distribution of the targeted area between network exchangers is unknown. Thus, to cost a network using Eq. (7.20), some area distribution must be assumed, the simplest being that all exchangers have the same area:

$$\text{Network capital cost} = N[a + b(A_{\text{NETWORK}}/N)^C] \quad (7.21)$$

where N is the number of units or shells, whichever is appropriate.

At first sight, the assumption of equal area exchangers used in Eq. (7.21) might seem crude. However, from the point of view of predicting capital cost, the assumption turns out to be a remarkably good one.[11] At the targeting stage, no given distribution can be judged consistently better than another, since the network is not yet known. The implications of these inaccuracies, together with others, will be discussed later.

If the problem is dominated by equipment with a single specification (i.e., a single material of construction, equipment type, and pressure rating), then the capital cost target can be calculated from Eq. (7.21) with the appropriate cost coefficients. However, if there is a mix of specifications, such as different streams requiring different materials of construction, then the approach must be modified.

Equation (7.21) uses a single cost function in conjunction with the targets for the number of units (or shells) and network area. Differences in cost can be accounted for either by introducing new cost functions or by adjusting the heat exchange area to reflect the cost differences.[13] This can be done by weighting the stream heat transfer coefficients in the calculation of network area with a factor ϕ to account for these differences in cost. If, for example, a corrosive stream requires more expensive materials of construction than the other streams, it has a greater contribution to the capital cost than a similar noncorrosive stream. This can be accounted for by artifically decreasing its heat transfer coefficient to increase the contribution the stream makes to the network area. This fictitious area, when turned into a capital cost using the cost function for the noncorrosive materials, returns a higher capital cost, reflecting the increased cost resulting from special materials.

Heat exchanger cost data usually can be manipulated such that the fixed costs, represented by the coefficient a in Eq. (7.20), do not vary with exchanger specification.[13] If this is done, then Eq. (7.6), as derived in App. F, can be modified to[13]

$$A_{\text{NETWORK}}^* = \sum_{k}^{\text{INTERVALS }K} \frac{1}{\Delta T_{LMk}} \times \left(\sum_{i}^{\text{HOT STREAMS }I} \frac{q_i}{\phi_i h_i} + \sum_{j}^{\text{COLD STREAMS }J} \frac{q_j}{\phi_j h_j} \right) \quad (7.22)$$

where
$$\phi = \left(\frac{b_1}{b_2}\right)^{1/c_1}\left(\frac{A_{\text{NETWORK}}}{N}\right)^{1-(c_2/c_1)} \tag{7.23}$$

where ϕ_i = cost weighting factor for hot stream i
ϕ_j = cost weighting factor for cold stream j
a_1, b_1, c_1 = cost law coefficients for the reference cost law
a_2, b_2, c_2 = cost law coefficients for the special law
N = number of units or shells, whichever is applicable

Heat exchanger cost laws often can be adjusted with little loss of accuracy such that the coefficient c is constant for different specifications, i.e., $c_1 = c_2 = c$. In this case, Eq. (7.23) simplifies to[13]

$$\phi = \left(\frac{b_1}{b_2}\right)^{1/c} \tag{7.24}$$

Thus, to calculate the capital cost target for a network comprising mixed exchanger specifications, the procedure is

1. Choose a reference cost law for the heat exchangers. Greatest accuracy results if the category of streams which makes the largest contribution to capital cost is chosen as reference.[13]

2. Calculate ϕ factors for those streams which require a specification different from that of the reference using Eq. (7.23) or Eq. (7.24). If Eq. (7.23) is to be used, then the actual network area A_{NETWORK} must first be calculated using either Eq. (7.6) or Eq. (7.19) and N_{UNITS} or N_{SHELLS}, whichever is appropriate.

3. Calculate the weighted network area A^*_{NETWORK} from Eq. (7.22). When the weighted h values (ϕh) vary appreciably, say, by more than one order of magnitude, an improved estimate of A^*_{NETWORK} can be evaluated by linear programming.[4]

4. Calculate the capital cost target for the mixed specification heat exchanger network from Eq. (7.21) using the cost law coefficients for the reference specification.

Example 7.4 For the process in Fig. 6.2, the stream and utility data are given in Table 7.1. Pure countercurrent (1-1) shell and tube heat exchangers are to be used.

a. Calculate the capital cost target if all individual heat exchangers can be costed by the relationship:

Heat exchanger capital cost = 40,000 + 500A ($)

where A is the heat transfer area in square meters.

b. Calculate the capital cost target if cold stream 3 from Table 7.1 required a more expensive material. Individual heat exchangers made entirely from this more expensive material can be costed by the relationship

$$\text{Heat exchanger capital cost (special)} = 40{,}000 + 1100A \; (\$)$$

where A is the heat transfer area in square meters.

Solution

a. The capital cost target of the network can be calculated from Eq. (7.21). To apply this equation requires the target for both the number of units (N_{UNITS}) and the heat exchange area (A_{NETWORK}). In Example 7.1 we calculated $N_{\text{UNITS}} = 7$ and in Example 7.2 $A_{\text{NETWORK}} = 7410 \text{ m}^2$. Thus:

$$\text{Network capital cost} = 7[40{,}000 + 500(7410/7)^1]$$

$$= 3.99 \times 10^6 \; \$$$

b. To calculate the capital cost target of the network with mixed materials of construction, a reference material is first chosen. In principle, either of the materials can be chosen as reference. However, greater accuracy is obtained if the reference is taken to be that category of streams which makes the largest contribution to capital cost. In this case, the reference should be taken to be the cheaper material of construction. Now calculate ϕ factors for those streams which require a specification different from the reference. In this problem it only applies to stream 3. Since the c constant is the same for both cost laws, Eq. (7.24) can be used:

$$\phi_3 = (500/1100)^{1/1}$$

$$= 1/2.2$$

$$\phi_3 h_3 = 0.0008/2.2$$

Now recalculate the network area target substituting $\phi_3 h_3$ for h_3 in Fig. 7.5. Table 7.2 is revised to the values shown in Table 7.3:

Thus the weighted network area A^*_{NETWORK} is 9546 m². Now calculate the network capital cost for mixed materials of construction by using A^*_{NETWORK}

TABLE 7.3 Area Target per Enthalpy Interval

Enthalpy interval	ΔT_{LMk}	Hot streams $\Sigma\,(q_i/h_i)_k$	Cold streams $\Sigma\,(q_j/h_j)_k$	A_k
1	17.38	1,500	4,125	323.6
2	25.30	2,650	21,037.5	936.3
3	28.65	5,850	16,087.5	765.4
4	14.43	23,125	46,333.3	4813.5
5	29.38	25,437.5	36,666.7	2113.8
6	59.86	6,937.5	6,666.7	227.3
7	34.60	6,000	6,666.7	366.1
			ΣA_k	9546.0

in conjunction with the cost coefficients for the reference material in Eq. (7.21):

$$\text{Network capital cost (mixed materials)} = 7[40{,}000 + 500(9546/7)^1]$$

$$= 5.05 \times 10^6 \ \$$$

Let us now address the question of how accurate the capital cost targets are likely to be. It was discussed earlier how the basic area targeting equation [Eq. (7.6) or Eq. (7.19)] represents a true minimum network area if all heat transfer coefficients are equal but is slightly above the true minimum if there are significant differences in heat transfer coefficients. Providing heat transfer coefficients vary by less than one order of magnitude, Eqs. (7.6) and (7.19) predict an area which is usually within 10 percent of the minimum. However, this does not turn into a 10 percent error in capital cost of the final design, since practical designs are almost invariably slightly above the minimum. There are also two errors inherent in the approach to capital cost targets:

- Total heat transfer area is assumed to be divided equally between exchangers. This tends to overestimate the capital cost.

- The area target is usually slightly less than the area observed in design.

These small positive and negative errors partially cancel each other. The result is that capital cost targets predicted by the methods described in this chapter are usually within 5 percent of the final design, providing heat transfer coefficients vary by less than one order of magnitude. If heat transfer coefficients vary by more than one order of magnitude, then a more sophisticated approach can sometimes be justified.[5]

If the network comprises mixed exchanger specification, then an additional degree of uncertainty is introduced into the capital cost target. Applying the ϕ-factor approach to a single exchanger, where both streams require the same specification, there is no error. In practice, there can be different specifications on two streams being matched, and $\phi_H \neq \phi_C$ for specifications involving shell-and-tube heat exchangers with different materials of construction and pressure rating. This does not present a problem, since the exchanger can be designed for different materials of construction or pressure rating on the shell side and the tube side of a heat exchanger. If, for example, there is a mix of streams, some requiring carbon steel and some stainless steel, then some of the matches involve a corrosive stream on one side of the exchanger and a noncorrosive stream on the other. The capital cost of such exchangers will lie somewhere between the cost based on the sole use of either material. This is what the capital cost target predicts. Thus introducing mixed specifications for materials of construction and pressure rating does not significantly decrease the accuracy of the capital cost predictions.

By contrast, this is not true of mixed exchanger types. For networks comprising different exchanger types (e.g., shell-and-tube, plate, spiral, etc.), it is not possible to mix types in a single unit. Although a cost weighting factor may be applied to one stream in targeting, this assumes that different exchanger types can be mixed. In practice, such a match is forced to be a special-type exchanger. Thus there may be some discrepancy between cost targets and design cost when dealing with mixed exchanger types.

Overall, the accuracy of the capital cost targets is more than good enough for the purposes for which they are used:

- Screening design options in the material and energy balance. For example, changes in reactor or separation system design can be screened effectively without performing repeated network design.

- Different utility options such as furnaces, gas turbines, and different steam levels can be assessed more easily and with greater confidence knowing the capital cost implications for the heat exchanger network.

- Preliminary process optimization is greatly simplified, as will be seen in the next chapter.

- The design of the heat exchanger network is greatly simplified if the design is initialized with an optimized value for $\Delta T_{\min}$.

7.5 Total Cost Targets

Increasing the chosen value of process energy consumption also increases all temperature differences available for heat recovery and hence decreases the necessary heat exchanger surface area (see Fig. 6.6). The network area can be distributed over the targeted number of units or shells to obtain a capital cost using Eq. (7.21). This capital cost can be annualized as detailed in App. A. The annualized capital cost can be traded off against the annual utility cost as shown in Fig. 6.6. The total cost shows a minimum at the optimal energy consumption.

Example 7.5 For the process in Fig. 6.2, determine the value of $\Delta T_{\min}$ and the total cost of the heat exchanger network at the optimal setting of the

TABLE 7.4 Variation of Annualized Costs with ΔT_{min}

ΔT_{min}	Q_{Hmin} (MW)	Annual hot utility cost (10^6 \$ yr^{-1})	Q_{Cmin} (MW)	Annual cold utility cost (10^6 \$ yr^{-1})	$A_{NETWORK}$ (m^2)	N_{UNITS}	Annualized capital cost (10^6 \$ yr^{-1})	Annualized total cost (10^6 \$ yr^{-1})
2	4.3	0.516	6.8	0.068	15,519	7	2.121	2.705
4	5.1	0.612	7.6	0.076	11,677	7	1.614	2.302
6	5.9	0.708	8.4	0.084	9,645	7	1.346	2.138
8	6.7	0.804	9.2	0.092	8,336	7	1.173	2.069
10	7.5	0.900	10.0	0.100	7,410	7	1.051	2.051
12	8.3	0.996	10.8	0.108	6,716	7	0.960	2.064
14	9.1	1.092	11.6	0.116	6,174	7	0.888	2.096

capital/energy tradeoff. The stream and utility data are given in Table 7.1. The utility costs are

$$\text{Steam cost} = 120{,}000\ (\$\ \text{MW}^{-1}\ \text{yr}^{-1})$$

$$\text{Cooling water cost} = 10{,}000\ (\$\ \text{MW}^{-1}\ \text{yr}^{-1})$$

The heat exchangers to be used are single tube and shell pass. The installed capital cost is given

$$\text{Heat exchanger capital cost} = 40{,}000 + 500A\ (\$)$$

where A is the heat transfer area in square meters. The capital cost is to be paid back over 5 years at 10 percent interest.

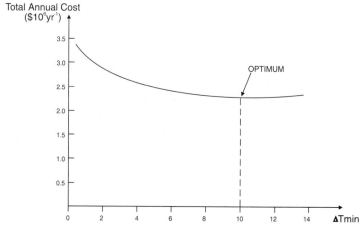

Figure 7.11 Optimization of the capital/energy tradeoff for Example 7.5.

Solution   From Eq. (A10) in App. A,

$$\text{Annualized heat exchanger capital cost} = \text{capital cost} \times \frac{i(1+i)^n}{(1+i)^n - 1}$$

where i = fractional interest rate per year
n = number of years

$$\text{Annualized heat exchanger capital cost} = (40,000 + 500A)$$

$$\times \frac{0.1(1+0.1)^5}{(1+0.1)^5 - 1}$$

$$= (40,000 + 500A)0.2638$$

$$= 10,552 + 131.9A$$

$$\text{Annualized network capital cost} = N_{\text{UNITS}}\left(10,552 + \frac{131.9 A_{\text{NETWORK}}}{N_{\text{UNITS}}}\right)$$

Now scan a range of values of $\Delta T_{\min}$ and calculate the targets for energy, number of units, and network area and combine these into a total cost. The results are given in Table 7.4.

The data from Table 7.4 are presented graphically in Fig. 7.11. The optimal $\Delta T_{\min}$ is at 10°C, confirming the initial value used for this problem in Chap. 6.

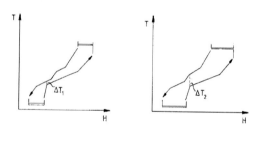

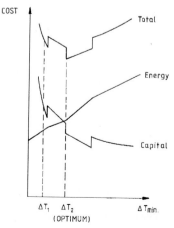

Figure 7.12 Energy and capital cost targets can be combined to optimize prior to design. *(From Smith and Linnhoff, Trans. IChemE, ChERD, 66: 195, 1988; reproduced by permission of the Institution of Chemical Engineers.)*

The total annualized cost at the optimal setting of the capital/energy tradeoff is 2.05×10^6 \$ yr^{-1}.

For more complex examples, total cost profiles return step changes such as shown in Fig. 7.12 (due to changes in N_{UNITS} and N_{SHELLS}). These step changes are easily located, prior to design, through simple software. Most important, extensive experience has shown that predicted overall costs are typically accurate within 5 percent or better.[5]

7.6 Heat Exchanger Network and Utilities Capital and Total Costs—Summary

There are parts of the flowsheet synthesis problem which can be "solved" without having to study actual designs. These are the layers of the process onion relating to the heat exchanger network and utilities. For these parts of the process design, targets can be set for energy costs and capital costs directly from the material and energy balance without having to resort to heat exchanger network design for evaluation.

Once a design is known for the first two layers of the onion (i.e., reactors and separators only), the overall total cost of this design for all four layers of the onion (i.e., reactors, separators, heat exchanger network, and utilities) is simply the total cost of all reactors and separators (evaluated explicitly) plus the total cost target for heat exchanger network and utilities.

7.7 References

1. Linnhoff, B., Mason, D. R., and Wardle, I., "Understanding Heat Exchanger Networks," *Computers Chem. Eng.,* 3: 295, 1979.
2. Townsend, D. W., and Linnhoff, B., "Surface Area Targets for Heat Exchanger Networks," IChemE Annual Research Meeting, Bath, U.K., 1984.
3. Linnhoff, B., and Ahmad, S., "Cost Optimum Heat Exchanger Networks: I. Minimum Energy and Capital Using Simple Models for Capital Cost," *Computers Chem. Eng.,* 14: 729, 1990.
4. Saboo, A. K., Morari, M., and Colberg, R. D., "RESHEX: An Interactive Software Package for the Synthesis and Analysis of Resilient Heat Exchanger Networks: II. Discussion of Area Targeting and Network Synthesis Algorithms," *Computers Chem. Eng.,* 10: 591, 1986.
5. Ahmad, S., Linnhoff, B., and Smith, R., "Cost Optimum Heat Exchanger Networks: II. Targets and Design for Detailed Capital Cost Models," *Computers Chem. Eng.,* 14: 751, 1990.
6. Perry, R. H., *Chemical Engineers Handbook,* 6th ed., McGraw-Hill, New York, 1984.
7. Polley, G. T., Panjeh Shahi, M. H., and Jegede, F. O., "Pressure Drop Considerations in the Retrofit of Heat Exchanger Networks," *Trans. IChemE,* part A, 68: 211, 1990.
8. Polley, G. T., and Panjeh Shahi, M. H., "Interfacing Heat Exchanger Network

Synthesis and Detailed Heat Exchanger Design," *Trans. IChemE,* part A, 69: 445, 1991.

9. Bowman, R. A., Mueller, A. C., and Nagle, W. M., "Mean Temperature Differences in Design," *Trans. ASME,* 62: 283, 1940.

10. Kern, D. Q., *Process Heat Transfer,* McGraw-Hill, New York, 1950.

11. Ahmad, S., Linnhoff, B., and Smith, R., "Design of Multipass Heat Exchangers: An Alternative Approach," *Trans. ASME, J. Heat Transfer,* 110: 304, 1988.

12. Ahmad, S., and Smith, R., "Targets and Design for Minimum Number of Shells in Heat Exchanger Networks," *IChemE, ChERD,* 67: 481, 1989.

13. Hall, S. G., Ahmad, S., and Smith, R., "Capital Cost Target for Heat Exchanger Networks Comprising Mixed Materials of Construction, Pressure Ratings and Exchanger Types," *Computers Chem. Eng.,* 14: 319, 1990.

8

Economic Tradeoffs

The basic structural decisions on the choice of reactor and choice of separation and recycle structure for the two inner layers of the onion have now been established. It also has been seen how targets can be set for the energy and capital costs of the heat exchanger network and utilities, the two outer layers. Before any thought is given to establishing a design for the heat exchanger network, there are important optimization variables to be explored. Let us now consider the more important optimization variables that will shape the final flowsheet.

8.1 Local and Global Tradeoffs

Consider again the simple process shown in Fig. 4.4d in which FEED is reacted to PRODUCT. If the process uses a distillation column as separator, there is a tradeoff between reflux ratio and the number of plates if the feed and products to the distillation column are fixed, as discussed in Chap. 3 (Fig. 3.7). This, of course, assumes that the reboiler and/or condenser are not heat integrated. If the reboiler and/or condenser are heat integrated, the tradeoff is quite different from that shown in Fig. 3.7, but we shall return to this point later in Chap. 14. The important thing to note for now is that if the reboiler and condenser are using external utilities, then the tradeoff between reflux ratio and the number of plates does not affect other operations in the flowsheet. It is a *local* tradeoff.

By contrast, if the reactor conversion is optimized, this is a *global* tradeoff, since changes in the reactor conversion affect operations throughout the flowsheet.[1] If the reactor conversion is changed so as to optimize its setting, then not only is the reactor affected in size but also the distillation column, since it now has a different separation task. The size of the recycle also will change. The heating and cooling associated with the reactor are different, as are the reboiler and condenser duties of the distillation column.

As the reactor conversion increases, the reactor volume increases and hence reactor capital cost increases. At the same time, the amount of unconverted feed needing to be separated decreases and hence the cost of recycling unconverted feed decreases, as shown in Fig. 8.1. Combining the reactor and recycle costs into a total cost indicates that there is an optimal reactor conversion.

It also should be noted in Fig. 4.4d that there is an impurity entering with the feed which is being removed via a purge. The concentration of impurity in the recycle can be varied as a degree of freedom. If the impurity is allowed to build up to a high concentration, then this reduces the loss of valuable raw materials in the

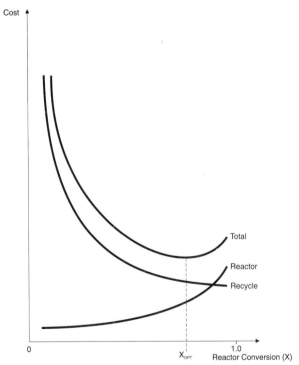

Figure 8.1 Overall cost tradeoffs as functions of reactor conversion.

purge. However, this decrease in raw materials cost is offset by an increase in the cost of recycling the additional impurity, together with increased capital cost of equipment in the recycle. Changes in the recycle inert concentration make changes throughout the flowsheet. This again is a global optimization variable. Thus the reactor/separation system involves both *local* and *global* tradeoffs.[1]

Caution should be exercised in judging whether an optimization is local if it involves the use of energy. If an operation uses energy, this might be supplied from an external utility or by heat integration. The apparent "local" tradeoff between reflux and plates in a distillation column might well turn out to have features of a global tradeoff if the reboiler and/or condenser are heat integrated with the rest of the process. If this is the case, then changing the reflux ratio affects the overall heat integration problem.

8.2   Optimization of Reactor Conversion for Single Reactions

Consider again the simple process in Fig. 4.1 in which FEED is reacted to PRODUCT via the reaction

$$\text{FEED} \longrightarrow \text{PRODUCT} \qquad (8.1)$$

The effluent from the reactor contains both PRODUCT and unreacted FEED which must be separated in a distillation column. Unreacted FEED is recycled to the reactor via a pump if the recycle is liquid or a compressor if the recycle is vapor.

Optimization of the system can be carried out by minimizing a cost function or maximizing economic potential EP defined by (see App. A)

$$EP = \text{value of products} - \text{raw materials costs}$$

$$- \text{annualized capital cost} - \text{energy cost} \qquad (8.2)$$

Plots of economic potential versus reactor conversion allow the optimal reactor conversion for a given flowsheet to be identified (Fig. 8.2). Although this approach allows the location of the optimum to be found, it does not give any indication of why the optimum occurs where it does.

Component costs for the simple process in Fig. 4.1 are decomposed according to the layers of the onion model in Fig. 8.3.[2] In Fig. 8.3, the annualized reactor cost (capital only) increases because high conversion requires a large volume and hence high capital cost. The annualized separation and recycle cost (capital only in this case) decreases with increasing reactor conversion because the amount of unreacted FEED to separate and recycle decreases. If the recycle had required a compressor, the capital and power costs of the compressor would have been included in the separation and recycle cost. The cost of the heat exchanger network and utilities is a combination of annualized energy cost and annualized capital cost of all exchangers, heaters, and coolers. Here the energy and capital cost targeting procedures developed in Chaps. 6 and 7 are used. Figure 8.3 shows the cost of heat exchanger network and utilities to decrease with increasing conversion, since the separation duty is decreased and the heating and cooling duties in the recycle decrease. Combining the reactor, separation and recycle, heat exchanger network and utility costs into a total annual cost (energy and capital) reveals the optimal reactor conversion. From Fig. 8.3 it can be seen that, for this example, heat integration and the cost of the heat exchanger network and utilities have a major influence on the optimal conversion. In other cases, the relative importance of the component costs will be different.

If the cost of the heat exchanger network changes, perhaps through a change in energy cost, then the optimal reactor conversion can change significantly. This change will likely dictate a different optimal reactor conversion and hence different separator design and process flow rates. However, such sensitivity is easily explored by changing the component cost for the heat exchanger network and utilities in Fig. 8.3 based on new targets.

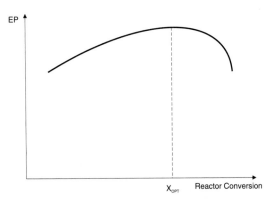

Figure 8.2 A plot of economic potential versus conversion shows where the optimum is, but why does the optimum occur where it is?

In Fig. 8.3, the only cost forcing the optimal conversion back from high values is that of the reactor. Hence, for such simple reaction systems, a high optimal conversion would be expected. This was the reason in Chap. 2 that an initial value of reactor conversion of 0.95 was chosen for simple reaction systems.

In Fig. 8.3, the curves were limited by a reactor conversion of 1. If the reaction had been reversible, then a similar picture would have been obtained. However, instead of being limited by a reactor conversion of 1, the curves would have been limited by the equilibrium conversion.

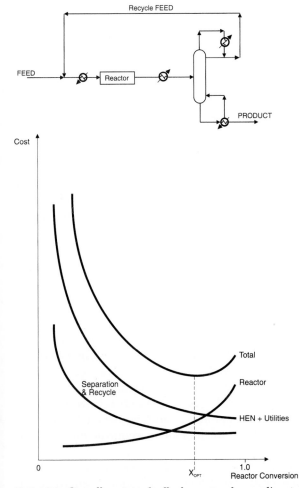

Figure 8.3 Overall cost tradeoffs decomposed according to the onion model.

8.3   Optimization of Reactor Conversion for Multiple Reactions Producing Byproducts

Consider the example of a process that involves the following multiple reactions:

$$\text{FEED} \longrightarrow \text{PRODUCT}$$

$$\text{PRODUCT} \longrightarrow \text{BYPRODUCT}$$

$$(8.3)$$

Because there is a mixture of FEED, PRODUCT, and BYPRODUCT in the reactor effluent, an additional separator is required.

The economic tradeoffs now become more complex, and a new cost must be added to the tradeoffs. This is a raw material efficiency cost due to byproduct formation. If the PRODUCT formation is kept constant despite varying levels of BYPRODUCT formation, then the cost can be defined to be[2]

Cost due to BYPRODUCT formation

= cost of FEED lost to BYPRODUCT

− value of BYPRODUCT   (8.4)

The value of PRODUCT formation and the raw materials cost of FEED that reacts to PRODUCT are constant. Alternatively, if the byproduct has no value, the cost of disposal should be included as

Cost due to BYPRODUCT formation

= cost of FEED lost to BYPRODUCT

+ cost of disposal of BYPRODUCT   (8.5)

By considering only those raw materials which undergo reaction to undesired byproduct, only the raw materials costs which are in principle avoidable are considered. Those raw materials costs which are inevitable (i.e., the stoichiometric requirements for FEED which converts into the desired PRODUCT) are not included. Raw materials costs which are in principle avoidable are distinguished from those which are inevitable from the stoichiometric requirements of the reaction.[2]

Figure 8.4 shows the cost tradeoffs for the present case. At high conversions, the raw materials costs due to byproduct formation are dominant. This is so because the reaction to the undesired

BYPRODUCT is series in nature, which leads to the selectivity becoming very poor at high conversions. We recall that in Chap. 2 the initial setting for reactor conversion was to be 0.5 for such reaction systems. Figure 8.4 shows why a high setting for reactor conversion would be inappropriate. The byproduct formation cost forces the optimum to lower values of conversion. Again, if the primary reaction had been reversible, then a similar picture would have been obtained. However, instead of being limited by a reactor conversion of 1, the curves would have been limited by the equilibrium conversion.

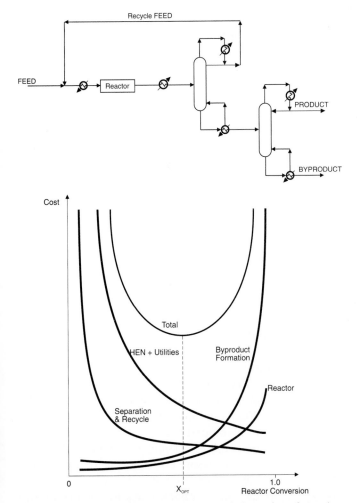

Figure 8.4 Cost tradeoffs for a process with byproduct formation. *(From Smith and Linnhoff, Trans. IChemE, ChERD, 66: 195, 1988; reproduced by permission of the Institution of Chemical Engineers.)*

Also, if there are two separators, the order of separation can change. The tradeoffs for these two alternative flowsheets will be different. The choice between different separation sequences can be made using the methods described in Chap. 5. However, we should be on guard to the fact that as the reactor conversion changes, the most appropriate sequence also can change. In other words, different separation system structures become appropriate for different reactor conversions.

8.4 Optimization of Processes Involving a Purge

As with the case of byproduct losses, another cost needs to be added to the tradeoffs when there is a purge. This is a raw materials efficiency cost due to purge losses. If the PRODUCT formation is constant, this cost can be defined to be[2]

$$\text{Cost of purge losses} = \text{cost of FEED lost to purge}$$

$$- \text{value of purge} \qquad (8.6)$$

The purge usually only has value in terms of its fuel value. Alternatively, if the purge must be disposed of by effluent treatment,

$$\text{Cost of purge losses} = \text{cost of FEED lost to purge}$$

$$+ \text{cost of disposal of purge} \qquad (8.7)$$

Again, as with the byproduct case, those raw materials costs which are in principle avoidable (i.e., the purge losses) are distinguished from those which are inevitable (i.e., the stoichiometric requirements for FEED entering the process which converts to the desired PRODUCT). Consider the tradeoffs for the reaction in Eq. (8.1), but now with IMPURITY entering with the FEED.

Now there are two global variables in the optimization. These are reactor conversion (as before) but now also the concentration of IMPURITY in the recycle. For each setting of the IMPURITY concentration in the recycle, a set of tradeoffs can be produced analogous to those shown in Figs. 8.3 and 8.4.

However, the concentration of impurity in the recycle is varied as shown in Fig. 8.5, so each component cost shows a family of curves when plotted against reactor conversion. Reactor cost (capital only) increases as before with increasing conversion (see Fig. 8.5a). Separation and recycle costs decrease as before (see Fig. 8.5b). Figure 8.5c shows the cost of the heat exchanger network and utilities to again decrease with increasing conversion. In Fig. 8.5d, the purge

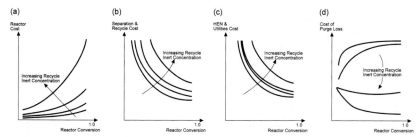

Figure 8.5 Cost tradeoffs for processes with a purge when reactor conversion and recycle inert concentration are allowed to vary. *(From Smith and Linnhoff, Trans. IChemE, ChERD, 66: 195, 1988; reproduced by permission of the Institution of Chemical Engineers.)*

losses show a complex pattern which depends on the relative costs of raw materials and fuel value of the purge and whether the reaction also produces IMPURITY as a byproduct.[2]

Figure 8.6 shows the component costs combined to give a total cost which varies with both reactor conversion and recycle inert concentration. Each setting of the recycle inert concentration shows a cost profile with an optimal reactor conversion. As the recycle inert concentration is increased, the total cost initially decreases but then

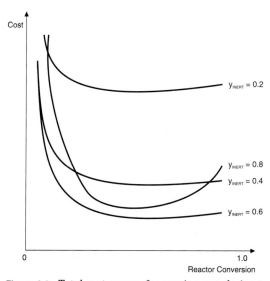

Figure 8.6 Total cost curves for varying recycle inert concentration. *(From Smith and Linnhoff, Trans. IChemE, ChERD, 66: 195, 1988; reproduced by permission of the Institution of Chemical Engineers.)*

increases for larger values of recycle inert concentration. The optimal conditions in Fig. 8.6 are in the region of $y_{\text{INERT}} = 0.6$ and $X = 0.5$.

An alternative way to view the tradeoffs shown in Fig. 8.6 is as a *contour diagram*. The contours in Fig. 8.7a show lines of constant total cost. The objective of the optimization is to find the lowest point. The shape of the contours dictates the optimization strategy. A naive strategy would fix the first variable, then optimize the second variable, and then fix the second variable and optimize the first. This is shown in Fig. 8.7b, where reactor conversion (X) is first fixed and inert concentration (y_{INERT}) optimized. Inert concentration is then fixed and conversion optimized. In this case, after two searches the solution is close to the optimum. Whether or not this is adequate depends on how flat the solution space is in the region of the optimum. Whether or not such a strategy will find the actual optimum depends on the shape of the solution space and the initialization for the optimization. In fact, such a strategy will only be sure to find the optimum if the contours are "circular." In Fig. 8.7, the contours are not circular; therefore, searching across each variable once will not be sure to identify the optimum. What would be required would be to repeat the strategy of fixing the first variable and optimizing the second and fixing the second and optimizing the first several times until the cost did not change significantly (see Fig. 8.7c). Much more efficient strategies for optimization based on the slope of the solution space can be developed.[3]

Obviously, the use of purges is not restricted to dealing with impurities. Purges can be used to deal with byproducts also.

8.5 Batch Processes

In Sec. 4.4 the possibility of using batch rather than continuous operations in the flowsheet was discussed. At that time, our only interest was the recycle structure of the flowsheet. There the approach was first to synthesize a flowsheet based on continuous

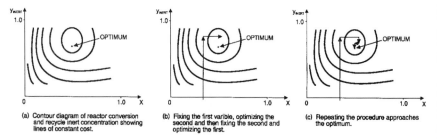

(a) Contour diagram of reactor conversion and recycle inert concentration showing lines of constant cost.

(b) Fixing the first varible, optimizing the second and then fixing the second and optimizing the first.

(c) Repeating the procedure approaches the optimum.

Figure 8.7 Optimization using a contour diagram.

operation and then, if required, change the process steps to batch. It was noted that if overlapping batch operation is used, recycling normally requires intermediate storage of the recycle, since, in general, the reaction and separation steps are carried out at different times.

It was also noted in Sec. 4.4 that the batch nature of a process can lead to less than full utilization of the equipment. Let us consider how utilization of equipment can be improved.

Consider again Example 4.5. Figure 4.15 shows the original time-event chart with overlapping batches. In Fig. 4.15 the reactor limits the batch cycle time; i.e., it has no "dead" time. On the other hand, the evaporator and stripper both have significant "dead" time. Figure 8.8 shows the time-event chart for an alternative arrangement with two reactors operating in parallel. With parallel operation, the reaction operations can overlap, allowing the evaporation and stripping operations to be carried out more frequently. This improves the overall utilization of equipment.

The batch cycle time has been reduced from 2.6 to 1.3 hours. This means that a greater number of batches can be processed, and hence, if there are two reactors each with the original capacity, the process capacity has increased. However, the increase in capacity has been achieved at the expense of increased capital cost for the second reactor. An economic assessment is required before we can judge whether the tradeoff is justified.

Perhaps the additional capacity might not be added. If it is not

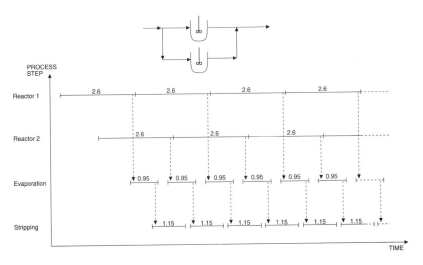

BATCH CYCLE TIME = 1.3hrs

Figure 8.8 Time-event chart for Example 4.4 with two reactors in parallel.

needed, then the size of the reactors, evaporator, and stripper can be reduced. Keeping the original process capacity with parallel operation of the reactors would mean a tradeoff between the increased capital cost of two (smaller) reactors versus the reduced capital cost of the evaporator and stripper. An economic comparison would be required to judge whether this would be beneficial.

Another option to improve utilization of equipment is, instead of adding a reactor in parallel, installing intermediate storage. Figure 8.9 shows the time-event chart with intermediate storage between the reactor and evaporator and between the evaporator and stripper. The evaporator step is no longer constrained to start on completion of the reaction step and start the stripping step on completion of the evaporation step. The individual steps can be decoupled via the intermediate storage. This maintains the original batch cycle time of 2.6 hours but allows, as shown in Fig. 8.9, the elimination of "dead" time in the evaporation and stripping steps. Now more evaporation and stripping steps can be carried out and the size of the evaporator and stripper reduced accordingly. This time the capital cost of intermediate storage is traded off against reduced capital cost of the evaporator and stripper. In Fig. 8.9, the intermediate storage between the reactor and evaporator has a significant effect on equipment utilization. The intermediate storage between the reactor and stripper has a less pronounced effect and would be more difficult to justify economically.

In general, utilization can be improved by

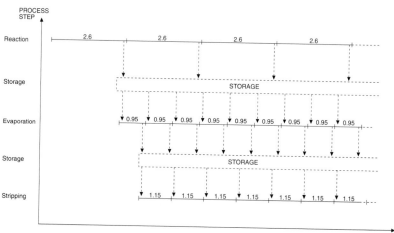

BATCH CYCLE TIME = 2.6hrs

Figure 8.9 Time-event chart for Example 4.4 with intermediate storage.

- Merging more than one operation into a single piece of equipment (e.g., feed preheating and reaction in the same vessel), providing these operations are not limiting the cycle time.
- Overlapping batches; i.e., more than one batch at different processing stages resides in the plant at any given time.
- Introducing parallel operations to the steps which limit the batch cycle time.
- Introducing multiple operations in series to the steps which limit the batch cycle time.
- Increasing the size of equipment in the steps which limit the batch cycle time to reduce the "dead" time for those steps which are not limiting.
- Decreasing the size of equipment for those steps which are not limiting to increase the time required for those steps which are not limiting and hence reduce the "dead" time for the nonlimiting steps.
- Introducing intermediate storage between batch steps.

Whether parallel operations, larger or smaller items of equipment, and intermediate storage should be used can only be judged on the basis of economic tradeoffs. However, this is still not the complete picture as far as the batch process tradeoffs are concerned. So far the batch size has not been varied. Batch size can be varied as a function of cycle time. Overall, the variables are

- Batch size
- Batch cycle time
- Number of batch units in parallel
- Number of batch units in series
- Size of equipment
- Intermediate storage

In addition to these variables which result from the batch nature of the process, there are still the variables considered earlier for continuous processes:

- Reactor conversion
- Recycle inert concentration

All these variables must be varied in order to minimize the total cost or maximize the economic potential (see App. A). This is a complex optimization problem involving both continuous variables (e.g., batch size) and integer variables (e.g., number of units in parallel) and is outside the scope of the present text.

8.6 Economic Tradeoffs—Summary

Two broad classes of tradeoffs can be identified when optimizing a given flowsheet structure. Local tradeoffs, when carried out around an operation, do not affect other operations in the flowsheet. Global tradeoffs cause changes throughout the flowsheet. Caution should be exercised in judging whether an optimization is local if it involves the use of energy. If an operation uses energy, this might be supplied from an external utility or by heat integration. The apparently "local" tradeoff between reflux and plates in a distillation column might well turn out to have features of a global tradeoff if the reboiler and/or condenser are heat integrated with the rest of the process. If this is the case, then changing the reflux ratio affects the overall heat integration problem.

Interactions between the reactor and the rest of the process are extremely important. Reactor conversion is the most significant optimization variable because it tends to influence most operations through the process.

When inerts are present in the recycle, the concentration is another important optimization variable, again influencing operations throughout the process.

In carrying out these optimizations, targets should be used for the energy and capital cost of the heat exchanger network. This is the only practical way to carry out these optimizations, since changes in reactor conversion and recycle inert concentration change the material and energy balance of the process, which, in turn, change the heat recovery problem. Each change in the material and energy balance in principle calls for a different heat exchanger network design. Furnishing a new heat exchanger network design for each setting of reactor conversion and recycle inert concentration is just not practical. On the other hand, targets for energy and capital cost of the heat exchanger network are, by comparison, easily generated.

We should be on guard for the fact that as the reactor conversion changes, the most appropriate separation sequence also can change. In other words, different separation system structures become appropriate for different reactor conversions.

In batch process optimization, one of the principal objectives is to improve equipment utilization through reduction in "dead" time. This requires both structural and parameter optimization, with many options available.

8.7 References

1. Douglas, J. M., *Conceptual Design of Chemical Processes,* McGraw-Hill, New York, 1988.
2. Smith, R., and Linnhoff, B., "The Design of Separators in the Context of Overall Processes," *Trans. IChemE, ChERD,* 66: 195, 1988.
3. Humphreys, K. K., *Jelen's Cost and Optimization Engineering,* 3d ed., McGraw-Hill, New York, 1991.

9

Safety and Health
Considerations

All too often safety and health (and environmental) considerations are left to the final stages of the design. Returning to the hierarchy of design illustrated by the onion diagram in Fig. 1.6, such considerations would add another layer in the diagram outside the utilities layer. This approach leaves much to be desired.

Early decisions made purely for process reasons often can lead to problems of safety and health (and environment) which require complex and often expensive solutions. It is far better to consider them early as the design progresses. Designs that avoid the need for hazardous materials, or use less of them, or use them at lower temperatures and pressures, or dilute them with inert materials will be *inherently safe* and will not require elaborate safety systems.[1]

Here we shall restrict consideration to safety and health considerations that can be built in while the design is developing rather than the detailed hazard and operability studies that take place in the later stages of design. The three major hazards in process plants are fire, explosion, and toxic release.[2]

9.1 Fire

The first major hazard in process plants is fire, which is usually regarded as having a disaster potential lower than both explosion or toxic release.[2] However, fire is still a major hazard and can, under the worst conditions, approach explosion in its disaster potential. It may, for example, give rise to toxic fumes. Let us start by examining the important factors in assessing fire as a hazard.

1. *Flash point.* The flash point of a liquid is the lowest temperature at which it gives off enough vapor to form an ignitable mixture with air. The flash point generally increases with increasing pressure.

2. *Autoignition temperature.* The autoignition temperature of a material is the temperature at which it will ignite spontaneously in air without any external source of ignition.

3. *Flammability limits.* A flammable gas will burn in air only over a limited range of composition. Below a certain concentration of the flammable gas, the *lower flammability limit,* the mixture is too "lean" to burn, i.e., lacks fuel. Above a certain concentration, the *upper flammability limit,* it is too "rich" to burn, i.e., lacks oxygen. Concentrations between these limits constitute the *flammable range.*

Combustion of a flammable gas-air mixture occurs if the composition of the mixture lies in the flammable range *and* if there is a source of ignition. Alternatively, combustion of the mixture occurs without a source of ignition if the mixture is heated up to its autoignition temperature.

The most flammable mixture usually approximates the stoichiometric mixture for combustion. It is often found that the concentrations of the lower and upper flammability limits are approximately one-half and twice that of the stoichiometric mixture, respectively.[2,3]

Flammability limits are affected by pressure. The effect of pressure changes is specific to each mixture. In some cases, decreasing the pressure can narrow the flammable range by raising the lower flammability limit and reducing the upper flammability limit until eventually the two limits coincide and the mixture becomes nonflammable. Conversely, an increase in pressure can widen the flammable range. However, in other cases, increasing the pressure has the opposite effect of narrowing the flammable range.[2,3]

Flammability limits are also affected by temperature. An increase in temperature usually widens the flammable range.[2,3]

Flammable liquids are potentially much more dangerous than flammable gas mixtures because of the greater mass which may be present. This is especially true if the liquids are processed or stored under pressure at a temperature above their atmospheric boiling point. Gases leak at a lower mass flow rate than liquids through an opening of a given size. Flashing liquids leak at about the same rate as a subcooled liquid but then turn into a mixture of vapor and spray. The spray, if fine, is just as hazardous as the vapor and can be spread as easily by the wind. Thus the leak of a flashing liquid through a hole of a given size produces a much greater hazard than the corresponding leak of gas.[1]

When synthesizing a flowsheet, it is more important to avoid the occurrence of flammable gas mixtures than to rely on elimination of sources of ignition. This may be achieved in the first instance by changing the process conditions such that mixtures are outside the flammable range. If this is not possible, inert material such as nitrogen, carbon dioxide, or steam should be introduced. The use of flammable liquids held under pressure above their atmospheric boiling points should be avoided. Adopting atmospheric subcooled conditions or vapor conditions in the process will be much safer. In addition, sources of ignition such as flames, electrical equipment, static electricity, etc. also should be eliminated wherever possible.

9.2 Explosion

The second of the major hazards is explosion, which has a disaster potential usually considered to be greater than fire but lower than toxic release.[2] Explosion is a sudden and violent release of energy.

The energy released in an explosion in a process plant is either chemical or physical:

1. *Chemical energy.* Chemical energy derives from a chemical reaction. The source of the chemical energy is exothermic chemical reactions, which includes combustion of flammable material. Explosions based on chemical energy can be either *uniform* or *propagating*. An explosion in a vessel will tend to be a uniform explosion, while an explosion in a long pipe will tend to be a propagating explosion.

2. *Physical energy.* Physical energy may be pressure energy in gases, thermal energy, strain energy in metals, or electrical energy. An example of an explosion caused by release of physical energy would be fracture of a vessel containing high-pressure gas. Thermal energy is generally important in creating the conditions for explosions rather than as a source of energy for the explosion itself. In particular, as already mentioned, superheat in a liquid under pressure causes flashing of the liquid if it is accidentally released to the atmosphere.

There are two basic kinds of explosions involving the release of chemical energy:

1. *Deflagration.* In a deflagration, the flame front travels through the flammable mixture relatively slowly, i.e., at subsonic velocity.

2. *Detonation.* In a detonation, the flame front travels as a shock wave, followed closely by a combustion wave, which releases the energy to sustain the shock wave. The detonation front travels with a velocity greater than the speed of sound in the unreacted medium.

A detonation generates greater pressures and is more destructive than a deflagration. Whereas the peak pressure caused by the deflagration of a hydrocarbon-air mixture in a closed vessel at atmospheric pressure is on the order of 8 bar, a detonation may give a peak pressure on the order of 20 bar. A deflagration may turn into a detonation, particularly if traveling down a long pipe.

Just as there are two basic kinds of explosions, they can occur in two different conditions:

1. *Confined explosions.* Confined explosions are those which occur within vessels, pipework, or buildings. The explosion of a flammable mixture in a process vessel or pipework may be a deflagration or a detonation. The conditions for a deflagration to occur are that the gas mixture is within the flammable range *and* there is a source of ignition. Alternatively, the deflagration can occur without a source of ignition if the mixture is heated to its autoignition temperature. An explosion starting as a deflagration can make the transition into a detonation. This transition can occur in a pipeline but is unlikely to happen in a vessel.

2. *Vapor cloud explosions.* Explosions which occur in the open air are vapor cloud explosions. A vapor cloud explosion is one of the most serious hazards in the process industries. Although a large toxic release may have a greater disaster potential, vapor cloud explosions tend to occur more frequently.[2] Most vapor cloud explosions have been the result of leaks of flashing flammable liquids.

The problem of explosion of a vapor cloud is not only that it is potentially very destructive but also that it may occur some distance from the point of vapor release and may thus threaten a considerable area. If the explosion occurs in an unconfined vapor cloud, the energy in the blast wave is generally only a small fraction of the energy theoretically available from the combustion of all the material that constitutes the cloud. The ratio of the actual energy released to that theoretically available from the heat of combustion is referred to as the *explosion efficiency.* Explosion efficiencies are typically in the range of 1 to 10 percent. A value of 3 percent is often assumed.

The hazard of an explosion should in general be minimized by avoiding flammable gas-air mixtures in the process. Again, this can be done either by changing process conditions or by adding an inert material. It is bad practice to rely solely on elimination of sources of ignition.

9.3 Toxic Release

The third of the major hazards and the one with the greatest disaster potential is the release of toxic chemicals.[2] The hazard posed by toxic release depends not only on the chemical species but also on the conditions of exposure. The high disaster potential from toxic release arises in situations where large numbers of people are briefly exposed to high concentrations of toxic material, i.e., *acute exposure.* However, the long-term health risks associated with prolonged exposure at low concentrations, i.e., *chronic exposure,* also present serious hazards.

For a chemical to affect health, a substance must come into contact with an exposed body surface. The three ways in which this happens are by inhalation, skin contact, and ingestion, the latter being rare.

In preliminary process design, the primary consideration is contact by inhalation. This happens either through accidental release of toxic material to the atmosphere or the *fugitive emissions* caused by slow leakage from pipe flanges, valve glands, and pump and compressor seals. Tank filling causes emissions when the rise in liquid level causes vapor in the tank to be released to the atmosphere.

The acceptable limits for toxic exposure depend on whether the exposure is brief or prolonged. *Lethal concentration* for airborne materials and *lethal dose* for nonairborne materials are measured by tests on animals. The limits for brief exposure to toxic materials which are airborne are usually measured by the concentration of toxicant which is lethal to 50 percent of the test group over a given exposure period, usually 4 hours. It is written as LC_{50} (lethal concentration for 50 percent of the test group). The test gives a comparison of the absolute toxicity of a compound in a single concentrated dose, i.e., acute exposure. For nonairborne materials, lethal dose LD_{50} is an index of the quantity of material administered which results in the death of 50 percent of the test group. It should be emphasized that it is extremely difficult to extrapolate tests on animals to human beings.

The limits for prolonged exposure are expressed as the *threshold limit values.* These are essentially acceptable concentrations in the workplace. There are three categories of threshold limit values:

1. *Time-weighted exposure.* This is the time-weighted average concentration for a normal 8-hour workday or 40-hour workweek to

which nearly all workers can be exposed, day after day, without adverse effects. Excursions above the limit are allowed if compensated by other excursions below the limit.

2. *Short-term exposure.* This is the maximum concentration to which workers can be exposed for a period of up to 15 minutes continuously without suffering from (a) intolerable irritation, (b) chronic or irreversible tissue change, or (c) narcosis of sufficient degree to increase accident proneness, impair self-rescue, or materially reduce efficiency, provided that no more than four excursions per day are permitted, with at least 60 minutes between exposure periods, and provided the daily time-weighted value is not exceeded.

3. *Ceiling exposure.* This is the concentration that should not be exceeded, even instantaneously.

When synthesizing a flowsheet, it is obviously best to try to avoid, where possible, the use of toxic materials altogether. However, this is often just not possible. In this case, the designer should take particular care to avoid, where possible, processing and storing toxic liquids under pressure at temperatures above their atmospheric boiling points. As with flammable materials, if a leak occurs, whether large or small, the mass flow rate through a hole of a given size is far greater for a liquid than for a gas. Release of a flashing liquid will result in higher levels of exposure than release of a subcooled liquid or a gas.

The best way to avoid fugitive emissions is by using leak-tight equipment (e.g., changing from packing to mechanical seals or even using sealless pumps, etc.). If this is not possible, then regular maintenance checks can reduce fugitive emissions. If all else, fails, the equipment can be enclosed and ventilated. The air would then be treated before finally passing to the atmosphere. Storage tanks should be prevented from breathing to atmosphere. There are three broad methods which allow this to be achieved:

1. *Vapor Treatment.* The vapors from the tank space can be sent to a treatment system (condenser, absorption, etc.) before venting. The system shown in Fig. 9.1 uses a vacuum-pressure relief valve which allows air in from the atmosphere when the liquid level falls (Fig. 9.1*a*) but forces the vapor through a treatment system when the tank is filled (Fig. 9.1*b*). If inert gas blanketing is required, because of the flammable nature of the material, then a similar system can be adopted which draws inert gas rather than air when the liquid level falls.

2. *Floating Roof.* Floating roofs can be used as shown in Fig. 9.2*a*.

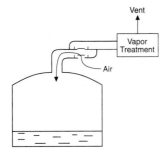

(a) As the liquid level falls, air is drawn in from atmosphere.

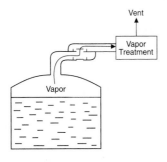

(b) As the liquid level rises vapor is forced through the vapor treatment system.

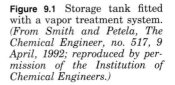

Figure 9.1 Storage tank fitted with a vapor treatment system. *(From Smith and Petela, The Chemical Engineer, no. 517, 9 April, 1992; reproduced by permission of the Institution of Chemical Engineers.)*

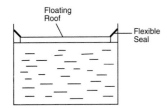

(a) Floating roof storage tank

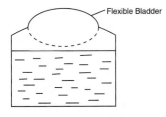

Figure 9.2 Floating roofs and flexible membranes can be used to prevent the release of material. *(From Smith and Petela, The Chemical Engineer, no. 517, 9 April, 1992; reproduced by permission of the Institution of Chemical Engineers.)*

(b) Flexible roof storage tank

This eliminates the vapor space but sealing the edge can be a problem. Double seals can help and sometimes a fixed roof is also added above the floating roof to help capture any leaks from the seal. However in this case, the space between the fixed and floating roof now breathes and an inert gas purge of this space would typically be used. The inert gas would be vented to atmosphere after treatment.

3. *Flexible membrane.* Another method to stop the vapor space breathing to atmosphere is to use a tank with a flexible membrane in the roof, Fig. 9.2b.

9.4   Intensification of Hazardous Materials

The best way to deal with a hazard in a flowsheet is to remove it completely. The provision of safety systems to control the hazard is much less satisfactory. One of the principal approaches to making a process inherently safe is to limit the inventory of hazardous material, called *intensification* of hazardous material. The inventories we wish to avoid most of all are flashing flammable liquids or flashing toxic liquids.

Once the process route has been chosen, it may be possible to synthesize flowsheets that do not require large inventories of materials in the process. The design of the reaction and separation system is particularly important in this respect, but heat transfer, storage, and pressure relief systems are also important.

1. *Reactors.* Perhaps the worst safety problem that can occur with reactors occurs when an exothermic reaction generates heat at a faster rate than the cooling can remove it. Such *runaway reactions* are usually caused by coolant failure, perhaps for a temporary period, or reduced cooling capacity due to perhaps a pump failure in the cooling water circuit. The runaway happens because the rate of reaction, and hence the rate of heat generation, increases exponentially with temperature, whereas the rate of cooling increases only linearly with temperature. Once heat generation exceeds available cooling capacity, the rate of temperature rise becomes progressively faster.[4] If the energy release is large enough, liquids will vaporize, and overpressurization of the reactor follows.

Clearly, the potential hazard from runaway reactions is reduced by reducing the inventory of material in the reactor. Batch operation requires a larger inventory than the corresponding continuous reactor. Thus there may be a safety incentive to change from batch to continuous operation. Alternatively, the batch operation can be changed to semi-batch in which one (or more) of the reactants is added over a period of time. The advantage of semi-batch operation is that the feed can be switched off in the event of a temperature (or pressure) excursion. This minimizes the chemical energy stored up for a subsequent exothermic release. For continuous reactors, plug-flow designs require smaller volumes and hence smaller inventories than continuous well-mixture designs for the same conversion.

Reaction rates often may be improved by using more extreme operating conditions. More extreme conditions may reduce inventory appreciably. However, more extreme conditions bring their own problems, as we shall discuss later. A very small reactor operating at a high temperature and pressure may be inherently safer than one operating at less extreme conditions because it contains a much lower inventory.[1] A large reactor operating close to atmospheric temperature and pressure may be safe for different reasons. Leaks are less likely, and if they do happen, the leak will be small because of the low pressure. Also, little vapor is produced from the leaking liquid because of the low temperature. A compromise solution employing moderate pressure and temperature and medium inventory may combine the worst features of the extremes.[1] The compromise solution may be such that the inventory is large enough for a serious explosion or serious toxic release if a leak occurs, the pressure will ensure that the leak is large, and the high temperature results in the evaporation of a large proportion of the leaking liquid.[1]

2. *Distillation.* There is a large inventory of boiling liquid, sometimes under pressure, in a distillation column, both in the base and held up in the column. If a sequence of columns is involved, then, as discussed in Chap. 5, the sequence can be chosen to minimize the inventory of hazardous material. If all materials are equally hazardous, then choosing the sequence that tends to minimize the flow rate of nonkey components also will tend to minimize the inventory. Use of the dividing-wall column shown in Fig. 5.17c will reduce considerably the inventory relative to two simple columns. Dividing-wall columns are inherently safer than conventional arrangements because they lower not only the inventory but also the number of items of equipment and hence lower the potential for leaks.

The column inventory also can be reduced by the use of low-holdup column internals, including the holdup in the column base. As the design progresses, other features can be included to reduce the inventory. Thermosyphon reboilers have a lower inventory than kettle reboilers. Peripheral equipment such as reboilers can be located inside the column.[1]

3. *Heat transfer operations.* Heat transfer fluids other than steam

and cooling water utilities are sometimes introduced into the design of the heat exchange system. These heat transfer media are sometimes liquid hydrocarbons used at high pressure. When possible, higher-boiling liquids should be used. Better still, the flammable material should be substituted by a nonflammable medium such as water or molten salt.

The use of an unnecessarily hot utility or heating medium should be avoided. This may have been a major factor that led to the runaway reaction at Seveso in Italy in 1976, which released toxic material over a wide area. The reactor was liquid phase and operated in a stirred tank (Fig. 9.3). It was left containing an uncompleted batch at around 160°C, well below the temperature at which a runaway reaction could start. The temperature required for a runaway reaction was around 230°C.[5]

In this accident, the steam was isolated from the reactor containing the unfinished batch and the agitator was switched off. The steam used to heat the reactor was the exhaust from a steam turbine at 190°C but which rose to about 300°C when the plant was shutdown. The reactor walls below the liquid level fell to the same temperature as the liquid, around 160°C. The reactor walls above the liquid level remained hotter because of the high-temperature steam at shutdown (but now isolated). Heat then passed by conduction and radiation from the walls to the top layer of the stagnant liquid, which became hot enough for a runaway reaction to start (see Fig. 9.3). Once started in the upper layer, the reaction then propagated throughout the reactor. If the steam had been cooler, say, 180°C, the runaway could not have occurred.[1]

Some operations need to be carried out at low temperature, which requires refrigeration. The refrigeration fluid might, for example, be propylene and present a major hazard. Operation of the process at a

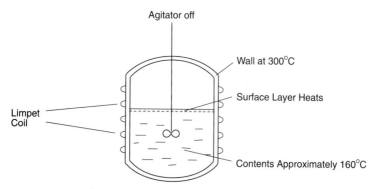

Figure 9.3 Schematic of the Seveso reaction system.

higher presssure on the one hand brings increased hazards in the process equipment but on the other hand might allow use of a less hazardous refrigeration fluid.

4. *Storage.* Some of the largest inventories of hazardous materials tend to be held up in the storage of raw materials and products and intermediate (buffer) storage. The most obvious way of reducing the inventory in storage is by locating producing and consuming plants near to each other so that hazardous intermediates do not have to be stored or transported.[1] It also may be possible to reduce storage requirements by making the design more flexible. Adjusting the capacity could then be used to cover delays in the arrival of raw materials, upsets in one part of the plant, etc. and thus reduce the need for storage.[1]

Large quantities of toxic gases such as chlorine and ammonia and flammable gases such as propane and ethylene oxide can be stored either under pressure or at atmospheric pressure under refrigerated conditions. If there is a leak from atmospheric refrigerated storage, the quantity of hazardous material that is discharged will be less than that from a corresponding pressurized storage at atmospheric temperature. For large storage tanks, refrigeration is safer. However, this might not be the case with small-scale storage, since the refrigeration equipment provides sources for leaks. Thus, in small-scale storage, pressurization may be safer.

5. *Relief systems.* Emergency discharge from relief valves can be dealt with in a number of ways:
a. Direct discharge to atmosphere under conditions leading to rapid dilution.
b. Total containment in a connected vessel, ultimate disposal being deferred.
c. Partial containment in which some of the discharge is separated either physically through gravitational, centrifugal means, etc. or chemically through absorption, etc. and contained.
d. Combustion in a flare. Flare systems might include a catchpot which collects liquids and passes gases to flare.

Relief systems are expensive and introduce considerable environmental problems. Sometimes it is possible to dispense with relief valves and all that comes after them by using stronger vessels, strong enough to withstand the highest pressures that can be reached. For example, if the vessel can withstand the pump delivery pressure, then a relief valve for overpressurization by the pump may not be needed. However, there may still be a need for a small relief device to guard against overpressurization in the event of a fire. It may be possible to avoid the need for a relief valve on a distillation column by making it strong enough to withstand the pressure developed if cooling is lost but heat input and feed pumping continue.[1]

At first sight, it might seem that making vessels strong enough to withstand the possible overpressurization would be an expensive option. However, we must not lose sight of the fact that we are not simply comparing one vessel with a thick wall versus one vessel with a thin wall protected by a relief valve. Material discharged through the relief valve might need to be partially contained, in which case the comparison might be between Fig. 9.4a and b.[1]

Similarly, instead of installing vacuum relief valves the vessels can be made strong enough to withstand vacuum. In addition, if the vessel contains flammable gas or vapor, vacuum relief valves will often need to admit nitrogen to avoid flammable mixtures. A stronger vessel often may be safer and cheaper.

6. *The overall inventory.* In the preceding chapter, the optimization of reactor conversion was considered. As the conversion increased, the size (and cost) of the reactor increased, but that of separation, recycle, and heat exchanger network systems decreased. The same also tends to occur with the inventory of material in these systems. The inventory in the reactor increases with increasing conversion, but the inventory in the other systems decreases. Thus, in some processes, it is possible to optimize for minimum overall inventory.[6] In the same way as reactor conversion can be varied to minimize the overall inventory, the recycle inert concentration also can be varied.

It might be possible to reduce the inventory significantly by changing reactor conversion and recycle inert concentration without a large cost penalty if the cost optimization profiles are fairly flat.

Intensification of hazardous materials results in a safer process. In

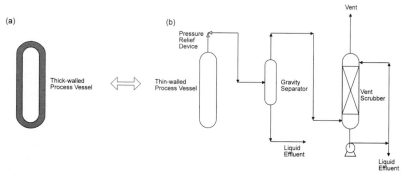

Figure 9.4 A thick-walled pressure vessel might be economical when compared with a thin-walled vessel and its relief and venting system.

the next chapter we shall see that it is also compatible with minimization of waste and can bring environmental benefits.

9.5 Attenuation of Hazardous Materials

So far the emphasis has been on substituting hazardous materials or using less, i.e., intensification. Let us now consider use of hazardous materials under less hazardous conditions, i.e. at less extreme temperatures or pressures or as a vapor rather than superheated liquid or diluted, in other words, *attenuation*.[1]

Operation at extremes of pressure and temperature brings a number of safety problems:

1. *High pressure.* Most process plant operates at pressures below 250 bar, but certain processes, such as high-pressure polyethylene plants, operate at pressures up to about 3000 bar. The use of high pressure greatly increases the stored energy in the plant. Although high pressures in themselves do not pose serious problems in materials of construction, the use of high temperatures, low temperatures, or corrosive chemicals together with high pressure does.[2] With high-pressure operation the problem of leaks becomes much more serious, since this increases the mass flow rate of fluid which can leak out through a given hole. This is particularly so when the fluid is a flashing liquid.

2. *Low pressure.* Low pressures are not in general as hazardous as the other extreme operating conditions. However, one particular hazard that does exist in low-pressure plants handling flammable materials is the possible ingress of air with the consequent formation of a flammable mixture.

3. *High temperature.* The use of high temperatures in combination with high pressures greatly increases stored energy in a plant. The heat required to obtain a high temperature is often provided by furnaces. These have a number of hazards, including possible rupture of the tubes carrying the process fluid and explosions in the radiant zone. There are also materials of construction problems associated with high-temperature operation. The main problem is creep, which is the gradual extension of a material which is under a steady tensile stress over a prolonged period of time.

4. *Low temperature.* Low-temperature process (below 0°C) can contain large amounts of fluids kept in the liquid state by pressure and/or low temperature. If for any reason it is not possible to keep them under pressure or keep them cold, then the liquids will begin to vaporize. If this happens, impurities in the fluids are liable to

precipitate from solution as solids, especially if equipment is allowed to boil dry. Deposited solids may not only be the cause of blockage but also in some cases the cause of explosions. It is necessary, therefore, to ensure that the fluids entering a low temperature plant are purified. A severe materials of construction problem in low temperature process is low-temperature embrittlement. Also, in low-temperature as in high-temperature operations, the equipment is subject to thermal stresses, especially during start-up and shutdown. Because of these dangers, large quantities of liquid stored at low temperature often use vessels with two skins, i.e., *double containment*.

When synthesizing a flowsheet, the designer should consider carefully the problems associated with operation under extreme conditions. Attenuation will result in a safer plant, providing the attenuation does not increase the inventory of hazardous materials. If the inventory does not increase, then attenuation not only will make the process safer but also will make it cheaper, since cheaper materials of construction and thinner vessel walls can be used and it is not necessary to add on so much protective equipment.

9.6   Quantitative Measures of Inherent Safety

It is easy to say that operation of a reactor at higher temperature *might* lead to a safer plant if the inventory can be reduced as a result, but how do we assess such changes quantitatively? Lowering the inventory makes the plant safer, but raising the temperature makes it less safe. Which effect is more significant?

Safety indices, such as the Dow index, have been suggested as measures of safety.[1,2] In these indices, the hazard associated with each material in the process is assessed and given a number based largely on judgment and experience. The numbers are weighted and combined to give an overall index for the process. The indices have no significance in an absolute sense but can be used to compare the relative hazards between two alternative designs. They are intended more for use in the later stages of design when more information is available. Detailed risk assessment, possibly including analysis of probabilities, also can be carried out in the later stages of design.[3]

However, in the early stages of design, decisions that have important safety implications must be made based on an incomplete picture. Let us explore simple quantitative measures which can be used to assist decision making in the early stages of design.

The major hazard from the release of flammable or toxic material arises from a material which, having been released to atmosphere, vaporizes. A simple quantitative measure of inherent safety for fire and explosion hazards is the energy released by combustion of a material that enters the vapor phase upon release from containment. The combustion energy releases associated with two process alternatives can be compared and some judgment made of the relative safety of the two options as far as fire and explosion hazards are concerned. However, the difficulty is one of defining the mode of release. We could, as a worst case, assume catastrophic failure involving release of all the materials and calculate the energy release from that part which would vaporize. On the other hand, the release could be assumed to occur from a standard-sized hole in the equipment. This would be a less hazardous scenario than catastrophic failure but more likely to occur. Comparing process options on the basis of these two alternative modes of release will not necessarily lead to the same conclusions when comparing two process alternatives. Judgment is required as to which mode of release is most appropriate.

On the other hand, if the hazard is toxicity, process alternatives can be compared by assessing the mass of toxic material that would enter the vapor phase on release from containment, weighting the components according to their lethal concentration.

Example 9.1 A process involves the use of benzene as a liquid under pressure. The temperature can be varied over a range. Compare the fire and explosion hazards of operating with a liquid process inventory of 1000 kmol at 100 and 150°C based on the theoretical combustion energy resulting from catastrophic failure of the equipment. The normal boiling point of benzene is 80°C, the latent heat of vaporization is 31,000 kJ kmol^{-1}, the specific heat capacity is 150 kJ kmol^{-1} °C^{-1}, and the heat of combustion is 3.2×10^6 kJ kmol^{-1}.

Solution The fraction of liquid vaporized on release is calculated from a heat balance.[3] The sensible heat above saturated conditions at atmospheric pressure provides the heat of vaporization. The sensible heat of the superheat is given by

$$mC_P(T_{\text{SUP}} - T_{\text{BT}})$$

where m = mass of liquid
C_P = heat capacity
T_{SUP} = temperature of the superheated liquid
T_{BT} = normal boiling point

If the mass of liquid vaporized is m_V, then

$$m_V = \frac{mC_P(T_{\text{SUP}} - T_{\text{BT}})}{\lambda}$$

where λ is the latent heat of vaporization. Thus the vapor fraction VF is given by

$$VF = \frac{m_V}{m} = \frac{C_P(T_{\text{SUP}} - T_{\text{BT}})}{\lambda}$$

For operation at 100°C,

$$VF = \frac{150(100 - 80)}{31{,}000}$$

$$= 0.097$$

$$m_V = 0.097 \times 1000$$

$$= 97 \, \text{kmol}$$

Theoretical combustion energy $= 97 \times 3.2 \times 10^6$

$$= 310 \times 10^6 \, \text{kJ}$$

For operation at 150°C,

$$VF = 0.339$$

$$m_V = 339 \, \text{kmol}$$

Theoretical combustion energy $= 1085 \times 10^6 \, \text{kJ}$

Thus, against this measure, the fire hazard is 3.5 times larger for operation at 150°C compared with operation at 100°C.

It is interesting to compare what might have been the conclusion if the inventory was in a reactor and not in a storage tank. If it is assumed, as an order of magnitude, that the reaction rate doubles for every 10°C rise in temperature, then the rate of reaction at 150°C would be 32 times faster than that at 100°C. For the same reactor conversion, this would mean that the inventory would be 32 times smaller. Thus operation at higher temperatures brings increased hazard as far as the fraction of released material that vaporizes is concerned but a lower hazard as far as the inventory required to give the same reactor conversion is concerned. Overall, operation of the reactor at higher temperature would be preferred against these measures. However, other factors would need to be taken into consideration in a detailed assessment.

In fact, the true fire load will be greater than the energy release calculated in Example 9.1. In practice, such a release of superheated liquid generates large amounts of fine spray in addition to the vapor. This can double the energy release based purely on vaporization.

If the material in two process alternatives is both flammable and highly toxic, then they can be compared on both bases separately. If the assessments of the relative flammability and toxicity are in conflict, then we can only resort to a safety index.

9.7 Safety and Health Considerations—Summary

Designs that avoid the need for hazardous materials, or use less of them, or use them at lower temperatures and pressures, or dilute them with inert materials will be inherently safe and will not require elaborate safety systems. When synthesizing a flowsheet, the occurrence of flammable gas mixtures should be avoided rather than relying on the elimination of sources of ignition.

One of the principal approaches to making a process inherently safe is to limit the inventory of hazardous materials. The inventories to avoid most of all are flashing flammable or toxic liquids, i.e., liquids under pressure above their atmospheric boiling points.

The following changes should be considered to improve safety:

Reactors

- Batch to continuous
- Batch to semi-batch
- Continuous well-mixed reactors to plug-flow
- Reduction of reactor inventory by increasing temperature or pressure, by changing catalyst, or by better mixing
- Lowering the temperature below the boiling point or diluting it with a safe solvent
- Substitute a hazardous solvent
- Externally heated/cooled to internally heated/cooled

Distillation

- Choose the distillation sequence to minimize the inventory of hazardous material.
- Use the divided wall column shown in Fig. 5.17c to reduce the inventory relative to two simple columns, and reduce the number of items of equipment and hence lower the potential for leaks.
- Use of low-holdup internals.

Heat transfer operations

- Use water or other nonflammable heat transfer medium.
- Use a lower-temperature utility or heat transfer medium.
- Use a heat transfer fluid below its atmospheric boiling point if flammable.
- If refrigeration is required, consider higher pressure process conditions if this allows a less hazardous refrigerant to be used.
- Use heat transfer equipment which requires a low inventory, such as plate heat exchangers.

Storage

- Locate producing and consuming plants near each other so that hazardous intermediates do not have to be stored and transported.

- Reduce storage by increasing design flexibility.

- Store in a safer form (less extreme pressure or temperature or in a different chemical form).

Relief systems. Relief systems are expensive and may bring significant environmental problems with them. Strengthening vessels may be a cheaper option.

Overall inventory. Consider changes to reactor conversion and recycle inert concentration to reduce the overall inventory. This might be possible without significant cost if the cost optimization profiles are fairly flat.

When synthesizing a flowsheet, the designer should consider carefully the problems associated with operation under extreme conditions. Attenuation will result in a safer plant, providing the attenuation does not increase the inventory of hazardous materials.

What you don't have, can't leak.[1] If we could design our plants so that they use safer raw materials and intermediates, or not so much of the hazardous ones, or use the hazardous ones at lower temperatures and pressures or diluted with inert materials, then many problems later in the design could be avoided.

As the design progresses, it is necessary to carry out hazard and operability studies.[2] These are generally only meaningful when the design has been progressed as far as the preparation of detailed flowsheets and are outside the scope of this text.

9.8   References

1. Kletz, T. A., "Cheaper, Safer Plants," *IChemE Hazard Workshop,* 2d., IChemE, Rugby, U.K., 1984.
2. Lees, F. P., *Loss Prevention in the Process Industries,* vol. 1, Butterworth, Reading, Mass., 1980.
3. Crowl, D. A., and Louvar, J. F., *Chemical Process Safety: Fundamentals with Applications,* Prentice-Hall, Englewood Cliffs, N.J., 1990.
4. Tharmalingam, S., "Assessing Runaway Reactions and Sizing Vents," *Chem. Eng.,* No. 463, Aug.: 33, 1989.
5. Cardillo, P., and Girelli, A., "The Seveso Runaway Reaction: A Thermoanalytical Study," *IChemE Symp. Series No. 68,* 3/N: 1, 1981.
6. Boccara, K., "Inherent Safety for Total Processes," M.Sc. dissertation, UMIST, U.K., 1992.

10

Waste Minimization

As with safety, environmental considerations are usually left to a late stage in the design. However, like safety, early decisions often can lead to difficult environmental problems which later require complex solutions. Again, it is better to consider effluent problems as the design progresses in order to avoid complex waste treatment systems.

The effects of pollution can be direct, such as toxic emissions providing a fatal dose of toxicant to fish, animal life, and even human beings. The effects also can be indirect. Toxic materials which are nonbiodegradable, such as waste from the manufacture of insecticides and pesticides, if released to the environment, are absorbed by bacteria and enter the food chain. These compounds can remain in the environment for long periods of time, slowly being concentrated at each stage in the food chain until ultimately they prove fatal, generally to predators at the top of the food chain such as fish or birds.

Thus emissions must not exceed levels at which they are considered harmful. There are two approaches to deal with emissions:

1. Treat the effluent using incineration, biological digestion, etc. to a form suitable for discharge to the environment, called *end-of-pipe treatment*.

2. Reduce or eliminate production of the effluent at the source by *waste minimization*.

The problem with relying on end-of-pipe treatment is that once waste has been created, it cannot be destroyed. The waste can be concentrated or diluted, its physical or chemical form can be changed, but it cannot be destroyed. Thus the problem with end-of-pipe effluent treatment systems is that they do not so much solve the problem as move it from one place to another. For example, aqueous solutions containing heavy metals can be treated by chemical precipitation to remove the metals. If the treatment system is designed and operated correctly, the aqueous stream can be passed on for further treatment or discharged to the receiving water. But what about the precipitated metallic sludge? This is usually disposed of to a landfill.[1]

The whole problem is best dealt with by not making the waste in the first place, i.e., waste minimization. If waste can be minimized at the source, this brings the dual benefit of reducing waste treatment costs *and* reducing raw materials costs.

Two classes of waste from chemical processes can be identified:[1]

- The two inner layers of the onion diagram in Fig. 1.6 (the reaction and separation and recycle systems) produce *process waste*. The process waste is waste byproducts, purges, etc.

- The outer layer of the onion diagram in Fig. 1.6 (the utility system) produces *utility waste*. The utility waste is composed of the products of fuel combustion, waste from the production of boiler feedwater for steam generation, etc. However, the design of the utility system is closely tied together with the design of the heat exchanger network. Hence, in practice, we should consider the two outer layers as being the source of utility waste.

There are three sources of process waste:[1]

1. *Reactors.* Waste is created in reactors through the formation of waste byproducts, etc.

2. *Separation and recycle systems.* Waste is produced from separation and recycle systems through the inadequate recovery and recycling of valuable materials from waste streams.

3. *Process operations.* The third source of process waste we can classify under the general category of process operations. Operations such as start-up and shutdown of continuous processes, product changeover, equipment cleaning for maintenance, tank filling, etc. all produce waste.

The principal sources of utility waste are associated with hot utilities (including cogeneration) and cold utilities. Furnaces, steam boilers, gas turbines, and diesel engines all produce waste as gaseous combustion products. These combustion products contain carbon dioxide, oxides of sulfur and nitrogen, and particulates which contribute in various ways to the greenhouse effect, acid rain, and the formation of smog (Fig. 10.1). In addition to gaseous waste, steam generation creates aqueous waste from boiler feedwater treatment, etc.

Let us look at how waste from each of these sources might be reduced before considering treatment methods in the next chapter. Since one of the themes running throughout the design philosophy presented here has been waste minimization through high process yields, elimination of extraneous materials, etc., much of the discussion to follow will summarize arguments already presented. However, this discussion shall go further and draw together the arguments into an overall philosophy of waste minimization. Since the reactor is at the heart of the process, this is where to start when considering waste minimization.[1] The separation and recycle system comes next, then process operations, and finally, utility waste.

10.1 Minimization of Waste from Reactors

Under normal operating conditions, waste is produced in reactors in six ways:[2]

1. If it is not possible for some reason to recycle unreacted feed

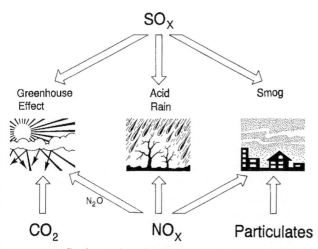

Figure 10.1 Products of combustion contribute in various ways to the greenhouse effect, acid rain, and smog. *(From Smith and Petela, Chem. Eng., 523: 32, 1992; reproduced by permission of the Institution of Chemical Engineers.)*

material to the reactor inlet, then low conversion will lead to waste of that unreacted feed.

2. The primary reaction can produce waste byproducts; for example,

$$\text{FEED 1} + \text{FEED 2} \longrightarrow \text{PRODUCT} + \text{WASTE BYPRODUCT}$$

$$(10.1)$$

3. Secondary reactions can produce waste byproducts; for example,

$$\text{FEED 1} + \text{FEED 2} \longrightarrow \text{PRODUCT}$$

$$\text{PRODUCT} \longrightarrow \text{WASTE BYPRODUCT}$$

$$(10.2)$$

In Chap. 2 the objective set was to maximize selectivity for a given conversion. This also will minimize waste generation in reactors for a given conversion.

4. Impurities in the feed materials can undergo reaction to produce waste byproducts.

5. If waste byproducts are formed, it may be possible to upgrade them by further reaction in a different reaction system.

6. Catalyst is either degraded and requires changing or is lost from the reactor and cannot be recycled.

Let us take each of these in turn and consider how reactor waste can be reduced.

   1. *Reducing waste when recycling is difficult.*
a. *Increasing conversion for single irreversible reactions.* If unreacted feed material is difficult to separate and recycle, it is necessary to force as high a conversion as possible. If the reaction is irreversible, then the low conversion can be forced to a higher conversion by longer residence time in the reactor, a higher temperature, higher pressure, or a combination of these.
b. *Increasing conversion for single reversible reactions.* The situation becomes worse if unreacted feed material is difficult to separate and recycle and this coincides with the reaction being reversible. Chapter 2 considered what can be done to increase equilibrium conversion:

   - *Excess reactants.* An excess of one of the reactants can be used, as shown in Fig. 2.9a.

- *Product removal during reaction.* Separation of the product before completion of the reaction can force a higher conversion, as discussed in Chap. 2. Figure 2.4 showed how this is done in sulfuric acid processes. Sometimes the product (or one of the products) can be removed continuously from the reactor as the reaction progresses, e.g., by allowing it to vaporize from a liquid phase reactor.

- *Inerts concentration.* The reaction might be carried out in the presence of an inert material. This could be a solvent in a liquid-phase reaction or an inert gas in a gas-phase reaction. Figure 2.9b shows that if the reaction involves an increase in the number of moles, then adding inert material will increase equilibrium conversion. On the other hand, if the reaction involves a decrease in the number of moles, then inert concentration should be decreased (see Fig. 2.9b). If there is no change in the number of moles during reaction, then inert material has no effect on equilibrium conversion.

- *Reaction temperature.* For endothermic reactions, Fig. 2.9c shows that the temperature should be set as high as possible consistent with materials-of-construction limitations, catalyst life, and safety. For exothermic reactions, the ideal temperature is continuously decreasing as conversion increases (see Fig. 2.9c).

- *Reactor pressure.* In Chap. 2 it was deduced that vapor-phase reactions involving a decrease in the number of moles should be set to as high a pressure as practicable, taking into account that the high pressure might be expensive to obtain through compressor power, mechanical construction might be expensive, and high pressure brings safety problems (see Fig. 2.9d). Reactions involving an increase in the number of moles ideally should have a pressure that is continuously decreasing as conversion increases (see Fig. 2.9d). Reduction in pressure can be brought about either by a reduction in the absolute pressure or by the introduction of an inert diluent.

If the separation and recycle of unreacted feed material is not a problem, then we don't need to worry too much about trying to squeeze extra conversion from the reactor.

2. *Reducing waste from primary reactions which produce waste byproducts.* If a waste byproduct is formed from the reaction, as in Eq. (10.1) above, then it can only be avoided by different reaction chemistry, i.e., a different reaction path.

3. *Reducing waste from multiple reactions producing waste byproducts.* In addition to the losses described above for single reactions,
multiple reaction systems lead to further waste through the formation of waste byproducts in secondary reactions. Let us briefly review
from Chap. 2 what can be done to minimize byproduct formation.

a. *Reactor type.* First, make sure that the correct reactor type has
   been chosen to maximize selectivity for a given conversion in
   accordance with the arguments presented in Chap. 2.

b. *Reactor concentration.* Selectivity often can be improved by one or
   more of the following actions:[2]

- Use an excess of one of the feeds when more than one feed is
  involved.

- Increase the concentration of inerts if the byproduct reaction is
  reversible and involves a decrease in the number of moles.

- Decrease the concentration of inerts if the byproduct reaction is
  reversible and involves an increase in the number of moles.

- Separate the product partway through the reaction before
  carrying out further reaction and separation.

- Recycle waste byproducts to the reactor if byproduct reactions
  are reversible.

Each of these measures can, in the appropriate circumstances,
minimize waste (see Fig. 2.10).

c. *Reactor temperature and pressure.* If there is a significant
   difference between the effect of temperature or pressure on
   primary and byproduct reactions, then temperature and pressure
   should be manipulated to improve selectivity and minimize the
   waste generated by byproduct formation.

d. *Catalysts.* Catalysts can have a major influence on selectivity.
   Changing the catalyst can change the relative influence on the
   primary and byproduct reactions.

4. *Reducing waste from feed impurities which undergo reaction.* If
feed impurities undergo reaction, this causes waste of feed material,
products, or both. Avoiding such waste is most readily achieved by
purifying the feed. Thus increased feed purification costs are traded
off against reduced raw materials, product separation, and waste
disposal costs (Fig. 10.2).

5. *Reducing waste by upgrading waste byproducts.* Waste byproducts can sometimes be upgraded to useful materials by subjecting

them to further reaction in a different reaction system. An example was given in Chap. 4 in which hydrogen chloride, which is a waste byproduct of chlorination reactions, can be upgraded to chlorine and then recycled to the chlorination reactor.

6. *Reducing catalyst waste.* Both homogeneous and heterogeneous catalysts are used. In general, heterogeneous catalysts should be used whenever possible because separation and recycling of homogeneous catalysts can be difficult, leading to waste.

Heterogeneous catalysts are more common. However, they degrade and need replacement. If contaminants in the feed material or recycle shorten catalyst life, then extra separation to remove these contaminants before the feed enters the reactor might be justified. If the cataylst is sensitive to extreme conditions, such as high temperature, then some measures can help to avoid local hot spots and extend catalyst life:

- Better flow distribution
- Better heat transfer
- Introduction of catalyst diluent
- Better instrumentation and control

Fluidized-bed catalytic reactors tend to generate loss of catalyst through attrition of the solid particles, causing fines to be generated,

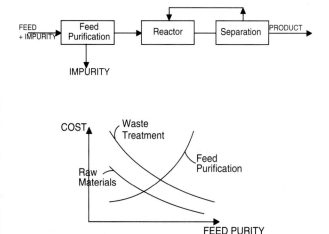

Figure 10.2 If feed impurity undergoes a reaction, then there is an optimal feed purity.

which are lost. More effective separation of catalyst fines from the reactor product and recycling of the fines will reduce catalyst waste up to a point. Improving the mechanical strength of the catalyst is likely to be the best solution in the long run.

Let us now turn our attention to losses from the separation and recycle system.

10.2 Minimization of Waste from the Separation and Recycle System

Waste also can be minimized if the separation system can be made more efficient such that useful materials can be separated and recycled more effectively.

Waste from the separation and recycle system can be minimized in five ways:[3]

- Recycle waste streams directly.
- Reduce feed impurities by purifying the feed.
- Eliminate extraneous materials used for separation.
- Employ additional separation of waste streams to allow increased recovery.
- Employ additional reaction and separation of waste streams to allow increased recovery.

Although this is generally the sequence in which the five actions would be considered, this sequence will not always be correct. The best sequence in which to consider the five actions will depend on the process. The magnitude of effect each action will have on waste minimization will vary for different processes.

1. *Recycle waste streams directly.* Sometimes waste can be reduced by recycling waste streams directly. If this can be done, it is clearly the simplest way to reduce waste and should be considered first. Most often, the waste streams that can be recycled directly are aqueous streams which, although contaminated, can substitute part of the freshwater feed to the process.

Figure 10.3*a* shows a simplified flowsheet for the production of isopropyl alcohol by the direct hydration of propylene.[3] Different reactor technologies are available for the process, and separation and recycle systems vary, but Fig. 10.3*a* is representative. Propylene containing propane as an impurity is reacted with water according to the reaction:

$$C_3H_6 + H_2O \longrightarrow (CH_3)_2CHOH$$

propylene   water   isopropyl alcohol

Some small amount of byproduct formation occurs. The principal byproduct is di-isopropyl ether. The reactor product is cooled, and a phase separation of the resulting vapor-liquid mixture produces a vapor containing predominantly propylene and propane and a liquid containing predominantly the other components. Unreacted propylene is recycled to the reactor, and a purge prevents the buildup of propane. The first distillation in Fig. 10.3a (column C1) removes

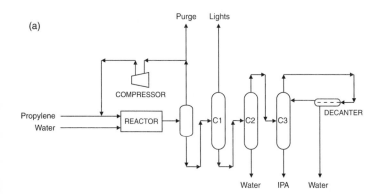

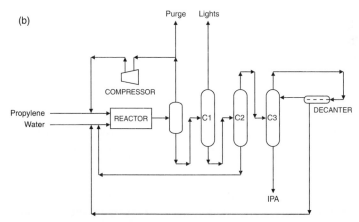

Figure 10.3 Outline flowsheet for the production of isopropyl alcohol by direct hydration of propylene. *(From Smith and Petela, Chem. Eng., 513: 24, 1991; reproduced by permission of the Institution of Chemical Engineers.)*

light ends (including the di-isopropyl ether). The second (column C2) removes as much water as possible to approach the azeotropic composition of the isopropyl alcohol–water mixture. The final column in Fig. 10.3a (column C3) is an azeotropic distillation using an entrainer. In this case, one of the materials already present in the process, di-isopropyl ether, can be used as the entrainer.

Wastewater leaves the process from the bottom of the second column and the decanter of the azeotropic distillation column. Although both these streams are essentially pure water, they will nevertheless contain small quantities of organics and must be treated before final discharge. This treatment can be avoided altogether by recycling the wastewater to the reactor inlet to substitute part of the freshwater feed (see Fig. 10.3b).

Sometimes waste streams can be recycled directly, but between different processes. Waste streams from one process can become the feedstock for another. The scope for such *waste exchanges* is often not fully realized, since it often means waste being transferred between different companies.

If waste streams can be recycled directly, this is clearly the simplest method for reducing waste. Most often, though, additional separation is required or a different separation method is needed to reduce waste.

2. *Feed purification.* Impurities that enter with the feed inevitably cause waste. If feed impurities undergo reaction, then this causes waste from the reactor, as already discussed. If the feed impurity does not undergo reaction, then it can be separated out from the process in a number of ways, as discussed in Sec. 4.1. The greatest source of waste occurs when we choose to use a purge. Impurity builds up in the recycle, and we would like it to build up to a high concentration to minimize waste of feed materials and product in the purge. However, two factors limit the extent to which the feed impurity can be allowed to build up:

a. High concentrations of inert material can have an adverse effect on reactor performance.

b. As more and more feed impurity is recycled, the cost of the recycle increases (e.g., through increased recycle gas compression costs, etc.) to the point where that increase outweighs the savings in raw materials lost in the purge.

In general, the best way to deal with a feed impurity is to purify the feed before it enters the process. Let us return to the isopropyl alcohol process from Fig. 10.3. Propylene is fed to the process containing propane as a feed impurity. In Fig. 10.3 the propane is removed from the process using a purge. This causes waste of propylene, together with a small amount of isopropyl alcohol. The purge can be virtually eliminated if the propylene is purified before entering the process. In this case the purification can be done by distillation. Examples of where similar schemes can be implemented are plentiful.

Many processes are based on an oxidation step for which air would be the first obvious source of oxygen. A partial list would include acetic acid, acetylene, acrylic acid, acrylonitrile, carbon black, ethylene oxide, formaldehyde, maleic anhydride, nitric acid, phenol, phthalic anhydride, sulfuric acid, titanium dioxide, vinyl acetate, and vinyl chloride.[5] Clearly, because the nitrogen in the air is not required by the reaction, it must be separated at some point. Because gaseous separations are difficult, the nitrogen is normally separated using a purge, or alternatively, the reactor is forced to as high a conversion as possible to avoid recycling. If a purge is used, the nitrogen will carry with it process materials, both feeds and products, and will probably require treatment before final discharge. If the air for the oxidation is substituted by pure oxygen, then, at worst, the purge will be very much smaller. At best, it can be eliminated altogether. Of course, this requires an air separation plant upstream of the process to provide the pure oxygen. However, despite this disadvantage, very significant benefits can be obtained, as the following example shows.

Consider vinyl chloride production (see Example 2.1). In the "oxychlorination" reaction step of the process, ethylene, hydrogen chloride, and oxygen are reacted to form dichloroethane:

$$C_2H_4 \ + \ 2HCl \ + 1/2O_2 \ \longrightarrow \ C_2H_4Cl_2 \ + H_2O$$
$$\text{ethylene} \quad \text{hydrogen} \quad \text{oxygen} \quad \quad \text{dichloroethane} \quad \text{water}$$
$$\text{chloride}$$

If air is used, then a single pass with respect to each feedstock is used and no recycle to the reactor (Fig. 10.4a). Thus the process operates at near stoichiometric feed rates to achieve high conversions. Typically, between 0.7 and 1.0 kg of vent gases are emitted per kilogram of dichloroethane produced.[6]

If the air is substituted by pure oxygen, then the problem of the large flow of inert gas is eliminated (see Fig. 10.4b). Unreacted gases can be recycled to the reactor. This allows oxygen-based processes to be operated with an excess of ethylene, thereby enhancing the HCl conversion without sacrificing ethylene yield. Unfortunately, this introduces a safety problem downstream of the reactor. Unconverted ethylene can create explosive mixtures with the oxygen. To avoid explosive mixtures, a small bleed of nitrogen is introduced.

Since nitrogen is drastically reduced in the feed and essentially all ethylene is recycled, only a small purge is required to be vented. This results in a 20- to 100-fold reduction in the size of the purge.[6]

3. *Eliminate extraneous materials for separation.* The third option is to eliminate extraneous materials added to the process to carry out separation. The most obvious example would be addition of a solvent, either organic or aqueous. Also, acids or alkalis are sometimes used to precipitate other materials from solution. If these extraneous materials used for separation can be recycled with a high efficiency, there is not a major problem. Sometimes, however, they cannot. If this is the case, then waste is created by discharge of that material. To reduce this waste, alternative methods of separation are needed, such as use of evaporation instead of precipitation.

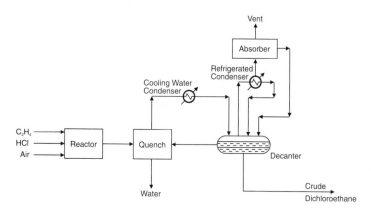

(a)   Air feed

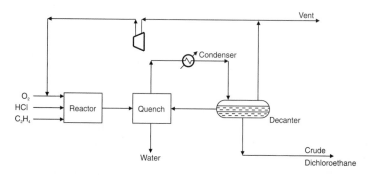

(b)   Oxygen feed

Figure 10.4 The oxychlorination step of the vinyl chloride process. *(From Smith and Petela, Chem. Eng., 513: 24, 1991; reproduced by permission of the Institution of Chemical Engineers.)*

As an example, consider again the manufacture of vinyl chloride. In the first step of this process, ethylene and chlorine are reacted to form dichloroethane:

$$\underset{\text{ethylene}}{C_2H_4} + \underset{\text{chlorine}}{Cl_2} \longrightarrow \underset{\text{dichloroethane}}{C_2H_4Cl_2}$$

A flowsheet for this part of the vinyl chloride process is shown in Fig. 10.5. The reactants, ethylene and chlorine, dissolve in circulating liquid dichloroethane and react in solution to form more dichloroethane. Temperature is maintained between 45 and 65°C, and a small amount of ferric chloride is present to catalyze the reaction. The reaction generates considerable heat.

In early designs, the reaction heat typically was removed by cooling water. Crude dichloroethane was withdrawn from the reactor as a liquid, acid-washed to remove ferric chloride, then neutralized with dilute caustic, and purified by distillation. The material used for separation of the ferric chloride can be recycled up to a point, but a purge must be done. This creates waste streams contaminated with chlorinated hydrocarbons which must be treated prior to disposal.

The problem with the flowsheet shown in Fig. 10.5 is that the ferric chloride catalyst is carried from the reactor with the product. This is separated by washing. If a reactor design can be found that prevents the ferric chloride leaving the reactor, the effluent problems created by the washing and neutralization are avoided. Because the ferric chloride is nonvolatile, one way to do this would be to allow the heat of reaction to raise the reaction mixture to the boiling point and remove the product as a vapor, leaving the ferric chloride in the reactor. Unfortunately, if the reaction mixture is allowed to boil, there are two problems:

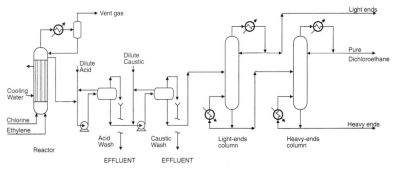

Figure 10.5 The direct chlorination step of the vinyl chloride process using a liquid phase reactor. *(From McNaughton, Chem. Engg., December 12, 1983, pp. 54–58; reproduced by permission.)*

- Ethylene and chlorine are stripped from the liquid phase, giving a low conversion.

- Excessive byproduct formation occurs.

This problem is solved in the reactor shown in Fig. 10.6. Ethylene and chlorine are introduced into circulating liquid dichloroethane. They dissolve and react to form more dichloroethane. No boiling takes place in the zone where the reactants are introduced or in the zone of reaction. As shown in Fig. 10.6, the reactor has a U-leg in which dichloroethane circulates as a result of gas lift and thermosyphon effects. Ethylene and chlorine are introduced at the bottom of the up-leg, which is under sufficient hydrostatic head to prevent boiling.

The reactants dissolve and immediately begin to react to form further dichloroethane. The reaction is essentially complete at a point only two-thirds up the rising leg. As the liquid continues to rise, boiling begins, and finally, the vapor-liquid mixture enters the disengagement drum. A very slight excess of ethylene ensures essentially 100 percent conversion of chlorine.

As shown in Fig. 10.6, the vapor from the reactor flows into the bottom of a distillation column, and high-purity dichloroethane is withdrawn as a sidestream several trays from the column top.[7] The design shown in Fig. 10.6 is elegant in that the heat of reaction is conserved to run the separation and no washing of the reactor

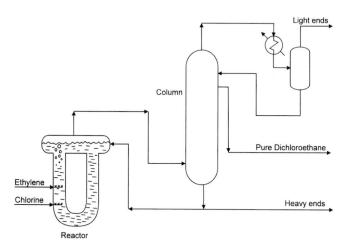

Figure 10.6 The direct chlorination step of the vinyl chloride process using a boiling reactor eliminates the washing and neutralization steps and the resulting effluents. *(From McNaughton, Chem. Engg., December 12, 1983, pp. 54–58; reproduced by permission.)*

products is required. This eliminates two aqueous waste streams which will inevitably carry organics with them, requiring treatment and causing loss of materials.

It is often possible to use the energy system inherent in the process to drive the separation system for us by improved heat recovery and in so doing carry out the separation at little or no increase in operating costs.

4. *Additional separation and recycling.* Once the possibilities for recycling streams directly, feed purification, and eliminating the use of extraneous materials for separation that cannot be recycled efficiently have been exhausted, attention is turned to the fourth option, the degree of material recovery from the waste streams that are left. One very important point which should not be forgotten is that once the waste stream is rejected, any valuable material turns into a liability as an effluent material. The level of recovery in such situations needs careful consideration. It may be economical to carry out additional separation of the valuable material with a view to recycling that additional recovered material, particularly when the cost of downstream effluent treatment is taken into consideration.

Perhaps the most extreme situation is encountered with purge streams. Purges are used to deal with both feed impurities and byproducts of reaction. In the preceding section we considered how the size of purges can be reduced in the case of feed impurities by purifying the feed. However, if it is impractical or uneconomical to reduce the purge by feed purification, or the purge is required to remove a byproduct of reaction, then the additional separation can be considered.

Figure 10.7 shows the basic tradeoff to be considered as additional feed and product materials are recovered from waste streams and recycled. As the fractional recovery increases, the cost of the separation and recycle increases. On the other hand, the cost of the lost materials decreases. It should be noted that the raw materials cost is a *net* cost, which means that the cost of lost materials should be adjusted to either

a. add the cost of waste treatment for unrecovered material, or

b. deduct the fuel value if the recovered material is to be burnt to provide useful heat in a furnace or boiler.

Figure 10.7 shows that the tradeoff between separation and net raw materials cost gives an economically optimal recovery. It is possible that significant changes in the degree of recovery can have a significant effect on costs other than those shown in Fig. 10.7 (e.g., reactor costs). If this is the case, then these also must be included in the tradeoffs.

It must be emphasized that any energy costs for the separation in the tradeoffs shown in Fig. 10.7 must be taken within the context of the overall heat integration problem. The separation might after all be driven by heat recovery.

5. *Additional reaction and separation of waste streams.* Sometimes it is possible to carry out further reaction as well as separation on waste streams. Some examples have already been discussed in Chap. 4.

10.3   Minimization of Waste from Process Operations

The third source of process waste after the reactor and separation and recycle systems is process operations.

1. *Sources of waste in process operations*
a. *Start-up/shutdown in continuous processes*

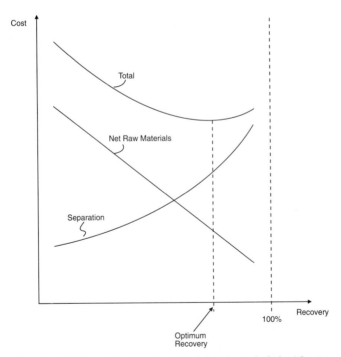

Figure 10.7 Effluent treatment costs should be included with raw materials costs when traded off against separation costs to obtain the optimal recovery. *(From Smith and Petela, Chem. Eng., 513: 24, 1991; reproduced by permission of the Institution of Chemical Engineers.)*

- Reactors give lower than design conversions.

- Reactors at nonoptimal conditions produce (additional) unwanted byproducts. Not only might this lead to loss of material through additional byproduct formation, but it also might prevent the recycling of material produced during the start-up.

- Separators working at unsteady conditions produce intermediates with compositions which do not allow them to be recycled. Alternatively, if the intermediate can be recycled, a nonoptimal recycle might produce (additional) unwanted byproducts in the reactor.

- Process intermediates are generated which, because the downstream process is not operational, cannot be processed further.

- When working at unsteady conditions, separators which normally split useful material from waste streams might lose material unnecessarily to the waste streams.

- Separators working at unsteady conditions produce products which do not meet the required sales specification.

b. *Product changeover*

- In continuous processes, all those sources of process waste associated with start-up and shutdown also apply to product changeover in multiproduct plants.

- In both batch and continuous processes, it may be necessary to clean equipment to prevent contamination of new product. Materials used for equipment cleaning often cannot be recycled, leading to waste.

c. *Equipment cleaning for maintenance, tank filling, and fugitive emissions*

- Equipment needs to be cleaned and made safe for maintenance.

- When process tanks, road tankers, or rail tank cars are filled, material in the vapor space is forced out of the tank and lost to atmosphere.

- Material transfer requires pipework, valves, pumps, and compressors. Fugitive emissions occur from pipe flanges, valve glands, and pump and compressor seals.

Let us now suggest what can be done, particularly in design, to overcome such waste.

2. *Process operation for waste minimization.* Many of the problems associated with waste from process operations can be mitigated if the

process is designed for low inventories of material in the process. This is also compatible with design for inherent safety. Other ways to minimize waste from process operations are

- Minimize the number of shutdowns by designing for high availability. Install more reliable equipment or standby equipment.

- Design continuous processes for flexible operation, e.g., high turndown rate rather than shutdown.

- Consider changing from batch to continuous operation. Batch processes, by their very nature, are always at unsteady state and thus are difficult to maintain at optimal conditions.

- Install enough intermediate storage to allow reworking of off-specification material.

- Changeover between products causes waste because equipment must be cleaned. Such waste can be minimized by scheduling operation to minimize product changeovers.

- Install a waste-collection system for equipment cleaning and sampling waste that allows waste to be segregated and recycled where possible. This normally requires separate sewers for organic and aqueous waste, collecting to sump tanks, and recycle or separate and recycle if possible. If equipment is steamed out during the cleaning process, the plant should allow collection and condensation of the vapors and recycling of materials where possible.

- Reduce losses from fugitive emissions and tank breathing as discussed under safety in Chap. 9.

There are many other sources of waste associated with process operations which can only be taken care of in the later stages of design or after the plant has been built and has become operational. For example, poor operating practice can mean that the process operates under conditions for which it was not designed, leading to waste. Such problems might be solved by an increased level of automation or better management of the process.[3] These considerations are outside the scope of this text.

10.4 Minimization of Utility Waste

1. *Utility systems as sources of waste.* The principal sources of utility waste are associated with hot utilities (including cogeneration systems) and cold utilities.[9] Furnaces, steam boilers, gas turbines, and diesel engines all produce waste from products of combustion. The principal problem here is the emission of carbon dioxide, oxides of sulfur and nitrogen, and particulates (metal oxides, unburnt

carbon, and hydrocarbon). As well as gaseous waste, the combustion of coal produces solid waste as ash. Steam systems and cooling water systems also produce aqueous waste, as we shall discuss later.

The waste streams created by utility systems tend, on the whole, to be less environmentally harmful than process waste. Unfortunately, complacency would be misplaced. Even though utility waste tends to be less harmful than process waste, the quantities of utility waste tend to be larger than those of process waste. This sheer volume can result in a greater environmental impact than process waste. Gaseous combustion products contribute in various ways to the greenhouse effect, acid rain, and can produce a direct health hazard due to the formation of smog (see Fig. 10.1). The aqueous waste generated by utility systems also can be a major problem if it is contaminated.

2. *Energy efficiency of the process.* If the process requires a furnace or steam boiler to provide a hot utility, then any excessive use of the hot utility will produce excessive utility waste through excessive generation of CO_2, NO_x, SO_x, particulates, etc. Improved heat recovery will reduce the overall demand for utilities and hence reduce utility waste.

3. *Local and global emissions.* When considering utility waste, it is tempting to consider only the *local* emissions from the process and its utility system (Fig. 10.8a). However, this only gives part of the picture. The emissions generated from central power generation are just as much part of the process as those emissions generated on-site (Fig. 10.8b). These emissions should be included in the assessment of utility waste. Thus *global* emissions are defined to be[10]

Global emissions = emissions from on-site utilities

> + emissions from central power generation corresponding with the amount of electricity imported

> − emissions saved at central power generation corresponding with the amount of electricity exported from the site

This is particularly important when considering the effect that combined heat and power generation (cogeneration) has on utility waste.

4. *Combined heat and power (cogeneration).* Combined heat and power generation can have a very significant effect on the generation of utility waste. However, great care must be taken to assess the effects on the correct basis.

Assessing only the local effects of combined heat and power is misleading. Combined heat and power generation increases the local utility emissions because, besides the fuel burnt to supply the heating demand, additional fuel must be burnt to generate the power. It is only when the emissions are viewed on a global basis, and the emissions from central power generation included, that the true picture is obtained. Once these are included, on-site combined heat and power generation can make major reductions in global utility waste. The reason for this is that even the most modern central power stations have a poor efficiency of power generation compared with a combined heat and power generation system. Once the other inefficiencies associated with centralized power generation are taken into account, such as distribution losses, the gap between the efficiency of combined heat and power systems and centralized power generation widens.

As an example, consider a process that requires a furnace to

(a) Local Emissions

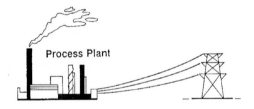

(b) Global Emissions

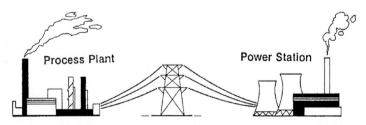

Figure 10.8 Local and global emissions. *(From Smith and Petela, Chem. Eng., 523: 32, 1992; reproduced by permission of the Institution of Chemical Engineers.)*

satisfy its hot utility requirements. Let us suppose it is a state-of-the-art furnace with a thermal efficiency of 90 percent producing 300 kg CO_2 per hour for each megawatt of heat delivered to the process. Power is being imported from centralized generation via the grid. If, instead of the furnace, a gas turbine is installed, this produces 500 kg CO_2 per hour for each megawatt of heat delivered to the process, an increase in local emissions of 200 kg CO_2 per hour per megawatt of heat. However, the gas turbine also generates 400 kW of power, replacing that much in centralized generation. If the same power was generated centrally to supplement the furnace, 450 kg CO_2 per hour would be released from centralized generation, giving a global emission of 750 kg CO_2 per hour for the furnace plus power from the grid.[9,10]

5. *Fuel switch.* The choice of fuel used in furnaces and steam boilers has a major effect on the gaseous utility waste from products of combustion. For example, a switch from coal to natural gas in a steam boiler can lead to a reduction in carbon dioxide emissions of typically 40 percent for the same heat released.[10] This results from the lower carbon content of natural gas. In addition, it is likely that a switch from coal to natural gas also will lead to a considerable reduction in both SO_x and NO_x emissions, as we shall discuss later.

Such a fuel switch, while being desirable in reducing emissions, might be expensive. If the problem is SO_x and NO_x emissions, there are other ways to combat these, which will be dealt with in the next chapter.

6. *Waste from steam systems.* If steam is used as a hot utility, then inefficiencies in the steam system itself cause utility waste. Figure 10.9 shows a schematic representation of a steam system. Raw water from a river or other source is fed to the steam system. This is

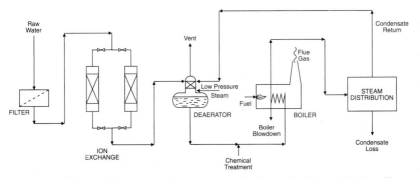

Figure 10.9 Schematic of a typical steam system. *(From Smith and Petela, Chem. Eng., 523: 32, 1992; reproduced by permission of the Institution of Chemical Engineers.)*

typically filtered, deionized in ion-exchange beds, and deaerated by steam stripping and chemical treatment before being fed to the boiler. Steam is then usually generated at high pressure to be let down through steam turbines to generate power before being used for steam heating. Not all the condensate from the heating duties will be returned to the boiler. This constant loss of condensate from the steam system means that there must be a constant makeup with fresh water.

This make up causes utility waste:

a. Wastewater is generated in the deionization process when the ion-exchange beds are regenerated with acid and alkaline solutions.

b. Wastewater is generated from *boiler blowdown*. This is a purge of the water in the boiler to prevent the buildup of dissolved solids which are present in the raw water but not removed in the deionization process. Boiler blowdown typically can range between 2 and 10 percent of the rate of steam generation in industrial boilers depending on the quality of the raw water, the water treatment process, and the percentage condensate return. The main problem with boiler blowdown is that it is contaminated with water treatment chemicals. Oxygen scavengers, phosphates (to precipitate calcium, magnesium, etc.), and polymer dispersants (to keep precipitates dispersed) are all added as treatment chemicals.

c. The lost condensate does not create a direct problem, since it is not likely to be contaminated except with perhaps a few parts per million of amines added to prevent corrosion in the condensate system. The major problems are indirect. The heat loss caused by the condensate loss ultimately must be made up by burning extra fuel and generating extra products of combustion.

These sources of waste from the steam system can be reduced by increasing the percentage of condensate returned (in addition to reducing steam generation by increased heat recovery).

7. *Waste from cooling systems.* Cooling water systems also give rise to wastewater generation. Most cooling water systems recirculate water rather than using "once through" arrangements. Water is lost from recirculating systems in the cooling tower mainly through evaporation but also, to a much smaller extent, through drift (wind carrying away water droplets). This loss is made up by raw water which contains solids. The evaporative losses from the cooling tower cause these solids to build up. The buildup of solids is prevented by a purge of water from the system, i.e., *cooling tower blowdown*. Cooling tower blowdown is the source of the largest volume of wastewater on many sites.

Cooling water systems are dosed with corrosion inhibitors, polymers to prevent solid deposition, and biocides to prevent the growth of microorganisms.

Cooling tower blowdown can be reduced by improving the energy efficiency of processes, thus reducing the thermal load on cooling towers. Alternatively, cooling water systems can be switched to air coolers, which eliminates the problem altogether.

10.5 Life-Cycle Analysis

When utility waste was considered, it was found that to obtain a true picture of the flue gas emissions associated with a process, both the local on-site emissions and those generated by centralized power generation corresponding to the amount of power imported (or exported) need to be included. In the limit, this basic idea can be extended to consider the total emissions (process and utility) associated with the manufacture of a given product in a *life-cycle analysis*.[11–13] In life-cycle analysis, a cradle-to-grave view of a particular product is taken. We start with the extraction of the initial raw materials from natural resources. The various transformations of the raw materials are followed through to the manufacture of the final consumer product, the distribution and use of the consumer product, recycling of the product, if this is possible, and finally, its eventual disposal. Each step in the life cycle creates waste. Waste generated by transportation and the manufacture and maintenance of processing equipment also should be included.

There are three components in a life-cycle analysis:[12]

1. The life cycle is first defined and the complete resource requirements (materials and energy) quantified. This allows the total environmental emissions associated with the life cycle to be quantified by putting together the individual parts. This defines the *life-cycle inventory*.

2. Once the life-cycle inventory has been quantified, we can attempt to characterize and assess the effects of the environmental emissions in a *life-cycle impact analysis*. While the life-cycle inventory can, in principle at least, be readily assessed, the resulting impact is far from straightforward to assess. Environmental impacts are usually not directly comparable. For example, how do we compare the production of a kilogram of heavy metal sludge waste with the production of a ton of contaminated aqueous waste? A comparision of two life cycles is required to pick the preferred life cycle.

3. Having attempted to quantify the life-cycle inventory and impact, a *life-cycle improvement analysis* suggests environmental improvements.

Life-cycle analysis, in principle, allows an objective and complete view of the impact of processes and products on the environment.[11] For a manufacturer, life-cycle analysis requires an acceptance of responsibility for the impact of manufacturing in total. This means not just the manufacturers' operations and the disposal of waste created by those operations but also those of raw materials suppliers and product users.

To the process designer, life-cycle analysis is useful because focusing exclusively on waste minimization at some point in the life cycle sometimes creates problems elsewhere in the cycle. The designer can often obtain useful insights by changing the boundaries of the system under consideration so that they are wider than those of the process being designed.

10.6   Waste Minimization in Practice

Knowing where waste is going is the key to reducing it. When reducing waste from process operations, a steady-state mass balance is not usually comprehensive enough. A balance that takes into account start-up, shutdown, and product changeovers is required.

Clearly, some of these measures to reduce waste in process operations—such as design for low process inventory—can be taken into consideration at the early stages of design, but many cannot. We should be aware of the problem and do whatever we can in the early stages to prevent problems later.

10.7   Waste Minimization—Summary

The best solution to effluent problems is not to produce the waste in the first place, i.e., *waste minimization*. If waste can be minimized at the source, this brings the dual benefit of reducing waste treatment costs *and* reducing raw materials costs.

There are three sources of process waste:

1. Reactor
2. Separation and recycle system
3. Process operations

Since the reactor is at the heart of the process, this is where to start when considering waste minimization. The separation and recycle system is next, and finally, process operations are considered.

Process waste minimization in general terms is a question of

- Changing the reaction path to reduce or eliminate the formation of unwanted byproducts.

- Increasing reactor conversion when separation and recycle of unreacted feed is difficult.

- Increasing process yields of raw materials through improved selectivity in the reactor.

- Reducing catalyst waste by changing from homogeneous to heterogeneous catalysts and protecting catalysts from contaminants and extreme conditions that will shorten their life.

- Increasing process yields through improved separation and recycling.

- Increasing process yields through feed purification to reduce losses in the reactor and separation and recycle system.

- Reducing the use of extraneous materials that cannot be recycled with high efficiency.

- Reducing process inventories.

- Allowing enough intermediate storage to rework off-specification material.

- Designing for a minimum number of shutdowns and product changeovers.

- Reducing the use of fluids (aqueous or organic) used for equipment cleaning.

- Segregating waste to maximize the potential for recycling.

- Reducing losses from fugitive emissions and tank breathing.

The utility system also creates waste through products of combustion from boilers and furnaces and wastewater from water treatment, boiler blowdown, etc. Utility waste minimization is in general terms a question of:

- Reducing products of combustion from furnaces, steam boilers, and gas turbines by making the process more energy efficient through improved heat recovery.

- Reducing wastewater associated with steam generation by both reducing steam use through improved heat recovery and by making the steam system itself more efficient.

- Reducing wastewater associated with cooling water systems.

10.8   References

1. Smith, R., and Petela, E., "Waste Minimization in the Process Industries: 1. The Problem," *Chem. Eng., 506:* 31, 1991.
2. Smith, R., and Petela, E., "Waste Minimization in the Process Industries: 2. Reactors," *Chem. Eng., 509/510:* 12, 1991.
3. Smith, R., and Petela, E., "Waste Minimization in the Process Industries: 3. Separation and Recycle Systems," *Chem. Eng., 513:* 13, 1991.
4. Redman, J., "Pollution is Waste," *Chem. Eng., 461:* 16 June 16, 1989.
5. Chowdhury, J., and Leward, R., "Oxygen Breathes More Life into CPI Processing," *Chem. Engg., 19:* 30, 1984.
6. Reich, P., "Air or Oxygen for VCM?" *Hydrocarbon Processing,* March, 85 1976.
7. McNaughton, K. J., "Ethylene Dichloride Process," *Chem. Engg., 12:* 54, 1983.
8. Smith, R., and Petela, E., "Waste Minimization in the Process Industries: 4. Process Operations," *Chem. Eng., 517:* 9, 1992.
9. Smith, R., and Petela, E., "Waste Minimization in the Process Industries: 5. Utility Waste," *Chem. Eng., 523:* 16, 1992.
10. Smith, R., and Delaby, O., "Targeting Flue Gas Emissions," *Trans. IChemE,* part A, 69: 492, 1991.
11. Hindle, P., and Payne, A. G., "Value-Impact Assessment," *Chem. Eng., 28:* 31, 1991.
12. Curran, M. A., "Broad-Based Environmental Life Cycle Assessment," *Environ. Sci. Technol.,* 27: 430 1993.
13. Guinee, J. B., Udo, H. A., and Huppes, G., "Quantitative Life Cycle Assessment of Products," *J. Cleaner Prod.,* 1: 3, 1993.

11

Effluent Treatment

Once waste minimization has been taken to an economic level, the treatment of the resulting emissions must be considered.

11.1 Incineration

Combustion in an incinerator is the only practical way to deal with many waste streams.[1,2] This is particularly true of solid and concentrated wastes and toxic wastes such as those containing halogenated hydrocarbons, pesticides, herbicides, etc. Many of the toxic substances encountered resist biological degradation and persist in the natural environment for a long period of time. Unless they are in dilute aqueous solution, the most effective treatment is usually incineration.

Incineration of toxic materials such as halogenated hydrocarbons, pesticides, herbicides, etc. requires a sustained temperature of 1100 to 1300°C in an excess of oxygen. The incinerator stack gases will contain acid gases such as hydrogen chloride, oxides of sulfur, and oxides of nitrogen depending on the waste being incinerated. These acid gases require scrubbers to treat the gaseous waste stream. This scrubbing, in turn, produces an aqueous effluent.

Depending on the mix of waste being burnt, the incinerator may or may not require auxiliary firing from fuel oil or natural gas.

There are five main classes of incinerators[1,2]

1. *Liquid injection incinerators.* This type of incinerator has a cylindrical refractory-lined combustion chamber mounted verti-

cally or horizontally. The waste material is fed to burners at one end of the incinerator. Although designed primarily to burn liquids, this type also can handle slurries and gases.

2. *Rotary kilns.* Rotary kilns involve a cylindrical refractory-lined shell mounted at small angle to horizontal and rotated at low speed. Solid material, sludges, and slurries are fed at the higher end and flow under gravity along the kiln. Liquids also can be incinerated. Rotary kilns are ideal for treating solid waste but have the disadvantage of high capital and maintenance costs.

3. *Hearth incinerators.* This type of incinerator is designed primarily to incinerate solid waste. Solids are moved through the combustion chamber mechanically using a rake.

4. *Fluidized-bed incinerators.* Liquid, solid, and gaseous wastes can be treated in fluidized-bed incinerators. Solid particles, however, must be small (typically around 1 mm). If low-temperature incineration (less than 900°C) is required, then sand can be used as the bed material. Higher temperatures result in fusion of the bed and require a particulate refractory material instead. If the waste can be incinerated at temperatures less than around 770°C, limestone can be used for the bed material, which absorbs the acid gases. Fluidized beds have two main advantages over other designs: good mixing allows lower excess air, which in turn leads to less auxiliary firing, and the heat capacity of the bed compensates for variations in feed material, giving more stable operation.

5. *Catalytic incinerators.* Catalytic incinerators allow oxidation of wastes at lower temperatures than conventional thermal incinerators. Operating temperatures are less than 550°C. Their advantages are lower fuel consumption if auxiliary fuel is required and less severe operating conditions for materials of construction. However, catalytic incinerators cannot handle solid waste, and catalyst fouling and aging are a problem. Catalysts are usually noble metals (such as platinum or rhodium) finely divided on a support such as alumina. Both fixed and fluidized beds are used. The most common applications for catalytic incinerators are dedicated devices to treat gaseous process vents, particularly purges.

The policy for waste heat recovery from the flue gas varies between incinerator operators. Incinerators located on the waste producer's site tend to be fitted with waste heat recovery systems, usually steam generation, which is fed into the site steam mains. Merchant incinerator operators, who incinerate other people's waste and operate in isolation remotely from the waste producers, tend not to fit heat recovery systems. Instead, the flue gases tend to be cooled prior to scrubbing and the waste heat is used to reheat the flue gases after scrubbing to avoid a visible steam plume from the stack.

While incineration is the preferred method of disposal for wastes containing high concentrations of organics, it becomes expensive for aqueous wastes with low concentrations of organics because auxiliary fuel is required, making the treatment expensive. Weak aqueous solutions of organics are better treated by wet oxidation (see Sec. 11.5).

11.2   Treatment of Solid Particle Emissions to Atmosphere

The selection of equipment for the treatment of solid particle emissions to the atmosphere depends on a number of factors[3]

- Size distribution of the particles to be separated
- Particle loading
- Gas throughput
- Permissible pressure drop
- Temperature

A wide range of equipment is available for the control of emissions of solid particles. These methods are classified in broad terms in Table 11.1.[3]

1. *Gravity settlers.* Gravity settlers were discussed in Chap. 3 and

TABLE 11.1   Methods to Control Emissions of Solid Particles

Equipment	Main particle separation mechanism	Approximate particle size range (μm)
Settling chambers	Gravity settling	>100
Inertial separators	Inertia	>50
Cyclones	Centrifugal settling	>5
Scrubbers	Inertia	>3
Venturi scrubbers	Inertia	>0.3
Bag filters	Sieving	>0.1
Electrostatic precipitators	Electrostatic migration	>0.001

illustrated in Figs. 3.1c and 3.3. They are used to collect coarse particles and may be used as prefilters. Only particles in excess of 100 μm can reasonably be removed.

2. *Inertial collectors.* In inertial collectors, an object is placed in the path of the gas. An example is shown in Fig. 11.1. While the gas passes around the shutters, particles with sufficiently high inertia impinge on them and are removed from the stream. Only particles in excess of 50 μm can reasonably be removed. Like gravity settlers, inertial collectors are widely used as prefilters.

3. *Cyclones.* Cyclones are also primarily used as prefilters. These also were discussed in Chap. 3 and illustrated in Fig. 3.4. The particle-laden gas enters tangentially and spins downward and inward, ultimately leaving the top of the unit. Particles are thrown radially outward to the wall by the centrifugal force and leave at the bottom.

Cyclones can be used under conditions of high particle loading. They are cheap, simple devices with low maintenance requirements. Problems occur when separating materials that have a tendency to stick to the cyclone walls.

4. *Scrubbers.* Scrubbers are designed to contact a liquid with the particle-laden gas and entrain the particles with the liquid. They offer the obvious advantage that they can be used to remove gaseous as well as particulate pollutants. The gas stream may need to be cooled before entering the scrubber. Some of the more common types of scrubbers are shown in Fig. 11.2.

Packed columns are widely used in gas absorption, but particulates are also removed in the process (see Fig. 11.2a). The main disadvan-

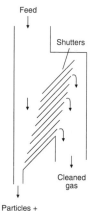

Figure 11.1 An inertial collector. *(Reproduced with permission from Stenhouse, "Pollution Control," in Teja, Chemical Engineering and the Environment, Blackwell Scientific Publications, Oxford, UK, 1981.)*

tage of packed columns is that the solid particles will often accumulate in the packing and require frequent cleaning. Some designs of packed column allow the packing to self-clear through inducing movement of the bed from the upflow of the gas.

In centrifugal scrubbers (Fig. 11.2*b*), an attempt is made to increase the relative velocity of particles and droplets by centrifuging the droplets in an outward direction.

A considerable reduction in particle size separation can be achieved at the expense of increased pressure drop using a Venturi scrubber (see Fig. 11.2*c*).

5. *Bag filters.* Bag filters, as discussed in Chap. 3 and illustrated in Fig. 3.6*b*, are probably the most common method of separating particulate materials from gases. A cloth or felt filter material is used that is impervious to the particles. Bag filters are suitable for use in very high dust load conditions. They have an extremely high efficiency, but they suffer from the disadvantage that the pressure drop across them may be high.[3]

6. *Electrostatic precipitators.* Electrostatic precipitators are used

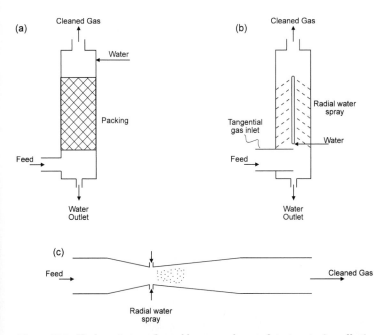

Figure 11.2 Various types of scrubbers can be used to treat air pollution from solid particles. *(Reproduced with permission from Stenhouse, "Pollution Control," in Teja, Chemical Engineering and the Environment, Blackwell Scientific Publications, Oxford, U.K., 1981.)*

where collection of fine particles at a high efficiency coupled with a low pressure drop is necessary. The arrangement is illustrated in Fig. 11.3. Particle-laden gas enters a number of tubes or passes between parallel plates. The particulates are charged and are deposited on the earthed plates or tube walls. The walls are mechanically *rapped* periodically to remove the accumulated dust layer.

11.3 Treatment of Gaseous Emissions to Atmosphere

The major gaseous pollutants arising from chemical processes are volatile organic compounds (VOCs), compounds containing sulfur (SO_x, H_2S), nitrogen (NO_x, NH_3), and halogens, and carbon dioxide and monoxide. The largest volume comes from the products of combustion, which will be dealt with in the next section. Here, we shall concentrate on the treatment of gaseous emissions other than those which arise from combustion processes.

1. *Condensation.* Condensation can be accomplished by increasing pressure or decreasing temperature. Generally, decreasing temperature is preferred, which usually means the use of refrigeration. Condensation is preferred when treating high concentrations. In many applications, recovered materials can be recycled.

2. *Absorption.* Absorption is used mainly for the removal of oxides

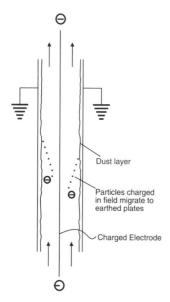

Figure 11.3 Electrostatic precipitation can be used to remove fine particles. *(Reproduced with permission from Stenhouse, "Pollution Control," in Teja, Chemical Engineering and the Environment," Blackwell Scientific Publications, Oxford, U.K., 1981.)*

of sulfur, hydrogen sulfide, oxides of nitrogen, ammonia, hydrogen fluoride, chlorine, and hydrogen chloride. Water is usually the first choice for solvent, with the aqueous solution being sent to aqueous waste disposal. However, many compounds have only a limited solubility in water, and for these, a reactive absorption is normally used. For example, acid gases are usually absorbed on sodium hydroxide solutions.

3. *Adsorption*. Adsorption of pollutants by passing the gas through a bed of solid is most suitable when the concentration of pollutant is very low. Materials such as activated carbon, silica gel, or alumina can be used as adsorbent. Activated carbon is the most common adsorbent for organics removal. The adsorptive capacity of the adsorbent (expressed as kilograms of pollutant removed per kilogram of adsorbent) depends not only on the properties of both the adsorbent and the pollutants but also on temperature and pressure. Adsorptive capacity decreases with increasing temperature and increases with increasing pressure.

When the bed is saturated, regeneration of the adsorbent is necessary. Carbon beds are typically regenerated with steam, hot air, or a combination of vacuum and hot gas.

4. *Flares*. Flares are used for the combustion of waste hydrocarbon gases in which the rates may vary over a wide range and for emergency releases. Steam injection is usually used to enhance mixing and the formation of a clean flame.

5. *Incineration*. Incinerators were discussed in Sec. 11.1. When incinerators are used to treat gaseous pollutants in relatively low concentration, auxiliary firing from fuel or other waste material normally will be necessary. The capital and operating costs may be high. In addition, long duct lines are often necessary.

11.4 Treatment of Combustion Product Emissions

The major products of combustion are CO_2, water, SO_x, and NO_x. The products of combustion are clearly best minimized by making the process efficient in its use of energy through improved heat recovery and avoiding unnecessary incineration through minimization of process waste.

1. *Dealing with CO_2 emissions*. If CO_2 emissions are of concern because of the greenhouse effect, then there are only two ways in which reductions can be made:

- *Increased energy efficiency.* Increasing energy efficiency and the introduction of cogeneration reduce CO_2 emissions. Remember that emissions should be viewed on a global basis, as discussed in Chap. 10.

- *Fuel switch.* Fuel switch from, say, coal to natural gas reduces the CO_2 emissions for the same heat release because of the lower carbon content of natural gas.

2. *Dealing with SO_x emissions.* If we need to reduce SO_x emissions, there are four ways in which this can be achieved:

- *Increase energy efficiency of the process.* Increasing the energy efficiency decreases the fuel burnt and hence decreases SO_x emissions at the source. Again, the emissions should be viewed on a global basis.

- *Fuel switch.* Switching to a low-sulfur fuel is an obvious solution.

- *Desulfurize the fuel.* Most types of fuel can be desulfurized. However, as we go from gaseous to liquid to solid fuels, the desulfurization process becomes increasingly difficult.

- *Desulfurize the flue gas.* A whole range of processes have been developed to remove SO_x from flue gases, such as injection of limestone into the furnace, absorption into wet limestone after the furnace, absorption into aqueous potassium sulfite after the furnace, and many others.[4] However, the byproducts from many of these desulfurization processes cause major disposal problems.

This is the order in which we should look to solve the problem of SO_x emissions. We should try to prevent creation of the waste, since treating the waste tends only to move the problem rather than solve it.

3. *Dealing with NO_x emissions.* There are two main reaction paths for NO_x formation[5]

a. Thermal NO_x from nitrogen in the combustion air:

$$N_2 + O_2 \rightleftharpoons 2NO$$

$$NO + 1/2O_2 \rightleftharpoons NO_2$$

Thermal NO_x is formed particularly at high temperatures.

b. Fuel-bound NO_x:

$$(\text{Fuel})\,N + 1/2O_2 \rightleftharpoons NO$$

$$NO + 1/2O_2 \rightleftharpoons NO_2$$

Fuel-bound NO_x is formed at low as well as high temperatures. However, part of the fuel nitrogen is directly reacted to N_2. Moreover, N_2O and N_2O_4 are also formed in various reactions and add to the complexity of the formation. It is virtually impossible to calculate a precise value for the NO_x emitted by a real combustion device. NO_x emissions depend not only on the type of combustion technology but also on its size and the type of fuel used.

If we need to reduce NO_x emissions, there are five ways in which this can be achieved:

- *Increased energy efficiency.* As with CO_2 and SO_x emissions, the less fuel burnt, the less NO_x that is produced. Yet again, the emissions should be viewed on a global basis.

- *Fuel switch.* Since NO_x formation is fuel-dependent, switching fuel can reduce formation. The general trend is that from solid to liquid to gaseous fuel, the NO_x formation decreases. However, it should be emphasized again that this is also very much dependent on the combustion device.

- *Low NO_x burners/staged combustion.* Low NO_x burners reduce the formation of NO_x by inducing fuel-air mixing patterns that lead to lower peak flame temperatures. Injection of the fuel in stages, injection of the air in stages, and steam injection are all possible. NO_x reductions on the order of 50 to 70 percent are possible by changing the burner design.

- *Flue gas recirculation.* Recirculation of part of the flue gas as shown in Fig. 11.4 lowers the peak flame temperature, thus reducing NO_x formation. There is clearly a limit to how much flue gas can be recirculated without affecting the stability of the flame.

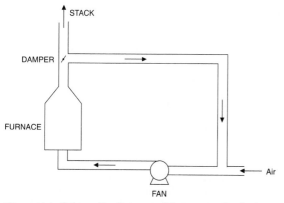

Figure 11.4 Schematic diagram of flue gas recirculation.

NO_x reductions on the order of 40 percent are possible by flue gas recirculation.

- *Chemical reduction.* The injection of ammonia reduces NO_x emissions by the reduction of NO_x to nitrogen and water. Although it can be used at higher temperatures without a catalyst, the most commonly used method injects the ammonia into the flue gas upstream of a catalyst bed (typically vanadium and/or tin on a silica support).

These techniques can be used in isolation or combination.

11.5   Treatment of Aqueous Emissions

Consider first aqueous emissions of organic waste material. When this is discharged to the receiving water, bacteria feed on the organic material. This organic material eventually will be oxidized to stable end products. Carbon molecules will be converted to CO_2, hydrogen to H_2O, nitrogen to NO_3, sulfur to SO_3, etc. As an example, consider the degradation of urea:

$$CH_4N_2O + 9/2O_2 \longrightarrow CO_2 + 2H_2O + 2NO_3$$
$$\text{urea} \qquad \text{oxygen} \qquad \text{carbon} \quad \text{water} \quad \text{nitrate}$$
$$\text{dioxide}$$

This equation indicates that every molecule of urea requires 9/2 molecules of oxygen for complete oxidation. The oxygen required for the reactions depletes the receiving water of oxygen, causing the death of aquatic life.

The amount of oxygen used in the degradation process is called the *biochemical oxygen demand* (BOD). A standard test has been devised to measure BOD in which the oxygen utilized by microorganisms in contact with the wastewater over a 5-day period at 20°C is measured.

While the BOD test gives a good indication of the effect the effluent will have on the environment, it requires 5 days to carry out. The *chemical oxygen demand* (COD) test has been developed to give a more rapid result. In the COD test, acidic oxidation with potassium dichromate is used. COD results are generally higher than BOD results because the COD test oxidizes materials that are only slowly biodegradable. Although the COD test provides a very strong oxidizing environment, certain compounds are not oxidized.

Another test is the *total oxygen demand* (TOD) test, which oxidizes the waste in the presence of a catalyst at 900°C in a stream of air. Under these harsh conditions, all the carbon is oxidized to CO_2. The oxygen demand is calculated from the difference in oxygen content of the air before and after oxidation. The resulting value of TOD embraces oxygen required to oxidize both organic and inorganic substances present.

The relationship between BOD, COD, and TOD for the same waste is in the order

$$BOD < COD < TOD$$

The process is designed from a knowledge of physical concentrations, whereas aqueous effluent treatment systems are designed from a knowledge of BOD and COD. Thus we need to somehow establish the relationship between BOD, COD, and the concentration of waste streams leaving the process. Without measurements, relationships can only be established approximately. The relationship between BOD and COD is not easy to establish, since different materials will oxidize at different rates. To compound the problem, many wastes contain complex mixtures of oxidizable materials, perhaps together with chemicals that inhibit the oxidation reactions.

If the composition of the waste stream is known, then the *theoretical oxygen demand* can be calculated from the appropriate stoichiometric equations. As a first level of approximation, we can assume that this theoretical oxygen demand would be equal to the COD. Then, experience with domestic sewage indicates that the average ratio of COD to BOD will be on the order 1.5 to 2. The following example will help to clarify these relationships.

Example 11.1 A process produces an aqueous waste stream containing 0.1 mol% acetone. Estimate the COD and BOD of the stream.

Solution First, calculate the theoretical oxygen demand from the equation that represents the overall oxidation of the acetone:

$$\underset{\text{acetone}}{(CH_3)_2CO} + \underset{\text{oxygen}}{4O_2} \longrightarrow \underset{\substack{\text{carbon} \\ \text{dioxide}}}{3CO_2} + \underset{\text{water}}{3H_2O}$$

Approximating the molar density of the waste stream to be that of pure water (i.e., $56 \, \text{kmol m}^{-3}$), then

$$\text{Theoretical oxygen demand} = 0.001 \times 56 \times 4 \, \text{kmol O}_2 \, \text{m}^{-3}$$

$$= 0.001 \times 56 \times 4 \times 32 \, \text{kg O}_2 \, \text{m}^{-3}$$

$$= 7.2 \, \text{kg O}_2 \, \text{m}^{-3}$$

Thus

$$COD \approx 7.2 \, \text{kg m}^{-3}$$

$$BOD \approx 7.2/1.5$$

$$\approx 4.8 \, \text{kg m}^{-3}$$

Wastewater treatment processes are generally classified in order as

- Primary (or pretreatment)
- Secondary (or biological)
- Tertiary (or polishing)

Primary or pretreatment of wastewater prior to biological treatment involves both physical and chemical treatment depending on the nature of the emission.

1. *Primary or pretreatment methods.* Primary or pretreatment processes serve two purposes:

- Recover useful material where possible.
- Prepare the aqueous waste for biological treatment by removing excessive load or components that will inhibit the biological processes.

The pretreatment processes may be most effective when applied to individual waste streams from particular processes or process steps before effluent streams are combined for biological treatment.

The capital cost of most aqueous waste treatment operations is proportional to the total flow of wastewater, and the operating cost increases with decreasing concentration for a given mass of contaminant to be removed. Thus, if two streams require different treatment operations, it makes no sense to mix them and treat both streams in both treatment operations. This will increase both capital and operating costs. Rather, the streams should be segregated and treated separately in a *distributed effluent treatment system.* Indeed, effective primary treatment might mean that some streams do not need biological treatment at all.

Let us briefly review the primary treatment methods used.[1,2,6–8] Pretreatment usually starts with phase separation if the effluent is a heterogeneous mixture.

a. Solids separation. Methods used include most of the commonly used techniques for the separation of solids from liquids:

- Screening
- Sedimentation
- Centrifuging
- Filtration

b. Coalescence. Immiscible liquid-liquid mixtures often can be separated by gravity in simple settling devices. The coalescence can be enhanced by the use of mesh pads and centrifugal forces, as discussed in Sec. 3.1.

c. *Flotation.* Flotation can be used to separate solid or immiscible liquid particles from the aqueous effluent. Fine gas bubbles (usually air) are introduced into the effluent which attach to the particles and rise to the surface. Once the particles have floated to the surface, they are collected by skimming. Flotation is particularly effective for the separation of very small or light particles. Direct air injection is usually not the most effective method of flotation. Dissolved-air flotation is usually superior. In dissolved-air flotation, air is dissolved in the effluent under a pressure of several atmospheres and then liberated in the flotation cell by reducing the pressure.

d. *Chemical precipitation.* Chemical precipitation followed by solids separation is particularly useful for separating heavy metals. The heavy metals of particular concern in the treatment of wastewaters include cadmium, chromium, copper, lead, mercury, nickel, and zinc. This is a particular problem in the manufacture of dyes and textiles and in metal processes such as pickling, galvanizing, and plating.

Heavy metals often can be removed effectively by chemical precipitation in the form of carbonates, hydroxides, or sulfides. Sodium carbonate, sodium bisulfite, sodium hydroxide, and calcium oxide are all used as precipitation agents. The solids precipitate as a floc containing a large amount of water in the structure. The precipitated solids need to be separated by thickening or filtration and recycled if possible. If recycling is not possible, then the solids are usually disposed of to a landfill.

The precipitation process tends to be complicated when a number of metals are present in solution. If this is the case, then the pH must be adjusted to precipitate out the individual metals, since the pH at which precipitation occurs depends on the metal concerned.

e. *Chemical oxidation.* Chemical oxidation can be used for the oxidation of organics which are difficult to treat biologically. When used before biological treatment, organic pollutants which are difficult to treat biologically can be oxidized to simpler, less refractory organic compounds. The most common agents are chlorine (as gaseous chlorine or hypochlorite ion), ozone, and hydrogen peroxide. The effectiveness of these oxidizing agents is enhanced by the presence of ultraviolet light. Chemical oxidation is also sometimes applied after biological treatment.

f. *Wet oxidation.* In wet oxidation, an aqueous mixture is heated under pressure in the presence of air or pure oxygen which oxidizes the organic material. The efficiency of the oxidation process depends on reaction time and pressure. Temperatures of

120 to 300°C are used together with pressures of 3 to 200 bar depending on the process and the nature of the waste being treated and whether a catalyst is used. The temperature and pressure required to treat a particular waste are less severe if a catalyst is used.

Wet oxidation is particularly effective in treating aqueous wastes containing organics with a COD up to 2 percent prior to biological treatment. COD can be reduced by up to 95 percent and organic halogen compounds by up to 95 percent also. Organic halogen compounds are particularly resistant to biological degradation.

g. *Adsorption.* Adsorption can be used for the removal of organics (including many toxic materials) and heavy metals (especially when complexed with organics). Activated carbon is used primarily as the adsorbent, although synthetic resins are also used. Obviously, as the adsorbent becomes saturated, regeneration is required. Activated carbon can be regenerated by steam stripping or heating in a furnace. Stripping allows recovery of material, whereas thermal regeneration destroys the organics. Thermal regeneration requires a furnace with temperatures above 800°C to oxidize the adsorbates. This causes a loss of carbon of perhaps 5 to 10 percent per regeneration cycle.

h. *Membrane processes.* In ultrafiltration, the effluent is passed across a semipermeable membrane. Water passes through the membrane, while submicron-sized particles and large molecules are entrained in the effluent and concentrated. The membrane is supported on a porous medium for strength. Configurations used include tubes, plate and frame arrangements, and spiral wound modules. Relatively low pressure differentials (1 to 10 bar) are used.

Reverse osmosis is a high-pressure membrane separation process (20 to 100 bar) which can be used to reject dissolved inorganic salt or heavy metals. The concentrated waste material produced by membrane process should be recycled if possible but might require further treatment or disposal.

i. *Ion exchange.* Ion exchange is used for selective ion removal and finds some application in the recovery of specific materials from wastewaters such as heavy metals. As with adsorption processes, regeneration of the medium is necessary. Resins are regenerated chemically, which produces a concentrated waste stream requiring further treatment or disposal.

j. *Solvent extraction.* The effluent is contacted with a solvent in which organic waste is more soluble. The waste is then separated from the solvent by evaporation or distillation and the solvent recycled. One common application of solvent extraction is the removal and recovery of phenol and compounds of phenol. While phenol is biodegradable, only limited levels can be treated biologically. Variations in phenol concentration are also a problem with biological treatment, since the biological processes take time to adjust to the variations.

k. *Stripping.* Volatile organics and inorganics can be stripped from wastewaters. The usual arrangement would involve the wastewater being fed down through a distillation column and the stripping agent (usually steam or air) fed to the bottom.

If steam is used as stripping agent, either live steam or a reboiler can be used. The use of live steam increases the effluent volume. The volatile organics are taken overhead, condensed, and recycled to the process, if possible. If recycling is not possible, then further treatment or disposal is necessary.

If air is used as stripping agent, further treatment of the stripped material will be necessary. The gas might be fed to an incinerator or some attempt made to recover material by use of adsorption.

l. *Evaporation.* If the wastewater is in low volume and the waste material involatile, then evaporation can be used to concentrate the waste. The relatively pure evaporated water might still require biological treatment after condensation. The concentrated waste can then be recycled or sent for further treatment or disposal. The cost of such operations can be prohibitively expensive unless the heat available in the evaporated water can be recovered.

m. *pH adjustment.* The pH of the wastewater often needs adjustment before biological treatment. The pH is normally adjusted to between 8 and 9 by dosing with acid or alkali.

2. *Biological treatment.* In *secondary,* or *biological treatment,* a concentrated mass of microorganisms is used to break down organic matter into stabilized wastes. Large chemical factories might require their own biological treatment processes. Smaller sites might rely on local municipal treatment processes which treat a mixture of industrial and domestic effluent.

There are two main types of biological reaction, aerobic and anaerobic:

a. *Aerobic.* Aerobic reactions take place only in the presence of free oxygen and produce stable, relatively inert end products such as carbon dioxide and water. Aerobic reactions are by far the most widely used, being capable of removing up to 95 percent of BOD.

b. Anaerobic. Anaerobic reactions function without the presence of free oxygen and derive their energy from organic compounds in the waste. Anaerobic reactions proceed relatively slowly and lead to end products which are unstable and contain considerable amounts of energy such as methane and hydrogen sulfide.

With wastewaters containing very high organic contents, the oxygen demand may be so high that it becomes very difficult and expensive to maintain aerobic conditions. In such circumstances, anaerobic processes can provide an efficient means of removing large quantities of organic material. Anaerobic processes tend to be used when BOD levels exceed 1000 mg/liter ($1 \, \text{kg m}^{-3}$). However, they are not capable of producing very high quality effluents, and further treatment is usually necessary.

The performance of anaerobic digestion processes varies according to the type of unit, throughput, and feed concentration, but such processes are typically capable of removing between 75 and 85 percent of COD.[8]

The inability to produce high-quality effluents is one significant disadvantage. Another disadvantage is that anaerobic processes must be maintained at temperatures between 35 and 40°C to get the best performance. If low-temperature waste heat is available from the production process, then this is not a problem.

One advantage of anaerobic reactions is that the methane produced can be a useful source of energy. This can be fed to steam boilers or burnt in a heat engine to produce power.

The degradable organic matter in the wastewater is used as food by the microorganisms. However, biological growth requires ample supplies of carbon, nitrogen, phosphorus, and inorganic ions such as calcium, magnesium, potassium, etc. Domestic sewage satisfies the requirements, but industrial wastewaters may lack nutrients, and this can inhibit biological growth. In such circumstances, nutrients may need to be added. As the waste treatment progresses, the microorganisms multiply, producing an excess of this *sludge* which cannot be recycled.

Various methods are used to contact the microorganisms with the wastewater. In a completely mixed system, the hydraulic residence time of the wastewater and the solids residence time of the microorganisms would be the same. Thus the minimum hydraulic residence time would be defined by the growth rate of the microorganisms. Since the crucial microorganisms can take several days to grow, this would lead to hydraulic residence times that would be prohibitively long. To overcome this, a number of methods have been developed to decouple the hydraulic and solids residence times.

a. Aerobic digestion. The *suspended growth* or *activated sludge method* is illustrated in Fig. 11.5*a*. Biological treatment takes place in a tank where the waste is mixed with a flocculated biological sludge. To maintain aerobic conditions, the tank must be aerated. Sludge separation from effluent is normally achieved by gravity sedimentation. Part of the sludge is recycled, and excess sludge is removed. The hydraulic flow pattern in the aeration tank can vary between extremes of well-mixed and plug-flow. For well-mixed reactors, the wastewater is rapidly dispersed throughout the reactor, and its concentration is reduced. This feature is advantageous at sites where periodic discharges of more concentrated waste are received. The rapid dilution of the waste means that the concentration of any toxic compounds present will be reduced, and thus the microorganisms within the reactor may not be affected by the toxicant. Thus well-mixed reactors produce an effluent of uniform quality in response to fluctuations in the feed.

Plug-flow reactors have a decreasing concentration gradient from inlet to outlet, which means that toxic compounds in the feed remain undiluted during their passage along the reactor, and this may inhibit or kill many of the microorganisms within the

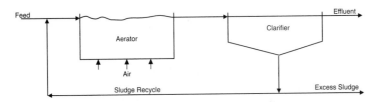

(a) Suspended growth aerobic digestion.

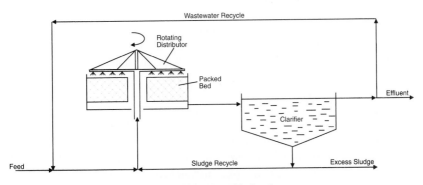

(b) Attached growth aerobic digestion.

Figure 11.5   Typical aerobic digestion processes.

reactor. The oxygen demand along the reactor also will vary. On the other hand, the increased concentration means that rates of reaction are increased, and for two reactors of identical volume and hydraulic retention time, a plug-flow reactor will show a greater degree of BOD removal than that shown by a well-mixed reactor.

The biochemical population can be adapted specifically to particular pollutants. However, in the majority of cases, a wide range of organics must be dealt with, and mixed cultures are used.

In aerobic processes, the mean sludge residence time is typically 5 to 10 days. The hydraulic residence time is typically 0.2 to 0.3 days. Suspended growth aerobic processes are capable of removing up to 95 percent of BOD.

In *attached growth (film) methods,* the wastewater is trickled over a packed bed through which air is allowed to percolate. A biological film or "slime" builds up on the packing under aerobic conditions. Oxygen from the air and biological matter from the wastewater diffuse into the slime. As the biological film grows, it eventually breaks its contact with the packing and is carried away with the water. Packing material varies from pieces of stone to preformed plastic packing.

Figure 11.5*b* shows a typical attached growth arrangement.

Attached growth processes are capable of removing up to 90 percent of BOD and are thus less effective than suspended growth methods.

b. Anaerobic digestion. One of the *suspended growth methods,* the *contact type of anaerobic digester* is similar to the activated sludge method of aerobic treatment. The feed and microorganisms are mixed in a tank (this time closed). Mechanical agitation is usually required, since there is no air injection. The sludge is separated from the effluent by sedimentation or filtration, part of the sludge is recycled, and excess sludge is removed.

Another suspended growth method is the *upward-flow anaerobic sludge blanket* illustrated in Fig. 11.6*a.* Here the sludge is contacted by upward flow of the feed at a velocity such that the sludge is not carried out of the top of the digester.

A third method of contact known as an *anaerobic filter* also uses upward flow but keeps the sludge in the digester by a physical barrier such as a grid.

In *attached growth (film) methods,* as with aerobic digestion, the microorganisms can be encouraged to grow attached to a support medium such as plastic packing or sand. In anaerobic digestion, the bed is usually fluidized rather than a fixed-bed arrangement (Fig. 11.6*b*). Anaerobic processes typically remove 75 to 85 percent of COD.[8]

In all types of biological processes, excess sludge is produced which must be disposed of. The treatment and disposal of sludge are major problems which can be costly to deal with. Anaerobic processes have the advantage here, since they produce considerably less sludge than aerobic processes (on the order of 5 percent of aerobic processes for the same throughput).

Sludge disposal typically can be responsible for 25 to 40 percent of the operating costs of a biological treatment system. Treatment of sludge is aimed primarily at reducing its volume. This is so because the sludge is usually 95 to 99 percent water and the cost of disposal

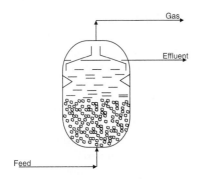

(a)   Upward flow anaerobic sludge blanket.

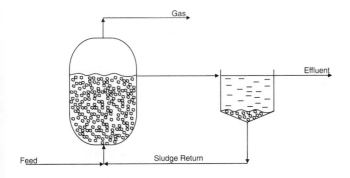

(b)   Fluidized anaerobic bed

Figure 11.6   Typical anaerobic digestion processes.

TABLE 11.2 Comparison of Aerobic and Anaerobic Wastewater Treatments

Aerobic	Anaerobic
BOD < 1000 mg/liter	BOD > 1000 mg/liter
Stable end products (CO_2, H_2O, etc.)	Unstable end products (CH_4, H_2S, etc.)
BOD removal up to 95%	BOD removal 75–85%
High sludge formation	Low sludge formation

is closely linked to its volume. The water is partly free, partly trapped in the flocs, and partly bound in the microorganisms. Anaerobic or aerobic digestion of the sludge can be used, followed by dewatering. The dewatering can be carried out by filtration or centrifuge. The resulting water content after these processes is reduced to typically 60 to 85 percent. The water content can be reduced to perhaps 10 percent by drying. The sludge finally may be used for agricultural purposes (albeit a poor fertilizer) or incinerated.

Table 11.2 provides a summary of the main features of aerobic and anaerobic wastewater treatment. Aerobic treatment processes are generally restricted to BOD < 1000 mg/liter unless pure oxygen is used for aeration.

3. *Tertiary treatment.* Tertiary or polishing treatment prepares the aqueous waste for final discharge. The final quality of the effluent depends on the nature and flow of the receiving water. Table 11.3 gives an indication of the final quality required.[9]

Aerobic digestion is normally capable of removing up to 95 percent of the BOD. Anaerobic digestion is capable of removing less, in the range 75 to 85 percent. With municipal treatment processes, which treat a mixture of domestic and industrial effluent, at minimum, some disinfection of the effluent is usually required to destroy any disease-causing organisms before discharge to the environment. Tertiary treatment processes vary, but they constitute the final stage of effluent treatment to ensure that the effluent meets specifications for disposal. Processes used include the following:

a. *Filtration.* Examples of such processes are microstrainers (a fine screen with openings ranging from 20 to 60 mm) and sand filters.

TABLE 11.3 Typical Effluent Quality for Various Receiving Waters[9]

	Probable effluent	
Receiving water	BOD (mg/liter)	Suspended solids (mg/liter)
Tidal estuary	150	150
Lowland river	20	30
Upland river	10	10
High-quality river with low dilution	5	5

They are designed to improve effluents from secondary treatment processes by removing suspended material and with it some of the remaining BOD.

b. *Ultrafiltration.* Ultrafiltration was described under pretreatment methods. It is used to remove finely divided suspended solids, and when used as a tertiary treatment, it can remove virtually all the BOD remaining after secondary treatment.

c. *Adsorption.* Some organics are not removed in biological systems operating under normal conditions. Removal of residual organics can be achieved by adsorption. Both activated carbon and synthetic resins are used. As described earlier under pretreatment methods, regeneration of the activated carbon in a furnace can cause carbon losses of perhaps 5 to 10 percent.

d. *Nitrogen and phosphorus removal.* Since nitrogen and phosphorus are essential for growth of the microorganisms, the effluent from secondary treatment will contain some nitrogen and phosphorus. The amount discharged to receiving waters can have a considerable effect on the growth of algae. If discharge is to a high-quality receiving water with low dilution rates, then removal may be necessary. Nitrogen principally occurs as ammonium (NH_4^+), nitrate (NO_3^-), and nitrite (NO_2^-). Phosphorus principally occurs as orthophosphate (PO_4^{3-}). A number of biological and chemical processes are available for the removal of nitrogen and phosphorous.[6–8] These processes produce extra biological and inorganic sludge that requires disposal.

e. *Disinfection.* Chlorine, as gaseous chlorine or as the hypochlorite ion, is widely used as a disinfectant. However, its use in some cases can lead to the formation of toxic organic chlorides, and the discharge of excess chlorine can be harmful. Ozone as an alternative disinfectant leads to products that have a lower toxic potential. Treatment is enhanced by ultraviolet light. Indeed, disinfection can be achieved by ultraviolet light on its own.

11.6   Effluent Treatment—Summary

When viewing effluent treatment methods, it is clear that the basic problem of disposing of waste material safely is, in many cases, not so much solved but moved from one place to another. The fundamental problem is that once waste has been created, it cannot be destroyed. The waste can be concentrated or diluted, its physical or chemical form can be changed, but it cannot be destroyed.

If a method of treatment can be used that allows material to be recycled, then the waste problem is truly solved. However, if the treatment simply concentrates the waste as a concentrated liquid, slurry, or solid in a form that cannot be recycled, then it will need to be disposed of. Landfill disposal of such waste is increasingly unacceptable, and incineration causes pollution through products of combustion and liquors from scrubbing systems.

It must be clear that the best method for dealing with effluent problems is to solve the problem at source, i.e., waste minimization.

11.7 References

1. Freeman, H. M., *Standard Handbook of Hazardous Waste Treatment and Disposal,* McGraw-Hill, New York, 1989.
2. Berkowitz, J. B., Funkhouser, J. T., and Stevens, J. I., *Unit Operations for Treatment of Hazardous Industrial Wastes,* Noyes Data Corporation, Park Ridge, N.J., 1978.
3. Stenhouse, J. I. T., "Pollution Control," in A. S. Teja (ed.), *Chemical Engineering and the Environment,* Blackwell Scientific Publications, Oxford, U.K., 1981.
4. Cynes, B. L., *Chemical Reactions as a Means of Separation—Sulfur Removal,* Marcel Dekker, New York, 1977.
5. Glassman, J., *Combustion,* 2d ed., Academic Press, New York, 1987.
6. Steel, E. W., and McGhee, T. J., *Water Supply and Sewage,* 5th ed., McGraw-Hill, New York, 1979.
7. Schroeder, E. D., *Water and Wastewater Treatment,* McGraw-Hill, New York, 1977.
8. Metcalf & Eddy, revised by Tchobanoglous, G., and Burton, F. L., *Wastewater Engineering: Treatment, Disposal and Reuse,* 3d ed., McGraw-Hill, 1991.
9. Tebbutt, T. H. Y., *BASIC Water and Wastewater Treatment,* Butterworth, London, 1990.

Chapter 12

Process Changes for Improved Heat Integration

In Chap. 10, modification of the process for reducing process waste was considered in detail. It also was concluded that to minimize utility waste, the single most effective measure would be improved heat recovery. The energy-targeting methods presented in Chaps. 6 and 7 maximize heat recovery for a given set of process conditions. However, the process conditions can be changed to improve the heat recovery further.

The insights developed in Chaps. 6 and 7 can now be directed toward changing process conditions to improve heat recovery. The basic decisions made for the reaction and separation and recycle systems will now be questioned. In other words, the strategy is go back in from the outer layers of the onion to the inner layers. The process changes will be considered using targets for the energy system. Only when we are satisfied with the targets will the heat exchanger network be designed.

12.1 The Plus/Minus Principle

Consider the composite curves in Fig. 12.1. Any process change[1] which

- increases the total hot stream heat duty above the pinch
- decreases the total cold stream heat duty above the pinch

- decreases the total hot stream heat duty below the pinch
- increases the total cold stream heat duty below the pinch

will bring about a decrease in utility requirements. This is known as the *plus/minus principle.*[1] These simple guidelines provide a definite reference for appropriate design changes to improve the targets. The changes apply throughout the process to reactors, recycle flow rates, distillation columns, etc.

If a process change, such as a change in distillation column pressure, allows shifting a hot stream from below the pinch to above, this has the effect of increasing the overall hot stream duty above the pinch and therefore decreasing the hot utility. Simultaneously, it decreases the overall hot stream duty below the pinch and decreases the cold utility. Shifting a cold stream from above the pinch to below decreases the overall cold stream duty above the pinch, decreasing the hot utility, and increases the overall cold stream duty below the pinch, reducing the cold utility. Thus one way to implement[2] the plus/minus principle is

- to shift hot streams from below to above the pinch, or
- to shift cold streams from above to below the pinch.

Another way to relate these principles is to remember that heat integration will always benefit by keeping hot streams hot and cold streams cold.[3]

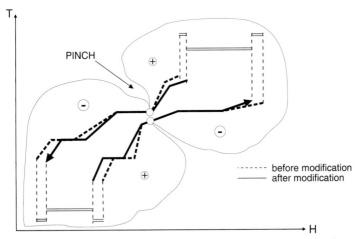

Figure 12.1 The plus/minus principle guides process changes to reduce utility consumption. *(From Smith and Linnhoff, ChERD, 66: 195, 1988; reproduced by permission of the Institution of Chemical Engineers.)*

12.2   The Tradeoffs Between Process Changes, Utility Selection, Energy Cost, and Capital Cost

Although the plus/minus principle is the ultimate reference in guiding process changes to reduce utility costs, it takes no account of capital costs. Process changes to reduce utility consumption normally will bring about a reduction in temperature driving forces, as indicated in Fig. 12.1. Thus the capital/energy tradeoff (and hence $\Delta T_{\min}$) should be readjusted after process changes.

Having to readjust the capital/energy tradeoff after every process change would be a real problem if it were not for the existence of the total cost targeting procedures discussed in Chap. 7.

In addition, the decrease in driving forces in Fig. 12.1 caused by the process changes also affects the potential for using multiple utilities. For example, as the driving forces above the pinch become smaller, the potential to switch duty from high-pressure to low-pressure steam, as discussed in Sec. 6.6, decreases. Process changes are competing with better choices of utility levels, heat engines, and heat pumps for available spare driving forces. Each time either a process change or a different choice of utilities is suggested, the capital/energy tradeoff should be readjusted. If multiple utilities are used, the optimization of the capital/energy tradeoff is not straightforward, since each pinch (process and utility) can have its own value of $\Delta T_{\min}$. The optimization thus becomes multidimensional.

12.3   Process Changes for Improved Heat Integration—Summary

The ultimate reference in guiding process changes to reduce utility costs and utility waste is the plus/minus principle. However, process changes so identified prompt changes in the capital/energy tradeoff and utility selection. Using the total cost targeting techniques described in Chaps. 6 and 7, it is possible to effectively screen a wide range of options using relatively simple computation. In the next three chapters we shall focus in detail on heat integration of reactors and heat-driven separators.

12.4   References

1. Linnhoff, B., and Vredeveld, D. R., "Pinch Technology Has Come of Age," *Chem. Eng. Progr., 80* July: 33, 1984
2. Linnhoff, B., and Parker, S. J., "Heat Exchanger Network with Process Modifications," IChemE Annual Research Meeting, Bath, U.K., April 1984.
3. Linnhoff, B., Townsend, D. W., and Boland, D., et al., *A Users Guide on Process Integration for the Efficient Use of Energy,* IChemE, Rugby, U.K., 1982.

Heat Integration of Reactors

13.1 The Heat Integration Characteristics of Reactors

The heat integration characteristics of reactors depend both on the decisions made for the removal or addition of heat and the reactor mixing characteristics. In the first instance, adiabatic operation is considered, since this gives the simplest design.

1. *Adiabatic operation.* If adiabatic operation leads to an acceptable temperature rise for exothermic reactors or an acceptable fall for endothermic reactors, then this is the option normally chosen. If this is the case, then the feed stream to the reactor requires heating and the effluent stream requires cooling. The heat integration characteristics are thus a cold stream (the reactor feed) and a hot stream (the reactor effluent). The heat of reaction appears as elevated temperature of the effluent stream in the case of exothermic reaction or reduced temperature in the case of endothermic reaction.

2. *Heat carriers.* If adiabatic operation produces an unacceptable rise or fall in temperature, then the option discussed in Chap. 2 is to introduce a heat carrier. The operation is still adiabatic, but an inert material is introduced with the reactor feed as a heat carrier. The heat integration characteristics are as before. The reactor feed is a cold stream and the reactor effluent a hot stream. The heat carrier serves to increase the heat capacity flow rate of both streams.

3. *Cold shot or hot shot.* Injection of cold fresh feed for exothermic reactions or preheated feed for endothermic reactions to inter-

mediate points in the reactor can be used to control the temperature in the reactor. The heat integration characteristics are similar to those of adiabatic operation. The feed is a cold stream and the product a hot stream. If any heat is provided to the cold-shot streams, then these are additional cold streams. In the case of hot shot, the preheated feed streams injected into the reactor are additional cold streams.

4. *Indirect heat transfer with the reactor.* Although indirect heat transfer with the reactor tends to bring about the most complex reactor design options, it is often preferable to the use of a heat carrier. A heat carrier creates complications elsewhere in the flowsheet. A number of options for indirect heat transfer were discussed earlier in Chap. 2.

The first distinction to be drawn, as far as heat transfer is concerned, is between the plug-flow and continuous well-mixed reactor. In the plug-flow reactor shown in Fig. 13.1, the heat transfer can take place over a range of temperatures. The shape of the profile depends on

- Inlet feed concentration

- Inlet temperature

- Inlet pressure and pressure drop (gas-phase reactions)

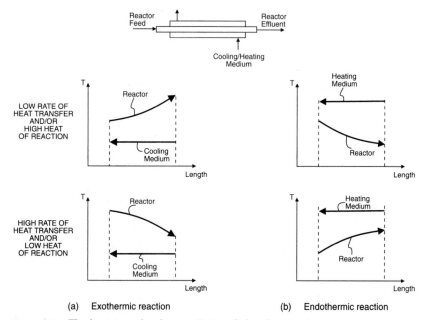

Figure 13.1 The heat transfer characteristics of plug-flow reactors.

- Conversion
- Byproduct formation
- Heat of reaction
- Rate of cooling/heating
- Presence of catalyst diluents or changes in catalyst through the reactor

Figure 13.1*a* shows two possible thermal profiles for exothermic plug-flow reactors. If the rate of heat removal is low and/or the heat of reaction is high, then the temperature of the reacting stream will increase along the length of the reactor. If the rate of heat removal is high and/or the heat of reaction is low, then the temperature will fall. Under conditions between the two profiles shown in Fig. 13.1*a,* a maximum can occur in the temperature at an intermediate point between the reactor inlet and exit.

Figure 13.1*b* shows two possible thermal profiles for endothermic plug-flow reactors. This time the temperature falls for low rates of heat addition and/or high heat of reaction. The temperature rises for the reverse conditions. Under conditions between the profiles shown in Fig. 13.1*b,* a minimum can occur in the temperature profile at an intermediate point between the inlet and exit.

The thermal profile through the reactor will in most circumstances be carefully optimized to maximize selectivity, extend catalyst life, etc. Because of this, direct heat integration with other process streams is almost never carried out. The heat transfer to or from the reactor is instead usually carried out by a heat transfer intermediate. For example, in exothermic reactions, cooling might occur by boiling water to raise steam, which, in turn, can be used to heat cold streams elsewhere in the process.

By contrast, if the reactor is continuous well-mixed, then the reactor is isothermal. This behavior is typical of stirred tanks used for liquid-phase reactions or fluidized-bed reactors used for gas-phase reactions. The mixing causes the temperature in the reactor to be effectively uniform.

For indirect heat transfer, the heat integration characteristics of the reactor can be broken down into three cases:

a. If the reactor can be matched with other process streams (which is unlikely), then the reactor profile should be included in the heat integration problem. This would be a hot stream in the case of an exothermic reaction or a cold stream in the case of an endothermic reaction.

b. If a heat transfer intermediate is to be used and the cooling/ heating medium is fixed, then the cooling/heating medium should be included and not the reactor profile itself. Once the cooling medium leaves an exothermic reactor, it is a hot stream requiring cooling before being returned to the reactor. Similarly, once the heating medium leaves an endothermic reactor, it is a cold stream requiring heating before being returned to the reactor.

c. If a heat transfer intermediate is to be used but the temperature of the cooling/heating medium is not fixed, then both the reactor profile and the cooling/heating medium should be included. The temperature of the heating/cooling medium can then be varied within the context of the overall heat integration problem to improve the targets.

In addition to the indirect cooling/heating within the reactor, the reactor feed is an additional cold stream and the reactor product an additional hot stream.

For the ideal batch reactor, the temperature is uniform throughout the reactor at any instant in time. Figure 13.2a shows the variation in temperature with time for an exothermic reaction in a batch reactor. A family of curves illustrates the effect of increasing the rate of heat removal and/or decreasing heat of reaction. Each individual curve assumes the rate of heat transfer to the cooling medium to be constant for that curve throughout the batch cycle. Figure 13.2b shows similar curves for endothermic reactions. Again, each individual curve in Fig. 13.2b assumes the rate of heat addition from the heating medium to be constant throughout the batch cycle.

Fixing the rate of heat transfer in a batch reactor is often not the best way to control the reaction. The heating or cooling characteristics can be varied with time to suit the characteristics of the reaction. Because of the complexity of batch operation and the fact that operation is usually small scale, it is rare for any attempt to be made

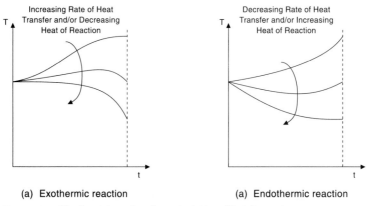

(a) Exothermic reaction   (a) Endothermic reaction

Figure 13.2 The heat transfer characteristics of batch reactors.

to recover heat from a batch reactor or supply heat by recovery. Instead, utilities are normally used.

The heat duty on the heating/cooling medium is given by

$$Q_{\text{REACT}} = -(\Delta H_{\text{STREAMS}} + \Delta H_{\text{REACT}}) \tag{13.1}$$

where Q_{REACT} = reactor heating or cooling required
$\Delta H_{\text{STREAMS}}$ = enthalpy change between feed and product streams
ΔH_{REACT} = reaction enthalpy (negative in the case of exothermic reactions)

5. *Quench.* As discussed in Chap. 2, the reactor effluent may need to be cooled rapidly (quenched). This can be by indirect heat transfer using conventional heat transfer equipment or by direct heat transfer by mixing with another fluid.

If indirect heat transfer is used with a large temperature difference to promote high rates of cooling, then the cooling fluid (e.g., boiling water) is fixed by process requirements. In this case, the heat of reaction is not available at the temperature of the reactor effluent. Rather, the heat of reaction becomes available at the temperature of the quench fluid. Thus the feed stream to the reactor is a cold stream, the quench fluid is a hot stream, and the reactor effluent after the quench is also a hot stream.

The reactor effluent might require cooling by direct heat transfer because the reaction needs to be stopped quickly, or a conventional exchanger would foul, or the reactor products are too hot or corrosive to pass to a conventional heat exchanger. The reactor product is mixed with a liquid that can be recycled, cooled product, or an inert material such as water. The liquid vaporizes partially or totally and cools the reactor effluent. Here, the reactor feed is a cold stream, and the vapor and any liquid from the quench are hot streams.

Now consider the placement of the reactor in terms of the overall heat integration problem.

13.2 Appropriate Placement of Reactors

In Chap. 12 it was discussed how the pinch takes on fundamental significance in improving heat integration. Let us now explore the consequences of placing reactors in different locations relative to the pinch.

Figure 13.3 shows a process represented simply as a heat sink and heat source divided by the pinch. Figure 13.3*a* shows the process with an exothermic reactor integrated above the pinch. The minimum hot utility can be reduced by the heat released by reaction, Q_{REACT}.

By comparison, Fig. 13.3*b* shows an exothermic reactor integrated below the pinch. Although heat is being recovered, it is being recovered into part of the process which is a heat source. The hot utility requirement cannot be reduced because the process above the pinch needs at least Q_{Hmin} to satisfy its enthalpy imbalance.

There is no obvious benefit from integrating an exothermic reactor below the pinch. The appropriate placement for exothermic reactors is above the pinch.[1]

Figure 13.4*a* shows an endothermic reactor integrated above the pinch. The endothermic reactor removes Q_{REACT} from the process above the pinch. The process above the pinch needs at least Q_{Hmin} to

(a)

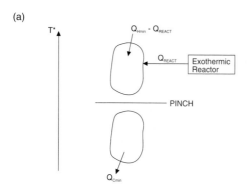

(b)

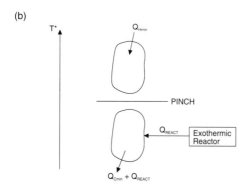

Figure 13.3 Appropriate placement of an exothermic reactor.

satisfy its enthalpy imbalance. Thus an extra Q_{REACT} must be imported from the hot utility to compensate, and there is no benefit by integrating an endothermic reactor above the pinch. Locally, it might seem that there is a benefit by running the reaction by recovery. However, additional hot utility must be imported elsewhere to compensate.

By contrast, Fig. 13.4b shows an endothermic reactor integrated below the pinch. The reactor imports Q_{REACT} from part of the process that needs to reject heat. Thus integration of the reactor serves to reduce the cold utility consumption by Q_{REACT}. There is an overall reduction in hot utility because, without integration, the process and reactor would require $(Q_{H\min} + Q_{\text{REACT}})$ from the utility.

There is no obvious benefit from integrating an endothermic reactor above the pinch. The appropriate placement for endothermic reactors is below the pinch.[1]

(a)

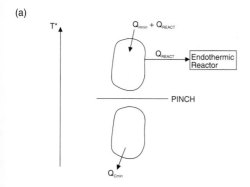

(b)

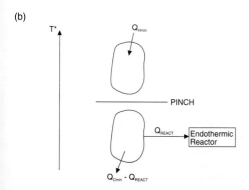

Figure 13.4 Appropriate placement of an endothermic reactor.

13.3   Use of the Grand Composite Curve for Heat Integration of Reactors

The preceding appropriate placement arguments assume that the process has the capacity to accept or give up the reactor heat duties at the given reactor temperature. A quantitative tool is needed to assess the capacity of the background process. For this purpose, the grand composite curve can be used and the reactor profile treated as if it was a utility, as explained in Chap. 6.

The problem with representing a reactor profile is that, unlike utility profiles, the reactor profile might involve several streams. The reactor profile involves not only streams such as those for indirect heat transfer shown in Fig. 13.1 but also the reactor feed and effluent streams, which can be an important feature of the reactor heating and cooling characteristics. The various streams associated with the reactor can be combined to form a grand composite curve for the reactor. This can then be matched against the grand composite curve for the rest of the process. The following example illustrates the approach.

Example 13.1   Phthalic anhydride is an important intermediate for the plastics industry. Manufacture is by the controlled oxidation of o-xylene or naphthalene. The most common route uses o-xylene via the reaction

$$\underset{\substack{o\text{-xylene}}}{C_8H_{10}} + \underset{\substack{\text{oxygen}}}{3O_2} \longrightarrow \underset{\substack{\text{phthalic}\\\text{anhydride}}}{C_8H_4O_3} + \underset{\substack{\text{water}}}{3H_2O}$$

A side reaction occurs in parallel:

$$\underset{\substack{o\text{-xylene}}}{C_8H_{10}} + \underset{\substack{\text{oxygen}}}{21/2O_2} \longrightarrow \underset{\substack{\text{carbon dioxide}}}{8CO_2} + \underset{\substack{\text{water}}}{5H_2O}$$

The reaction uses a fixed-bed vanadium pentoxide–titanium dioxide catalyst which gives good selectivity for phthalic anhydride, providing temperature is controlled within relatively narrow limits. The reaction is carried out in the vapor phase with reactor temperatures typically in the range 380 to 400°C.

The reaction is exothermic, and multitubular reactors are employed with indirect cooling of the reactor via a heat transfer medium. A number of heat transfer media have been proposed to carry out the reactor cooling, such as hot oil circuits, water, sulfur, mercury, etc. However, the favored heat transfer medium is usually a molten heat transfer salt which is a eutectic mixture of sodium–potassium nitrate–nitrite.

Figure 13.5 shows a flowsheet for the manufacture of phthalic anhydride by the oxidation of o-xylene. Air and o-xylene are heated and mixed in a Venturi, where the o-xylene vaporizes. The reaction mixture enters a tubular catalytic reactor. The heat of reaction is removed from the reactor by recirculation of molten salt. The temperature control in the reactor would be difficult to maintain by methods other than molten salt.

The gaseous reactor product is cooled first by boiler feedwater before entering a cooling water condenser. The cooling duty provided by the boiler

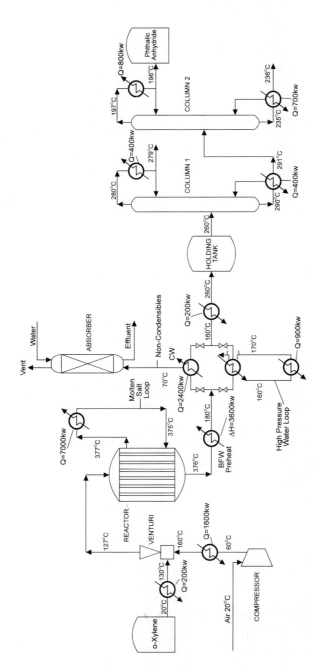

Figure 13.5 Outline of phthalic anhydride flowsheet.

feedwater has been fixed to avoid condensation. The phthalic anhydride in fact forms a solid on the tube walls in the cooling water condenser and is cooled to 70°C. Periodically, the on-line condenser is taken off-line and the phthalic anhydride melted off the surfaces by recirculation of high-pressure hot water. Two condensers are used in parallel, one on-line performing the condensation duty and one off-line recovering the phthalic anhydride. The heat duties for condensation and melting shown in Fig. 13.5 are time-averaged values. The noncondensible gases contain small quantities of byproducts and traces of phthalic anhydride and are scrubbed before being vented to the atmosphere.

The crude phthalic anhydride is heated and held at 260°C to allow some byproduct reactions to go to completion. Purification is by continuous distillation in two columns. In the first column, maleic anhydride and benzoic and toluic acids are removed overhead. In the second column, pure phthalic anhydride is removed overhead. High boiling residues are removed from the bottom of the second column.

There are two existing steam mains. These are high-pressure steam at 41 bar superheated to 270°C and medium-pressure steam at 10 bar saturated at 180°C. Boiler feedwater is available at 80°C and cooling water at 25°C to be returned at 30°C.

a. Extract the data from the flowsheet.

b. Examine the placement of the reactor relative to the rest of the process.

c. Determine the utility requirements of the process.

Solution

a. From the flowsheet in Fig. 13.5, the stream data for the heat recovery problem are presented in Table 13.1. A number of points should be noted about the *data extraction* from the flowsheet:

(1) The reactor is highly exothermic, and the data have been extracted as the molten salt being a hot stream. The basis of this is that it is assumed that the molten salt circuit is an essential feature of the reactor design. Thereafter, there is freedom within reason to choose how the molten salt is cooled.

(2) The product sublimation and melting are both carried out on a noncontinuous basis. Thus time-averaged values have been taken.

TABLE 13.1 Stream Data for the Process in Fig. 13.5

	Stream		T_S	T_T	ΔH
No.	Name	Type	(°C)	(°C)	(kW)
1	Reactor cooling	Hot	377	375	−7000
2	Reactor product cooling	Hot	376	180	−3600
3	Product sublimation	Hot	180	70	−2400
4	Column 1 condenser	Hot	280	279	−400
5	Column 2 condenser	Hot	197	196	−800
6	Air feed	Cold	60	160	1600
7	o-Xylene feed	Cold	20	130	200
8	Product melting	Cold	70	160	900
9	Holding tank feed	Cold	160	260	200
10	Column 1 reboiler	Cold	290	291	400
11	Column 2 reboiler	Cold	235	236	700

(3) The product sublimation and product melting imply a linear change in enthalpy over a relatively large change in temperature. However, changes of phase normally take place with a relatively small change in temperature. Thus the product sublimation might involve desuperheating over a relatively large range of temperature, change of phase over a relatively small change in temperature, and subcooling over a relatively large range in temperature. Product melting might involve heating to melting point over a relatively large range of temperature followed by melting over a relatively small change in temperature. Thus representation of the product sublimation and product melting as a linear change in enthalpy seems to be inappropriate. To overcome this, these two streams could be broken down into linear *segments* to represent this nonlinear temperature-enthalpy behavior. Here, for the sake of simplicity, the streams will be assumed to have a linear temperature-enthalpy behavior.

(4) The air starts at 20°C, but it is heated to 60°C in the compressor by the increase in pressure. If the compressor is an essential feature of the process, then the heating between 20 and 60°C is serviced by the compressor and should not be included in the heat recovery problem.

(5) The air and o-xylene are mixed at unequal temperatures in the Venturi, where the o-xylene vaporizes. Mixing at unequal temperatures provides heat transfer by direct contact and might in principle be direct contact heat transfer across the pinch, the location of which is as yet unknown.[2] Thus accepting the direct contact heat transfer might lead to unnecessarily high energy targets if the mixing causes heat transfer across the pinch. The problem is avoided in targeting by mixing streams, where possible, at the same temperature, thus avoiding any direct contact heat transfer.[2] Of course, once the targets have been established and the location of the pinch known, streams can then be mixed at unequal temperatures in the design away from the pinch in the knowledge that there is no cross-pinch heat transfer. In this case, the process conditions will be accepted, initially at least, because of the vaporization occurring in the mixing.

b. Figure 13.6a shows the composite curves for the process. The problem is clearly threshold in nature, requiring only cooling, with a threshold value of ΔT_{min} of 86°C. Figure 13.6b shows the grand composite curve for $\Delta T_{min} = 10$°C. The reason ΔT_{min} has been taken to be 10°C and not 86°C is that the cooling will be supplied by the introduction of steam generation, which will turn the threshold problem into a pinched problem, as discussed in Chap. 6, for which the value of ΔT_{min} is 10°C for this problem.

The stream data in Fig. 13.6 include those associated with the reactor and those for the rest of the process. If the placement of the reactor relative to the rest of the process is to be examined, those streams associated with the reactor need to be separated from the rest of the process. Figure 13.7 shows the grand composite curves for the two parts of the process. Figure 13.7b is based on streams 1, 2, 6, and 7 from Table 13.1, and Fig. 13.7c is based on streams 3, 4, 5, 8, 9, 10, and 11.

In Fig. 13.7d, the grand composite curve for the reactor and that for the rest of the process are superimposed. To obtain maximum overlap, one of the curves must be taken as a mirror image. It can be seen in Fig. 13.7d that the reactor is appropriately placed relative to the rest of the process. Had the reactor not been appropriately placed, it would have been extremely unlikely that we would have changed the reactor to make it so. Rather, to obtain appropriate placement of the reactor, the rest of the process would more likely have been changed.

c. Figure 13.8 shows the grand composite curve for all the streams with a steam generation profile matched against it. The process cooling demand is satisfied by the generation of high-pressure (41 bar) steam from boiler feedwater which is superheated to 270°C. High-pressure steam generation is preferable to low-pressure generation. There is apparently no need for cooling water.

A greater amount of steam would be generated if the noncondensible vent was treated using catalytic incineration rather than absorption. The

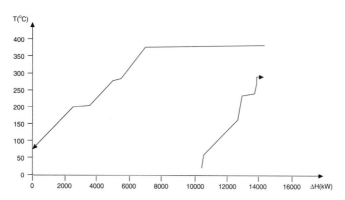

(a) The composite curves for the process show it to be a threshold problem.

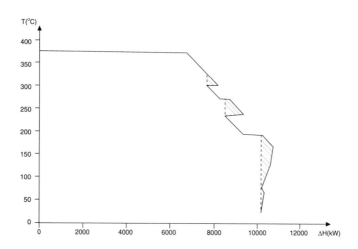

(b) The grand composite curve for the process for $\Delta T_{min} = 10°C$.

Figure 13.6 The composite curves and grand composite curve for the phthalic anhydride process.

exotherm from catalytic incineration would create an extra hot stream for steam generation.

13.4 Evolving Reactor Design to Improve Heat Integration

If the reactor proves to be inappropriately placed, then the process changes might be possible to correct this. One option would be to change the reactor conditions to bring this about. Most often,

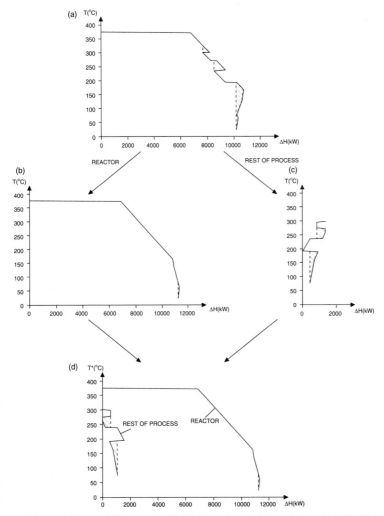

Figure 13.7 The problem can be divided into two parts, one associated with the reactor and the other with the rest of the process ($\Delta T_{min} = 10°C$), and then superimposed.

however, the reactor conditions will probably have been optimized for selectivity, catalyst performance, etc., which, taken together with safety, materials-of-construction constraints, control, etc., makes it unlikely that the reactor conditions would be changed to improve heat integration. Rather, to obtain appropriate placement of the reactor, the rest of the process would most likely be changed.

If changes to the reactor design are possible, then the simple criteria introduced in Chap. 12 can be used to direct those changes. Heat integration will always benefit by making hot streams hotter and cold streams colder. This applies whether the heat integration is carried out directly between process streams or through an intermediate such as steam. For example, consider the exothermic reactions in Fig. 13.1a. Allowing the reactor to work at higher temperature improves the heat integration potential if this does not interfere with selectivity or catalyst life or introduce safety and control problems, etc. However, if the reactor must work with a fixed intermediate cooling fluid, such as steam generation, then the only benefit will be a reduced heat transfer area in the reactor. The steam becomes a hot stream available for heat integration after leaving the reactor. If the pressure of steam generation can be increased, making a hot stream hotter, then there may be energy or area benefits when it is integrated with the rest of the process.

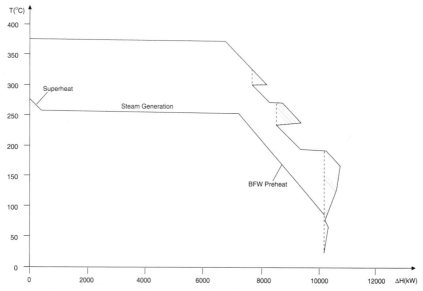

Figure 13.8 The grand composite curve for the whole process apparently requires only high-pressure steam generation from boiler feedwater.

Care should be taken when preheating reactor feeds within the reactor using the heat of reaction. This is achieved in practice simply by passing the cold feeds directly to the reactor and allowing them to be preheated by mixing with hot materials within the reactor. However, if the exothermic reactor is appropriately placed above the pinch and the feeds start below the pinch, then the preheating within the reactor is cross-pinch heat transfer. Rather, feeds should be preheated by recovery using streams below the pinch before being fed to the reactor. This increases the heat generated within the reactor available for recovery.

13.5 Heat Integration of Reactors—Summary

The appropriate placement of reactors, as far as heat integration is concerned, is that exothermic reactors should be integrated above the pinch and endothermic reactors below the pinch. Care should be taken when reactor feeds are preheated by heat of reaction within the reactor for exothermic reactions. This can constitute cross-pinch heat transfer. The feeds should be preheated to pinch temperature by heat recovery before being fed to the reactor.

Appropriate placement can be assessed quantitatively using the grand composite curve. The streams associated with the reactor can be represented as a grand composite curve for the reactor and then matched against the grand composite curve for the rest of the process.

If the reactor is not appropriately placed, then it is more likely that the rest of the process would be changed to bring about appropriate placement rather than changing the reactor. If changes to the reactor design are possible, then the simple criterion of making hot streams hotter and cold streams colder can be used to bring about beneficial changes.

13.6 References

1. Glavic, P., Kravanja, Z., and Homsak, M., "Heat Integration of Reactors: I. Criteria for the Placement of Reactors into Process Flowsheet," *Chem. Eng. Sci.,* 43(3): 593, 1988.
2. Linnhoff, B., Townsend, D. W., and Boland, D., et al., *A Users Guide on Process Integration for the Efficient Use of Energy,* IChemE, Rugby, U.K., 1982.

14

Heat Integration of Distillation Columns

14.1 The Heat Integration Characteristics of Distillation

The dominant heating and cooling duties associated with a distillation column are the reboiler and condenser duties. In general, however, there will be other duties associated with heating and cooling of feed and product streams. These sensible heat duties usually will be small in comparison with the latent heat changes in reboilers and condensers.

Both the reboiling and condensing processes normally take place over a range of temperature. Practical considerations, however, usually dictate that the heat to the reboiler must be supplied at a temperature above the dew point of the vapor leaving the reboiler and that the heat removed in the condenser must be removed at a temperature lower than the bubble point of the liquid. Hence, in preliminary design at least, both reboiling and condensing can be assumed to take place at constant temperatures.[1]

14.2 Appropriate Placement of Distillation

The consequences of placing distillation columns in different locations relative to the pinch will now be explored. There are two possible ways in which the distillation column can be integrated. The reboiler and condenser can be integrated either across the pinch or not across the pinch.

1. *Distillation across the pinch.* This arrangement is shown in

Fig. 14.1a. The background process (which does not include the reboiler and condenser) is represented simply as a heat sink and heat source divided by the pinch. Heat Q_{REB} is taken into the reboiler above pinch temperature and rejected from the condenser at a lower temperature, which is in this case below pinch temperature. Because the process sink above the pinch requires at least $Q_{H\text{min}}$ to satisfy its

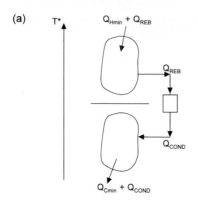

(a)

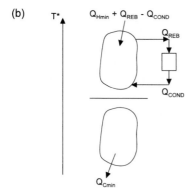

(b)

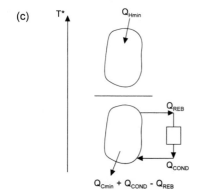

(c)

Figure 14.1 The appropriate placement of distillation columns. *(From Smith and Linnhoff, Trans. IChemE, ChERD, 66: 195, 1988; reproduced by permission of the Institution of Chemical Engineers.)*

enthalpy imbalance, the removal of Q_{REB} into the reboiler must be compensated by introducing an extra Q_{REB} from hot utility. Below the pinch, the process needs to reject Q_{Cmin} anyway, and an extra heat load Q_{COND} has been introduced from the condenser.

By integrating the separator with the process and by looking at the reboiler, it might be thought that energy has been saved. The reboiler is being run by heat recovery. However, as soon as the overall situation is considered, it becomes clear that heat is being transferred across the pinch through the distillation column and that the consumption of hot and cold utility in the process must increase correspondingly. There are fundamentally no energy savings from the integration of a distillation column across the pinch.[1]

2. *Distillation not across the pinch.* Here, the situation is somewhat different. Figure 14.1*b* shows a distillation column entirely above the pinch. The distillation column takes heat Q_{REB} from the process and returns Q_{COND} at a temperature above the pinch. The hot utility consumption changes by $(Q_{REB} - Q_{COND})$. The cold utility assumption is unchanged. Usually, Q_{REB} and Q_{COND} have a similar magnitude. If $Q_{REB} = Q_{COND}$, then the hot utility consumption is Q_{Hmin}, and there is no additional hot utility required to run the column. It takes a "free ride" from the process. Heat integration below the pinch is illustrated in Fig. 14.1*c*. Now the hot utility is unchanged, but the cold utility consumption changes by $(Q_{COND} - Q_{REB})$. Again, given that Q_{REB} and Q_{COND} usually have similar magnitude, the result is similar to heat integration above the pinch.

All these arguments can be summarized by a simple statement: *The appropriate placement for distillation is not across the pinch.*[1]

If both reboiler and condenser are integrated with the process, this can make the column difficult to start up and control. However, when the integration is considered more closely, it becomes clear that both the reboiler and condenser do not need to be integrated. Above the pinch the reboiler can be serviced directly from hot utility with the condenser integrated above the pinch. In this case the overall utility consumption will be the same as that shown in Fig. 14.1*b*. Below the pinch the condenser can be serviced directly by cold utility with the reboiler integrated below the pinch. Now the overall utility consumption will be the same as that shown in Fig. 14.1*c*.

14.3   Use of the Grand Composite Curve for Heat Integration of Distillation

The appropriate placement principle can only be applied if the process has the capacity to give up or accept the required heat duties. A quantitative tool is therefore needed to assess the source and sink capacities of any given background process. For this purpose, the grand composite curve is used. The grand composite curve would contain all heating and cooling duties for the process, including those associated with column feed heating and product cooling but excluding reboiler and condenser loads.

Let us now consider a few examples for the use of this simple representation. A grand composite curve is shown in Fig. 14.2. The distillation column reboiler and condenser duties are shown separately and are matched against it. Neither of the distillation columns in Fig. 14.2 fits. The column in Fig. 14.2a is clearly across the pinch. The distillation column in Fig. 14.2b does not fit, despite the fact that both reboiler and condenser temperatures are above the pinch. Strictly speaking, it is not appropriately placed, and yet some energy can be saved. By contrast, the distillation shown in Fig. 14.3a fits. The reboiler duty can be supplied by the hot utility. The condenser duty must be integrated with the rest of the process. Another example is shown in Fig. 14.3b. This distillation also fits. The reboiler duty must be supplied by integration with the process. Part of the condenser duty must be integrated, but the remainder of the condenser duty can be rejected to the cold utility.

14.4   Evolving the Design of Simple Distillation Columns to Improve Heat Integration

Starting with an inappropriately placed distillation, if it is shifted above the pinch by increasing its pressure, the condensing stream,

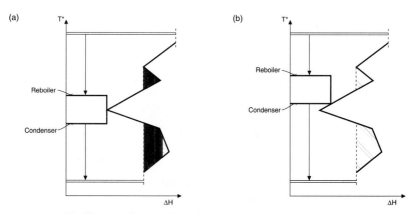

Figure 14.2 Distillation columns which do not fit against the grand composite curve. *(From Smith and Linnhoff, Trans. IChemE, ChERD, 66: 195, 1988; reproduced by permission of the Institution of Chemical Engineers.)*

which is a hot stream, is shifted from below to above the pinch. The reboiling stream, which is a cold stream, stays above the pinch. If the inappropriately placed column is shifted below the pinch by decreasing its pressure, then the reboiling stream, which is a cold stream, is shifted from above to below the pinch. The condensing stream stays below the pinch. Thus appropriate placement is a particular case of shifting streams, which, in turn, is a particular case of the plus/minus principle.[3]

If a distillation column is inappropriately placed across the pinch, it may be possible to change its pressure to achieve appropriate placement. Of course, as the pressure is changed, the shape of the "box" is also changed, since not only do the reboiler and condenser temperatures change but also the difference between them. The relative volatility also will be affected. Thus both the height and the width of the box will change as the pressure changes. Changes in pressure also affect the heating and cooling duties for column feed and products. These streams normally would be included in the background process. Hence the shape of the grand composite also will change as the column pressure changes. However, as pointed out previously, it is likely that these effects will not be significant in most processes, since the sensible heat loads involved usually will be small in comparison with the latent heat changes in condensers and reboilers.

If the distillation column will not fit either above or below the pinch, then other design options can be considered. One possibility is

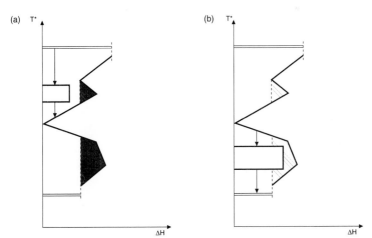

Figure 14.3 Distillation columns which fit against the grand composite curve. (*From Smith and Linnhoff, Trans. IChemE, ChERD, 66: 195, 1988; reproduced by permission of the Institution of Chemical Engineers.*)

double effecting the column, as shown in Fig. 14.4. The column feed is split and fed to two separate parallel columns. The relative pressures in the columns are chosen such that the two columns can each be appropriately placed.[3] Obviously, the capital cost of such a scheme will be higher than that of a single column, but it may be justified in favorable circumstances.

Another design option that can be considered if a column will not fit is use of an intermediate reboiler or condenser. An intermediate condenser is illustrated in Fig. 14.5. The shape of the "box" is now altered because the intermediate condenser changes the heat flow through the column. The particular design shown in Fig. 14.5 would require that at least part of the heat rejected from the intermediate condenser be passed to the process. An analogous approach can be used to evaluate the possibilities for use of intermediate reboilers. Flower and Jackson,[4] Kayihan,[5] and Dhole and Linnhoff[6] have presented procedures for the location of intermediate reboilers and condensers.

14.5   Heat Pumping in Distillation

Various heat pumping schemes have been proposed as methods for saving energy in distillation. Of these schemes, use of the column overhead vapor as the heat pumping fluid is usually the most economically attractive. This is the *vapor recompression scheme* shown in outline in Fig. 14.6.

For heat pumping to be economical on a stand-alone basis, the

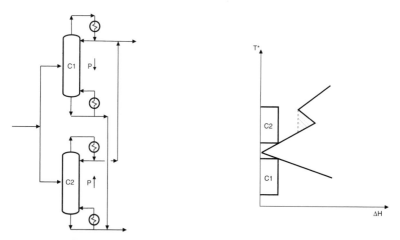

Figure 14.4 Double-effect distillation. *(From Smith and Linnhoff, Trans. IChemE, ChERD, 66: 195, 1988; reproduced by permission of the Institution of Chemical Engineers.)*

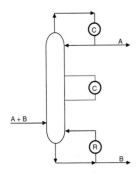

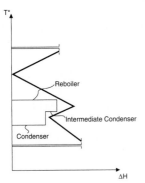

Figure 14.5 Distillation column with intermediate condenser. The profile can be designed to fit the background process. *(From Smith and Linnhoff, Trans. IChemE, ChERD, 66: 195, 1988; reproduced by permission of the Institution of Chemical Engineers.)*

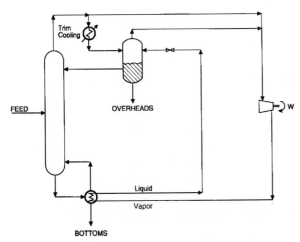

Figure 14.6 Heat pumping in distillation. A vapor recompression scheme. *(From Smith and Linnhoff, Trans. IChemE, ChERD, 66: 195, 1988; reproduced by permission of the Institution of Chemical Engineers.)*

heat pump must work across a small temperature difference, which for distillation means close-boiling mixtures. In addition, use of the scheme is only going to make sense at all if the column is constrained to operate across the pinch; otherwise, integration would be a much better option. However, distillation columns with close-boiling mixtures will be those which are easiest to move away from the pinch, since only a small pressure shift is required. Vapor recompression therefore only makes sense for the distillation of close-boiling mixtures in highly constrained situations.[3]

14.6   Evolving the Distillation Sequence

In Chap. 5 it was suggested that the sequencing of simple columns could be decoupled from their heat integration when synthesizing a design, providing there are no constraints severely restricting heat integration. The scope for heat integrating simple distillation columns is often limited. Practical constraints often prevent integration of distillation columns with the rest of the process. If the column cannot be integrated with the rest of the process, or if the potential for integration is limited by the heat flows in the background process, then attention must be turned back to the distillation operation itself and more complex arrangements considered. These were described in detail in Chap. 5. Prefractionator and thermally coupled arrangements such as the ones illustrated in Fig. 5.17 can reduce energy consumption by 20 to 30 percent when compared with a conventional arrangement of simple columns.

The following steps are necessary in establishing a heat integrated distillation sequence

1. *Establish simple sequences.* Using methods described in Chap. 5, sequences of simple columns with low overall vapor load are established. Consideration should not be restricted to the single sequence with the lowest overall vapor load, since many factors need to be considered in finally arriving at the best design.

2. *Establish the heat integration potential of simple columns.* Introduce heat recovery between reboilers, intermediate reboilers, condensers, intermediate condensers, and other process streams. Shift the distillation column pressures to allow integration, where possible, using the grand composite curve to assess the heat integration potential.

3. *Introduce complex distillation configurations.* Introduce prefractionation arrangements (with or without thermal coupling), side-rectifiers, and side-strippers to the extent that operability can be

maintained. This is done by replacing direct and indirect pairs from the simple column sequence with complex arrangements:

- Prefractionator arrangements (both with and without thermal coupling) can be used to replace either direct or indirect pairings.
- Direct sequence pairings should be replaced by side-rectifiers.
- Indirect sequence pairings should be replaced by side-strippers.

The logic behind these three evolutions is explained in Chap. 5.

As pointed out in Chap. 5, replacing simple columns by complex columns tends to reduce the vapor (and heat) load but requires more of the heat to be added or removed at extreme levels. This means that the introduction of complex columns in the design might prejudice heat integration opportunities. Thus the introduction of complex distillation arrangements needs to be considered simultaneously with the heat integration. This can be carried out manually with some trial and error or using an automated procedure such as that of Kakhu and Flower.[7]

14.7 Capital Cost Considerations

The design changes suggested so far for distillation columns have been motivated by the requirement to reduce energy costs by more effective integration between the distillation column and the rest of the process. There are, however, capital cost implications when distillation columns are integrated. These implications fall into two broad categories, i.e., changes in column capital cost and changes in heat exchanger network capital cost. Obviously, these capital cost changes should be considered together with the energy cost changes in order to achieve an optimal tradeoff between capital and energy costs.

1. *Distillation capital costs.* The classic optimization in distillation is to tradeoff capital cost of the column against energy cost for the distillation, as shown in Fig. 3.7. This would be carried out with distillation columns operating on utilities and not integrated with the rest of the process. Typically, the optimal ratio of actual to minimum reflux ratio lies in the range 1.05 to 1.1. Practical considerations often prevent a ratio of less than 1.1 being used, as discussed in Chap. 3.

If, however, the column is appropriately integrated, then the reflux ratio often can be increased without changing the overall energy consumption, as shown in Fig. 14.7a. Increasing the heat flow through the column decreases the requirement for plates in the column but increases the vapor rate. In designs initialized by traditional rules of thumb, this would have the effect of decreasing the capital cost of the column. However, the corresponding decrease in heat flow through the process shown in Fig. 14.7b will have the effect of decreasing temperature driving forces and increasing the capital cost of the heat exchanger network. Thus the tradeoff for an appropriately integrated distillation column becomes one between the capital cost of the column and the capital cost of the heat exchanger network[1,3] (see Fig. 14.7c).

Thus the optimal reflux ratio for an appropriately integrated distillation column will be problem-specific and is likely to be quite different from that for a stand-alone column.

If complex distillation columns are being considered, then these also can bring about significant reductions in capital cost. The dividing-wall column shown in Fig. 5.17 not only requires typically 20 to 30 percent less energy than a conventional arrangement but also can be typically 30 percent lower in capital cost than a conventional two-column arrangement.[8,9]

2. *Heat exchanger network capital costs.* It is easy for the designer to become carried away with the elegance of "packing boxes" into space around the grand composite curve. However, the full implications of integration are only clear by considering the composite curves of the process with the distillation column. Temperature driving forces become smaller throughout the process as a result of integration. This means that the capital/energy tradeoff should be readjusted, and a larger ΔT_{min} might be required. The optimization of the capital/energy tradeoff might undo part of the savings achieved by appropriate integration.

14.8   A Case Study

Figure 14.8a shows a simplified flowsheet for the manufacture of acetic anhydride as presented by Jeffries.[10] Acetone feed is cracked in a furnace to ketene and the byproduct methane. The methane is used as furnace fuel. A second reactor forms acetic anhydride by the reaction between ketene from the first reaction and acetic acid.

The composite curves for this flowsheet are shown in Fig. 14.8b. The composite curves are dominated by the reboilers and condensers of the two distillation columns and the feed vaporizer for the acetone feed. It is immediately apparent that the two distillation columns are both inappropriately placed across the pinch. Linnhoff and Parker[11]

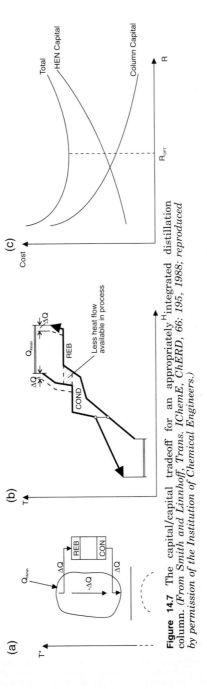

Figure 14.7 The capital/capital tradeoff for an appropriately integrated distillation column. (*From Smith and Linnhoff, Trans. IChemE, ChERD, 66: 195, 1988; reproduced by permission of the Institution of Chemical Engineers.*)

have discussed possible modifications for this process. Figure 14.9*a* presents one of the many schemes possible by manipulating column pressures.

The pressure in distillation column 1 has been increased to allow feed vaporization by heat recovery (from the distillation column condenser). Inspection of the new curves in Fig. 14.9*a* raises further possibilities. With the proposed modification, the overheads from the

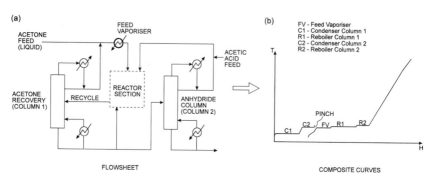

Figure 14.8 Simplified flowsheet for the acetic anhydride process. The composite curves show both distillation columns to be across the pinch. *(From Smith and Linnhoff, Trans. IChemE, ChERD, 66: 195, 1988; reproduced by permission of the Institution of Chemical Engineers.)*

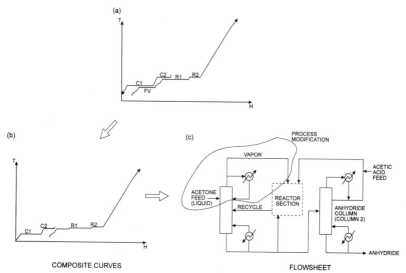

Figure 14.9 The composite curves after process modifications shows a much reduced energy target. *(From Smith and Linnhoff, Trans. IChemE, ChERD, 66: 195, 1988; reproduced by permission of the Institution of Chemical Engineers.)*

acetone recycle column would be heat exchanged with the feed vaporizer (i.e., acetone vaporizer), and a further modification becomes apparent in which the overheads from the column need only be partially condensed. The rest can be fed as vapor to the reactor. The missing reflux to the column can be provided by fresh process feed. The composite curves for this option are shown in Fig. 14.9b. The final flowsheet is shown in Fig. 14.9c.

Starting from the original flowsheet, Linnhoff and Parker[11] have shown that it is possible by a combination of distillation modifications and network improvements to reduce the energy consumption of this process by approximately 60 percent.

14.9   Heat Integration of Distillation Columns—Summary

The appropriate placement of distillation columns when heat integrated is not across the pinch. The grand composite curve can be used as a quantitative tool to assess integration opportunities.

The scope for integrating conventional distillation columns into an overall process is often limited. Practical constraints often prevent integration of columns with the rest of the process. If the column cannot be integrated with the rest of the process, or if the potential for integration is limited by the heat flows in the background process, then attention must be turned back to the distillation operation itself and complex arrangements considered.

Once the distillation is integrated, then driving forces between the composite curves become smaller. This in turn means the capital/energy tradeoff for the heat exchanger network should be adjusted accordingly.

Unfortunately, the overall design problem is even more complex in practice. Spare driving forces in the process could be exploited equally well to allow the use of moderate utilities or the integration of heat engines, heat pumps, etc. in preference to distillation integration.

14.10   References

1. Linnhoff, B., Dunford, H., and Smith, R., "Heat Integration of Distillation Columns into Overall Processes," *Chem. Eng. Sci.,* 38: 1175, 1983.
2. Andrecovich, M. J., and Westerberg, A. W., "A Simple Synthesis Method Based on Utility Bounding for Heat-Integrated Distillation Sequences," *AIChEJ,* 31(3): 363, 1985.
3. Smith, R., and Linnhoff, B., "The Design of Separators in the Context of Overall Processes," *Trans. IChemE, ChERD,* 66: 195, 1988.
4. Flower, J. R., and Jackson, M. A., "Energy Requirements in the Separation of Mixture by Distillation," *Trans. IChemE,* 42: T249, 1964.

5. Kayihan, F., "Optimum Distribution of Heat Load in Distillation Columns Using Intermediate Condensers and Reboilers," *AIChE Symp. Ser.*, 192(76): 1, 1980.
6. Dhole, V. R., and Linnhoff, B., "Distillation Column Targets," *Computers Chem. Eng.*, 17: 549, 1993.
7. Kakhu, A. I., and Flower, J. R., "Synthesizing Heat-Integrated Distillation Systems Using Mixed Integer Programming," *Trans. IChemE, ChERD*, 66: 241, 1988.
8. Kaibel, G., "Distillation Column Arrangements with Low Energy Consumption," *IChemE Symp. Ser.*, 109: 43, 1988.
9. Triantafyllou, C., and Smith, R., "The Design and Optimization of Fully Thermally Coupled Distillation Columns," *Trans. IChemE,* part A, 70: 118, 1992.
10. Jeffries, G. V., *The Manufacture of Acetic Anhydride,* IChemE, Rugby, U.K., 1961.
11. Linnhoff, B., and Parker, S., "Heat Exchanger Networks with Process Modifications," IChemE Annual Research Meeting, Bath, U.K., 1984.

15

Heat Integration of Evaporators and Dryers

15.1 The Heat Integration Characteristics of Evaporators

Evaporation processes usually separate a single component (typically water) from a nonvolatile material. As such, it is good enough in most cases to assume that the vaporization and condensation processes take place at constant temperatures.

As with distillation, the dominant heating and cooling duties associated with an evaporator are the vaporization and condensation duties. As with distillation, there will be other duties associated with the evaporator for heating or cooling of feed, product, and condensate streams. These sensible heat duties will usually be small in comparison with the latent heat changes.

Figure 15.1a shows a single-stage evaporator represented on both actual and shifted temperature scales. Note that in shifted temperature scale, the evaporation and condensation duties are shown at different temperatures even though they are at the same actual temperature. Figure 15.1b shows a similar plot for a three-stage evaporator.

Like distillation, evaporation can be represented as a box. This again assumes that any heating or cooling required by the feed and concentrate will be included with the other process streams in the grand composite curve.

15.2 Appropriate Placement of Evaporators

The concept of the appropriate placement of distillation columns was developed in the preceding chapter. The principle also clearly applies to evaporators. The heat integration characteristics of distillation columns and evaporators are very similar. Thus evaporator placement should be not across the pinch.[2]

15.3 Evolving Evaporator Design to Improve Heat Integration

The thermodynamic profile of an evaporator also can be manipulated. The approach is similar to that used for distillation columns, but the degrees of freedom are obviously different.[1,2]

In the first instance, temperature levels can be changed by manipulating the operating pressure in the same way as distillation (see Fig. 3.13). The ΔT between stages can change by varying the heat transfer area (see Fig. 3.13). The other basic degree of freedom shown in Fig. 3.13 is to change the number of stages. Further possibilities become apparent once the evaporator is considered in the process context.

Consider the three-stage evaporator against a background process

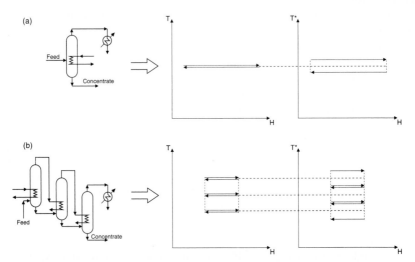

Figure 15.1 The representation of evaporators in shifted temperatures. *(Reprinted from Smith, R., and Jones, P. S., "The Optimal Design of Integrated Evaporation Systems," Heat Recovery Systems and CHP, 10: 341, 1990, with permission from Elsevier Science Ltd.)*

as shown in Fig. 15.2a. At the chosen pressure, the evaporator will not fit against the grand composite curve. The most obvious possibility is to first try an increase in pressure to allow appropriate placement above the pinch (Fig. 15.2b). Suppose, however, that the required increase in pressure would cause unacceptably high levels of decomposition and fouling in the evaporator.

The possibility of increasing the number of stages from three to, say, six could now be considered in order to allow a fit to the grand composite just above the pinch (Fig. 15.3a). The evaporator fits, but there is still a problem of product degradation because of high temperatures. However, it is not necessary for all evaporator stages to be linked thermally with each other. Instead, Fig. 15.3b shows a six-stage system, three stages appropriately placed above and three

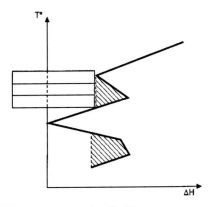

(a) A three stage evaporator will not fit.

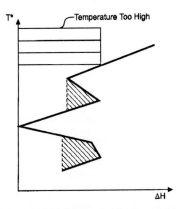

(b) Increasing the evaporator pressure allows appropriate. placement but the temperature is too high.

Figure 15.2 Integration of a three-stage evaporator.

below the pinch. This could be either a conventional six-stage system in which the first three and last three stages are not linked thermally. Alternatively, it could be two parallel three-stage systems, analogous to double effecting in distillation.

Yet another design option is shown in Fig. 15.3c, in which the heat flow (and hence mass flow) is changed between stages in the evaporator. Figure 15.3c shows an arrangement in which part of the vapor from the second stage is used for process heating rather than evaporation in the third stage. This means that more evaporation is taking place in the first two stages than in the third and other

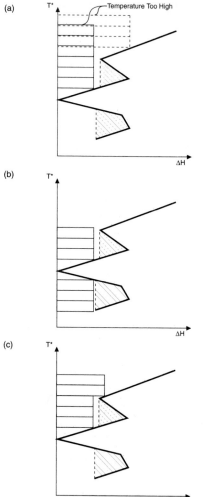

Figure 15.3 Evaporator design with the help of the grand composite curve. *(From Smith and Linnhoff, Trans. IChemE, ChERD, 66: 195, 1988; reproduced by permission of the Institution of Chemical Engineers.)*

stages. Note that even if the heat flow through the multistage evaporator is constant, the rate of evaporation will decrease because the latent heat increases as the pressure decreases.

15.4   The Heat Integration Characteristics of Dryers

The heat input to dryers is to a gas and as such takes place over a range of temperatures. Moreover, the gas is heated to a temperature higher than the boiling point of the liquid to be evaporated. The exhaust gases from the dryer will be at a lower temperature than the inlet, but again, the heat available in the exhaust will be available over a range of temperatures. The thermal characteristics of dryers tend to be design-specific and quite different in nature from both distillation and evaporation.

15.5   Evolving Dryer Design to Improve Heat Integration

It was noted earlier that dryers are quite different in character from both distillation and evaporation. However, heat is still taken in at a high temperature to be rejected in the dryer exhaust. The appropriate placement principle as applied to distillation columns and evaporators also applies to dryers. The plus/minus principle from Chap. 12 provides a general tool that can be used to understand the integration of dryers in the overall process context. If the designer has the freedom to manipulate drying temperature and gas flow rates, then these can be changed in accordance with the plus/minus principle in order to reduce overall utility costs.

15.6   A Case Study

Figure 15.4 shows a plant for the production of animal feed from spent grains in a whisky distillery. The plant has two feeds, one of low- and one of high-concentration solids. Water is removed from the low-concentration feed by an evaporator followed by a rotary dryer. Water is removed from the high-concentration feed by a centrifuge followed by two stages of drying in rotary dryers. As is usual with this type of plant, the evaporators and dryers have been designed on a stand-alone basis without consideration of the process context. Optimization of the evaporator on a stand-alone basis has indicated that heat pumping using a mechanical vapor recompression system would be economical.

Figure 15.4 shows the grand composite curve for this process and

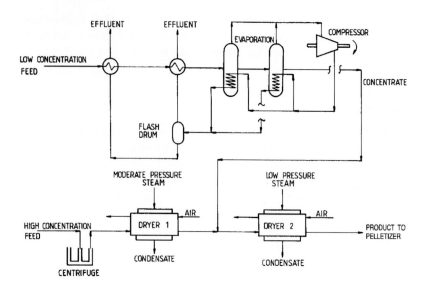

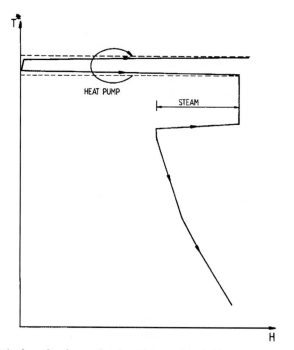

Figure 15.4 A plant for the production of animal feed. The heat pump encroaches into a "pocket" in the grand composite curve. *(From Smith and Linnhoff, Trans. IChemE, ChERD, 66: 195, 1988; reproduced by permission of the Institution of Chemical Engineers.)*

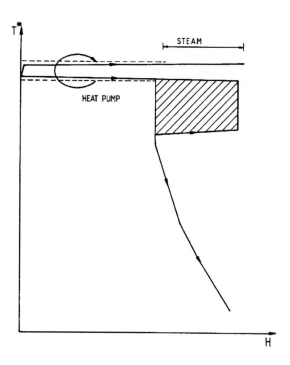

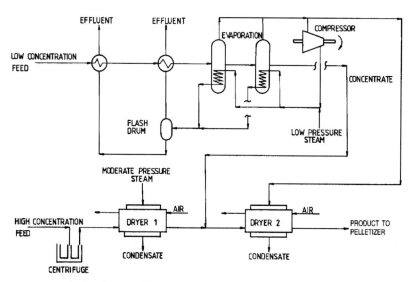

Figure 15.5 A simple modification reduces the load on the heat pump, saving electricity. *(From Smith and Linnhoff, Trans. IChemE, ChERD, 66: 195, 1988; reproduced by permission of the Institution of Chemical Engineers.)*

the location of the evaporator heat pump. The heating duty for the first dryer has been omitted from the grand composite because the required temperature is too high to allow integration with the rest of the process. The heat pump can be seen to be appropriately placed across the pinch. However, the cold side, below the pinch, encroaches into a pocket in the grand composite curve. If the design of the heat pump is changed so as not to encroach into the pocket, the result is shown in Fig. 15.5. The resulting steam consumption is virtually unchanged, but energy costs now will be lower. This results from the reduced load on the heat pump leading to a reduction in electricity demand.

15.7   Heat Integration of Evaporators and Dryers—Summary

Like distillation, the appropriate placement of evaporators and dryers is that they should be above the pinch, below the pinch, but not across the pinch. The grand composite curve can be used to assess appropriate placement quantitatively.

Also like distillation, the thermal profile of evaporators can be manipulated by changing the pressure. However, the degrees of freedom in evaporator design open up more options.

Dryers are different in characteristic from distillation columns and evaporators in that the heat is added and rejected over a large range of temperature. Changes to dryer design can be directed by the plus/minus principle.

15.8   References

1. Smith, R., and Jones, P. S., "The Optimal Design of Integrated Evaporation Systems," *J. Heat Recovery Sys. CHP,* 10: 341, 1990.
2. Smith, R., and Linnhoff, B., "The Design of Separators in the Context of Overall Processes," *Trans. IChemE, ChERD,* 66: 195, 1988.

16

Heat Exchanger Network Design

The structure of the reaction-separation system has now been fixed, and some optimization of the major design variables (reactor conversion, recycle inert concentration, etc.) has been carried out. This optimization has been carried out using only targets for the heat exchanger network and utilities. Minimization of process waste has been considered. Utility waste and cost have been minimized by improving heat integration. Again, the process changes to improve heat integration have been carried out using targets for the heat exchanger network and utilities.

Having explored the major degrees of freedom, the material and energy balance is now fixed, and hence the hot and cold streams which contribute to the heat exchanger network are firmly defined. The remaining task is to complete the design of the heat exchanger network.

16.1 The Pinch Design Method

In Chap. 6 the capital/energy tradeoff in the heat exchanger network was discussed. The relative position of the composite curves was changed by varying $\Delta T_{\min}$ (see Fig. 6.6). As $\Delta T_{\min}$ is changed from a small to a large value, the capital cost decreases, but the energy cost increases. When the two costs are combined to obtain a total cost, the optimal point in the capital/energy tradeoff is identified, corresponding with an optimal value of $\Delta T_{\min}$ (see Fig. 6.6). As pointed out in Chap. 6, the tradeoff between energy and capital suggests that, on average, individual exchangers should have a temperature difference no smaller than the $\Delta T_{\min}$ between the composite curves. It was suggested that a good initialization would be to assume that no individual exchanger should have a temperature difference smaller than ΔT_{min}.

Having decided that no exchanger should have a temperature difference smaller than ΔT_{min}, two rules were deduced. If the energy target set by the composite curves (or the problem table algorithm) is to be achieved, there must be no heat transfer across the pinch by

- process-to-process heat transfer
- inappropriate use of utilities

These rules are both necessary and sufficient for the design to achieve the energy target, given that no individual exchanger should have a temperature difference smaller than ΔT_{min}. To comply with these two rules, the process should therefore be divided at the pinch. As pointed out in Chap. 6, this is most clearly done by representing the stream data in the grid diagram. Figure 16.1 shows the stream data from Table 6.2 in grid form with the pinch marked. Above the pinch, steam can be used (up to Q_{Hmin}), and below the pinch, cooling water is used (up to Q_{Cmin}). But what strategy should be adopted for the design? A number of simple criteria can be developed to help.[1]

1. *Start at the pinch.* The pinch is the most constrained region of the problem. At the pinch, ΔT_{min} exists between all hot and cold streams. As a result, the number of feasible matches in this region is severely restricted. Quite often there are essential matches to be made. If such matches are not made, the result will be either use of temperature differences smaller than ΔT_{min} or excessive use of a utility resulting from heat transfer across the pinch. If the design

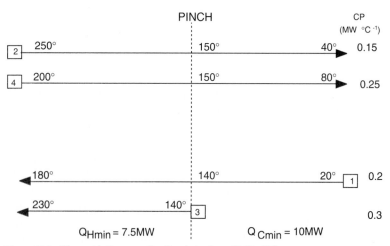

Figure 16.1 The grid diagram for the data from Table 6.2.

was started away from the pinch, at the hot end or cold end of the problem, then initial matches likely would need follow-up matches as the pinch is approached, which violate the pinch or the ΔT_{min} criterion. Putting the argument the other way around, if the design is started at the pinch, initial decisions are made in the most constrained part of the problem. This is much less likely to lead to difficulties later.

2. *The CP inequality for individual matches.* Figure 16.2*a* shows the temperature profile for an individual exchanger at the pinch, above the pinch.[1,2] Moving away from the pinch, temperature differences must increase. Figure 16.2*a* shows a match between a hot stream and a cold stream which has a *CP* smaller than the hot stream. At the pinch, the match starts with a temperature difference equal to ΔT_{min}. The relative slopes of the temperature-enthalpy profiles of the two streams mean that the temperature differences become smaller moving away from the pinch, which is infeasible. On the other hand, Fig. 16.2*b* shows a match involving the same hot stream but with a cold stream that has a larger *CP*. The relative slopes of the temperature-enthalpy profiles now cause the temperature differences to become larger moving away from the pinch, which is feasible. Thus, starting with ΔT_{min} at the pinch, for temperature differences to increase moving away from the pinch,[1,2]

$$CP_H \leq CP_C \quad \text{(above pinch)} \tag{16.1}$$

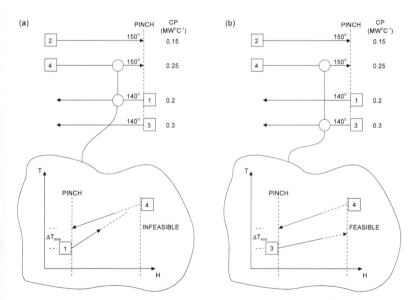

Figure 16.2 Criteria for pinch matches above the pinch.

Figure 16.3 shows the situation below the pinch at the pinch. If a cold stream is matched with a hot stream with a smaller CP, as shown in Fig. 16.3a (i.e., a steeper slope), then the temperature differences become smaller (which is infeasible). If the same cold stream is matched with a hot stream with a larger CP (i.e., a less steep slope), as shown in Fig. 16.3b, then temperature differences become larger (which is feasible). Thus, starting with ΔT_{min} at the pinch, for temperature differences to increase moving away from the pinch,[1,2]

$$CP_H \geq CP_C \quad \text{(below pinch)} \tag{16.2}$$

Note that the CP inequalities given by Eqs. (16.1) and (16.2) apply only at the pinch when both ends of the match are at pinch conditions.

3. *The CP table.* Identification of the essential matches in the region of the pinch is clarified by use of a *CP table*.[1,2] In a CP table, the CP values of the hot and cold streams at the pinch are listed in descending order.

Figure 16.4a shows the grid diagram with a CP table for design above the pinch. Cold utility must not be used above the pinch, which means that hot streams must be cooled to pinch temperature by recovery. Hot utility can be used, if necessary, on the cold streams above the pinch. Thus it is essential to match hot streams above the pinch with a cold partner. In addition, if the hot stream is at pinch conditions, the cold stream it is to be matched with must also be at

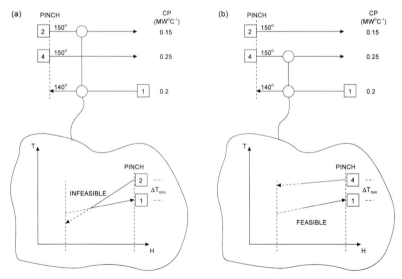

Figure 16.3 Criteria for pinch matches below the pinch.

pinch conditions; otherwise, the $\Delta T_{\min}$ constraint will be violated. Figure 16.4a shows a feasible design arrangement above the pinch that does not use temperature differences smaller than $\Delta T_{\min}$. Note again that the CP inequality applies only when a match is made between two streams that are both at the pinch. Away from the pinch, temperature differences increase, and it is no longer essential to obey the CP inequalities.

Figure 16.4b shows the grid diagram with a CP table for design below the pinch. Hot utility must not be used below the pinch, which means that cold streams must be heated to pinch temperature by recovery. Cold utility can be used, if necessary, on the hot streams below the pinch. Thus it is essential to match cold streams below the pinch with a hot partner. In addition, if the cold stream is at pinch conditions, the hot stream it is to be matched with also must be at pinch conditions; otherwise, the $\Delta T_{\min}$ constraint will be violated. Figure 16.4b shows a design arrangement below the pinch that does not use temperature differences smaller than $\Delta T_{\min}$.

Having decided that some essential matches need to be made around the pinch, the next question is how big should the matches be?

4. *The "tick-off" heuristic.* Once the matches around the pinch have been chosen to satisfy the criteria for minimum energy, the design should be continued in such a manner as to keep capital costs to a minimum. One important criterion in the capital cost is the number of units (there are others, of course, which shall be addressed later). Keeping the number of units to a minimum can be achieved using the *"tick-off" heuristic.*[1] To tick off a stream, individual units are made as

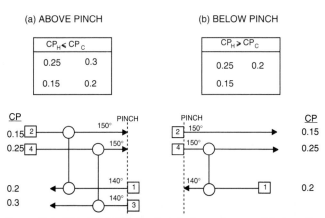

Figure 16.4 The CP table for the designs above and below the pinch for the problem from Table 6.2.

large as possible, i.e., the smaller of the two heat duties on the streams being matched.

Figure 16.5a shows the matches around the pinch from Fig. 16.4a with their duties maximized to tick-off streams. It should be emphasized that the tick-off heuristic is only a heuristic and can occasionally penalize the design. Methods will be developed later which allow such penalties to be identified as the design proceeds.

The design in Fig. 16.5a can now be completed by satisfying the heating and cooling duties away from the pinch. Cooling water must not be used above the pinch; therefore, if there are hot streams above the pinch for which the duties are not satisfied by the pinch matches, additional process-to-process heat recovery must be used. Figure 16.5b shows an additional match to satisfy the residual cooling of the hot streams above the pinch. Again, the duty on the unit is maximized. Finally, above the pinch the residual heating duty on the cold streams must be satisfied. Since there are no hot streams left above the pinch, hot utility must be used (Fig. 16.5c).

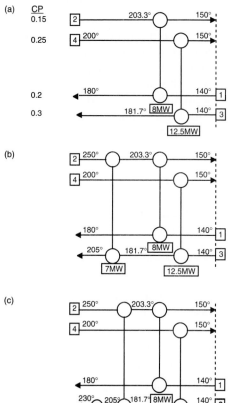

Figure 16.5 Sizing the units above the pinch using the tick-off heuristic.

Turning now to the cold-end design, Fig. 16.6*a* shows the pinch design with the streams ticked off. If there are any cold streams below the pinch for which the duties are not satisfied by the pinch matches, additional process-to-process heat recovery must be used, since hot utility must not be used. Figure 16.6*b* shows an additional match to satisfy the residual heating of the cold streams below the pinch. Again, the duty on the unit is maximized. Finally, below the pinch the residual cooling duty on the hot streams must be satisfied. Since there are no cold streams left below the pinch, cold utility must be used (Fig. 16.6*c*).

The final design shown in Fig. 16.7 amalgamates the hot-end design from Fig. 16.5*c* and the cold-end design from Fig. 16.6*c*. The duty on hot utility of 7.5 MW agrees with Q_{Hmin} and the duty on the cold utility of 10 MW agrees with Q_{Cmin} predicted by the composite curves and the problem table algorithm.

Note one further point from Fig. 16.7. The number of units is 7 in total (including the heater and cooler). Referring back to Example 7.1, the target for the minimum number of units was calculated to be 7. It therefore appears that there was something in our procedure

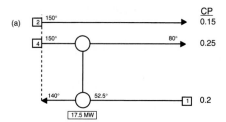

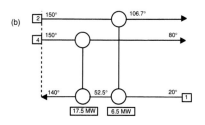

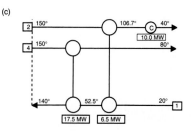

Figure 16.6 Sizing the units below the pinch using the tick-off heuristic.

which naturally steered us to a design that achieved the target for the minimum number of units.

It is in fact the tick-off heuristic that steered the design toward the minimum number of units.[1,2] The target for the minimum number of units was given by Eq. (7.2):

$$N_{\text{UNITS}} = S - 1 \qquad (7.2)$$

Before any matches are placed, the target indicates that the number of units needed is equal to the number of streams (including utility streams) minus one. The tick-off heuristic satisfied the heat duty on one stream every time one of the units was used. The stream that has been ticked off is no longer part of the remaining design problem. The tick-off heuristic ensures that having placed a unit (and used up one of our available units), a stream is removed from the problem. Thus Eq. (7.2) is satisfied if every match satisfies the heat duty on a stream or a utility.

This design procedure is known as the *pinch design method* and can be summarized in five steps:[1]

- Divide the problem at the pinch into separate problems.

- The design for the separate problems is started at the pinch, moving away.

- Temperature feasibility requires constraints on the *CP* values to be satisfied for matches between streams at the pinch.

- The loads on individual units are determined using the tick-off heuristic to minimize the number of units. Occasionally, the heuristic causes problems.

- Away from the pinch there is usually more freedom in the choice of matches. In this case, the designer can discriminate on the basis of judgment and process knowledge.

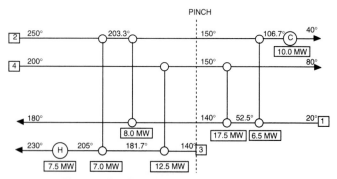

Figure 16.7 The completed design for the data from Table 6.2.

Example 16.1 The process stream data for a heat recovery network problem are given in Table 16.1. A problem table analysis on these data reveals that the minimum hot utility requirement for the process is 15 MW and the minimum cold utility requirement is 26 MW for a minimum allowable temperature difference of 20°C. The analysis also reveals that the pinch is located at a temperature of 120°C for hot streams and 100°C for cold streams. Design a heat exchanger network for maximum energy recovery in the minimum number of units.

Solution Figure 16.8a shows the hot-end design with the CP table. Above the pinch, adjacent to the pinch, $CP_H \leq CP_C$. The duty on the units has been maximized according to the tick-off heuristic.

Figure 16.8b shows the cold-end design with the CP table. Below the pinch, adjacent to the pinch, $CP_H \geq CP_C$. Again, the duty on units has been maximized according to the tick-off heuristic.

The completed design is shown in Fig. 16.8c. The minimum number of units for this problem is given by

$$N_{\text{UNITS}} = (S - 1)_{\text{ABOVE PINCH}} + (S - 1)_{\text{BELOW PINCH}}$$

$$= (5 - 1) + (4 - 1)$$

$$= 7$$

The design in Fig. 16.8 is seen to achieve the minimum number of units target.

16.2 Design for Threshold Problems

In Sec. 6.3 it was mentioned that some problems, known as *threshold problems,* do not have a pinch. They need either hot utility or cold utility but not both. How should the approach be modified to deal with the design of threshold problems?

The philosophy in the pinch design method was to start the design where it was most constrained. If the design is pinched, the problem is most constrained at the pinch. If there is no pinch, where is the design most constrained? Figure 16.9a shows a threshold problem that requires no hot utility, just cold utility. The most constrained part of this problem is the no-utility end.[2] This is where temperature differences are smallest, and there may be constraints, as shown in Fig. 16.9b, where the target temperatures on some of the cold

TABLE 16.1 Stream Data

Stream		Supply temperature (°C)	Target temperature (°C)	Heat capacity flow rate (MW °C⁻¹)
No.	Type			
1	Hot	400	60	0.3
2	Hot	210	40	0.5
3	Cold	20	160	0.4
4	Cold	100	300	0.6

streams can only be satisfied by very specific matches. Also, if individual matches are required to have a temperature difference no smaller than the threshold $\Delta T_{\min}$, the CP inequalities described in the pinch design method must be applied. For the most part, problems similar to that in Fig. 16.9a are treated as one-half of a pinched problem.

Figure 16.10 shows another threshold problem that requires only hot utility. This problem is different in characteristic from the one in Fig. 16.9. Now the minimum temperature difference is in the middle of the problem, causing a pseudopinch. The best strategy to deal with this type of threshold problem is to treat it as a pinched problem. For the problem in Fig. 16.10, the problem is divided into two parts at the pseudopinch, and the pinch design method is followed. The only complication in applying the pinch design method for such problems is that one-half of the problem (the cold end in Fig. 16.10) will not feature the flexibility offered by matching against utility.

16.3 Stream Splitting

The pinch design method developed earlier followed several rules and guidelines to allow design for minimum utility (or maximum energy recovery) in the minimum number of units. Occasionally, it appears not to be possible to create the appropriate matches because one or other of the design criteria cannot be satisfied.

Consider Fig. 16.11a, which shows the above-pinch part of a design. Cold utility must not be used above the pinch, which means

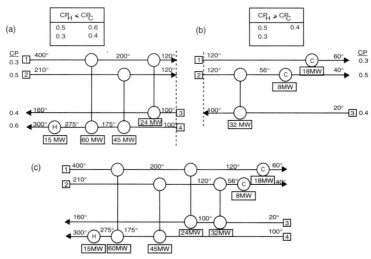

Figure 16.8 Maximum energy recovery design for Example 16.1.

that all hot streams must be cooled to pinch temperature by heat recovery. There are three hot streams and two cold streams in Fig. 16.11*a*. Thus, regardless of the *CP* values of the streams, one of the hot streams cannot be cooled to pinch temperature without some violation of the $\Delta T_{\min}$ constraint. The problem can only be resolved by splitting a cold stream into two parallel branches, as shown in Fig. 16.11*b*. Now each hot stream has a cold partner with which to match and capable of cooling it to pinch temperature. Thus, in addition to the *CP* inequality criteria introduced earlier, there is a stream number criterion above the pinch such that[1,2]

$$S_H \leq S_C \quad \text{(above pinch)} \qquad (16.3)$$

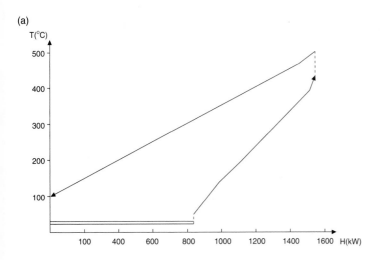

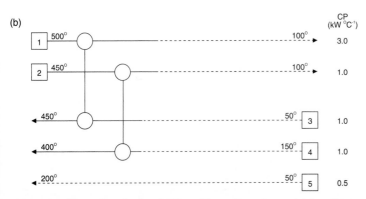

Figure 16.9 Even though threshold problems have large driving forces, there are still often essential matches to be made, especially at the no-utility end.

where S_H = number of hot streams at the pinch (including branches)

S_C = number of cold streams at the pinch (including branches)

If there had been more cold streams than hot streams in the design above the pinch, this would not have created a problem, since hot utility can be used above the pinch.

By contrast, now consider part of a design below the pinch (Fig. 16.12a). Here, hot utility must not be used, which means that all cold streams must be heated to pinch temperature by heat recovery. There are now three cold streams and two hot streams in Fig. 16.12a. Again, regardless of the CP values, one of the cold streams cannot be heated to pinch temperature without some violation of the $\Delta T_{\min}$ constraint. The problem can only be resolved by splitting a hot

(a)

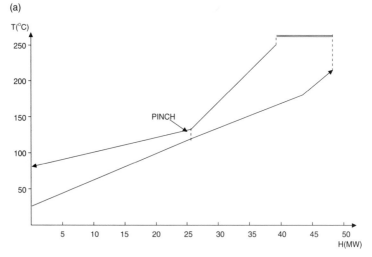

(b)

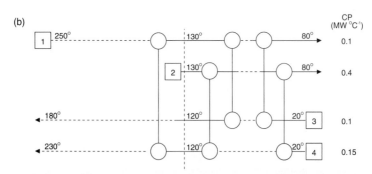

Figure 16.10 Some threshold problems must be treated as pinched problems requiring essential matches at both the no-utility end and the pinch.

stream into two parallel branches, as shown in Fig. 16.12*b*. Now each cold stream has a hot partner with which to match and capable of heating it to pinch temperature. Thus there is a stream number criterion below the pinch such that[1,2]

$$S_H \geq S_C \quad \text{(below pinch)} \qquad (16.4)$$

Had there been more hot streams than cold below the pinch, this would not have created a problem, since cold utility can be used below the pinch.

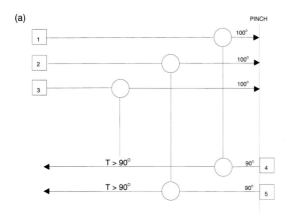

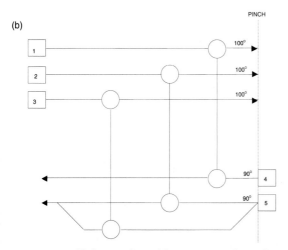

Figure 16.11 If the number of hot streams above the pinch at the pinch is greater than the number of cold streams, then stream splitting of the cold streams is required.

It is not only the stream number that creates the need to split streams at the pinch. Sometimes the *CP* inequality criteria [Eqs. (16.1) and (16.2)] cannot be met at the pinch without a stream split. Consider the above-pinch part of a problem in Fig. 16.13*a*. The number of hot streams is less than the number of cold, and hence Eq. (16.3) is satisfied. However, the *CP* inequality also must be satisfied, i.e., Eq. (16.1). Neither of the two cold streams has a large enough *CP*. The hot stream can be made smaller by splitting it into two parallel branches (Fig. 16.13*b*).

Figure 16.14*a* shows the below-pinch part of a problem. The

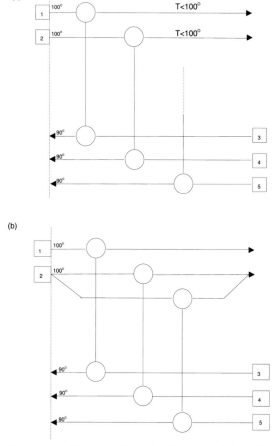

Figure 16.12 If the number of cold streams below the pinch at the pinch is greater than the number of hot streams, then stream splitting of the hot streams is required.

number of hot streams is greater than the number of cold, and hence Eq. (16.4) is satisfied. However, neither of the two hot streams has a large enough CP to satisfy the CP inequality, i.e., Eq. (16.2). The cold stream can be made smaller by splitting it into two parallel branches (Fig. 16.14b).

Clearly, in designs different from those in Figs. 16.13 and 16.14 when streams are split to satisfy the CP inequality, this might create a problem with the number of streams at the pinch such that Eqs. (16.3) and (16.4) are no longer satisfied. This would then require further stream splits to satisfy the stream number criterion. Figure 16.15 presents algorithms for the overall approach.[1]

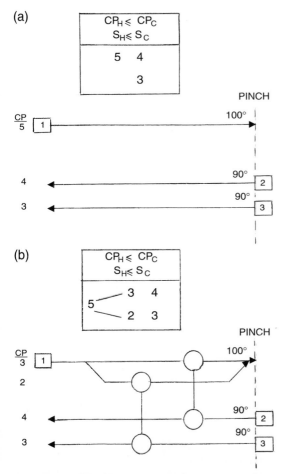

Figure 16.13 The CP inequality rules can necessitate stream splitting above the pinch.

One further important point needs to be made regarding stream splitting. In Fig. 16.13 the hot stream is split into two branches with CP values of 3 and 2 to satisfy the CP inequality criteria. However, a different split could have been chosen. For example, the split could have been into branch CP values of 4 and 1, or 2.5 and 2.5, or 2 and 3 (or any setting between 4 and 1 and 2 and 3). Each of these also would have satisfied the CP inequalities. Thus there is a degree of freedom in the design to choose the branch flow rates. By fixing the heat duties on the two units in Fig. 6.13b and changing the branch flow rates, the temperature differences across each unit are changed. The best choice can only be made by sizing and costing the various

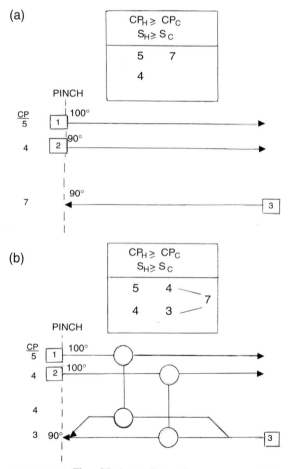

Figure 16.14 The CP inequality rules can necessitate stream splitting below the pinch.

units in the completed network for different branch flow rates. This is an important degree of freedom when the network is optimized later. Similar arguments could be made regarding the cold-end design in Fig. 16.14*b*.

Example 16.2 A problem table analysis of a petrochemicals process reveals that for a minimum temperature difference of 50°C the process requires 9.2 MW of hot utility, 6.4 MW of cold utility, and the pinch is located at 550°C

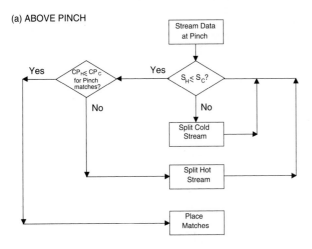

(a) ABOVE PINCH

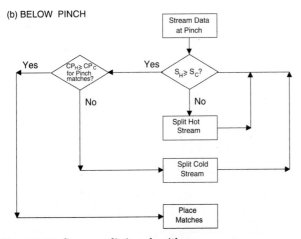

(b) BELOW PINCH

Figure 16.15 Stream-splitting algorithms.

for hot streams and 500°C for cold streams. The process stream data are given in Table 16.2. Design a heat exchanger network for maximum energy recovery which features the minimum number of units.

Solution Figure 16.16a shows the stream grid with the CP tables for the above- and below-pinch designs. Following the algorithms in Fig. 16.15, the

TABLE 16.2 Stream data

Stream		Supply temperature (°C)	Target temperature (°C)	Heat capacity flow rate (MW °C^{-1})
No.	Type			
1	Hot	750	350	0.045
2	Hot	550	250	0.04
3	Cold	300	900	0.043
4	Cold	200	550	0.02

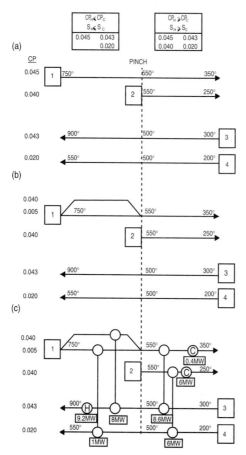

Figure 16.16 Maximum energy recovery design for Example 16.2.

hot stream must be split above the pinch to satisfy the CP inequality (see Fig. 16.16b). Thereafter, the design is straightforward, and the final design is shown in Fig. 16.16c.

The target for the minimum number of units is given by

$$N_{\text{UNITS}} = (S - 1)_{\text{ABOVE PINCH}} + (S - 1)_{\text{BELOW PINCH}}$$

$$= (4 - 1) + (5 - 1)$$

$$= 7$$

The design in Fig. 16.16c is seen to achieve the minimum units target.

16.4   Design for Multiple Pinches

In Chap. 6 it was discussed how the use of multiple utilities can give rise to multiple pinches. For example, the process from Fig. 6.2 could have used either a single hot utility or two steam levels, as shown in Fig. 6.26a. The targeting indicated that instead of using 7.5 MW of high-pressure steam at 240°C, 3 MW of this could be substituted with low-pressure steam at 180°C. Where the low-pressure steam touches the grand composite curve in Fig. 6.26a results in a utility pinch. Figure 16.17a shows the grid diagram when two steam levels are used with the utility pinch dividing the process into three parts.

Following the pinch rules, there should be no heat transfer across either the process pinch or the utility pinch by process-to-process heat exchange. Also, there must be no use of inappropriate utilities. This means that above the utility pinch in Fig. 16.17a, high-pressure steam should be used and no low-pressure steam or cooling water. Between the utility pinch and the process pinch, low-pressure steam should be used and no high-pressure steam or cooling water. Below the process pinch in Fig. 16.17, only cooling water should be used. The appropriate utility streams have been included with the process streams in Fig. 16.17a.

The network can now be designed using the pinch design method.[1,2] The philosophy of the pinch design method is to start at the pinch and move away. At the pinch, the rules for the CP inequality and the number of streams must be obeyed. Above the utility pinch and below the process pinch in Fig. 16.17, there is no problem in applying this philosophy. However, between the two pinches, there is a problem, since designing away from both pinches could lead to a clash where both meet.

More careful examination of Fig. 16.17a reveals that between the two pinches, one is more constrained than the other. Below the utility pinch, $CP_H \geq CP_C$ is required, and low-pressure steam is available as a hot stream with an extremely large CP. In fact, if it is assumed that the steam condenses isothermally, it will have a CP that is infinite. Thus, following the philosophy of starting the design in the most constrained region, the design between the pinches in Fig. 16.17a should be started at the most constrained pinch, which is the process pinch.

Following this approach, the design is straightforward, and the final design is shown in Fig. 16.17b. It achieves the energy targets

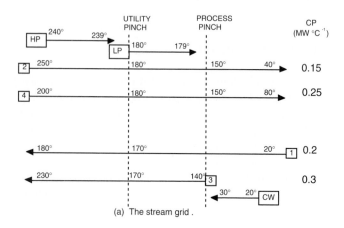

(a) The stream grid .

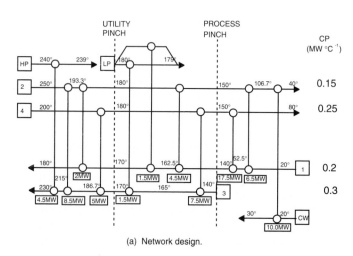

(a) Network design.

Figure 16.17 Network design for the process from Fig. 6.2 using two steam levels.

set in Example 6.3 in the minimum number of units. Remember that, in this case, to calculate the minimum number of units, the stream count must be performed separately in the three parts of the problem. Note that the stream split on the low-pressure steam in Fig. 16.17b is not strictly necessary but is made for practical reasons. Without the stream split, steam would have to be partially condensed in one unit and the steam-condensate mixture transferred to the next unit. The stream split allows two conventional steam heaters on low-pressure steam to be used. It is clear from Fig. 16.17b that the use of two steam levels has increased the complexity of the design considerably. However, the complexity can be reduced later when the structure is subjected to optimization. This optimization can remove units which are uneconomic.

It is rare for there to be two process pinches in a problem. Multiple pinches usually arise from the introduction of additional utilities causing utility pinches. However, cases such as that shown in Fig. 16.18 are not uncommon, where there is, strictly speaking, only one pinch (one place where $\Delta T_{\min}$ occurs), but there is a near-pinch. This near-pinch is a point in the process where the temperature difference becomes small enough to be effectively another pinch, even though the temperature difference is slightly larger than $\Delta T_{\min}$. Because the region around the near-pinch will be almost as constrained as the pinch, the best strategy is often to treat the near-pinch as if it was another pinch and divide the problem into three parts as shown in Fig. 16.18. The initial design would therefore avoid heat transfer across the near-pinch and the process pinch and use hot utility only above the near-pinch and cold utility only below the pinch. The

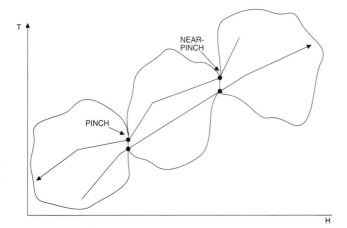

Figure 16.18 A near-pinch might require the design to be treated as if it had multiple pinches.

designer could then exploit the small amount of freedom around the near-pinch. Some heat transfer across the near-pinch is possible without causing an energy penalty. Exploiting this freedom allows the design to be simplified slightly.

Example 16.3 The stream data for a process are given in Table 16.3. It has been decided to integrate a gas turbine exhaust with the process. The exhaust temperature of the gas turbine is 400°C with $CP = 0.05 \, \text{MW} \, °\text{C}^{-1}$. Ambient temperature is 10°C.

a. Calculate the problem table cascade for $\Delta T_{min} = 20°\text{C}$.

b. Steam is to be generated by the process at a high-pressure level of 250°C and low-pressure level of 140°C. The generation of the higher-pressure steam is to be maximized. How much steam can be generated at the two levels assuming that boiler feedwater and final steam condition are both saturated?

c. Design a network for maximum energy recovery for $\Delta T_{min} = 20°\text{C}$ which generates steam at these two levels.

d. What is the residual cooling demand?

Solution

a. The problem table cascade is given in Table 16.4 for $\Delta T_{min} = 20°\text{C}$.

b. For high-pressure steam, $T^* = 260°\text{C}$. For low-pressure steam, $T^* = 150°\text{C}$.

TABLE 16.3 Stream Data

Stream		Supply temperature (°C)	Target temperature (°C)	Heat capacity flow rate (MW °C^{-1})
No.	Type			
1	Hot	635	155	0.044
2	Cold	10	615	0.023
3	Cold	85	250	0.020
4	Cold	250	615	0.020

TABLE 16.4 Problem Table Cascade

Interval temperature (°C)	Cascade heat flow (MW)
625	0
390	0.235
260	6.865
145	12.730
95	13.080
20	15.105
0	16.105

Figure 16.19 shows the grand composite curve plotted from the problem table cascade. The two levels of steam generation are shown.

$$\text{Duty on high-pressure steam generation} = 6.865\text{ MW}$$

By interpolation from the problem table cascade, at $T^* = 150°C$,

$$\text{Heat flow} = 6.865 + \frac{(260 - 150)}{(260 - 145)} \times (12.73 - 6.865)$$

$$= 12.475\text{ MW}$$

$$\text{Duty on low-pressure steam generation} = 12.475 - 6.865$$

$$= 5.61\text{ MW}$$

c. The use of two levels of steam generation in Fig. 16.19 creates two utility pinches. Thus the stream grid needs to be divided into three parts. Figure 16.20 shows the final design, which achieves the targets set for both high- and low-pressure steam generation.

d. There is a nominal cooling demand of 3.63 MW required on the gas turbine exhaust which can be satisfied by venting to the atmosphere.

16.5   Remaining Problem Analysis

The considerations addressed so far in network design have been restricted to those of energy performance and number of units. In addition, the problems have all been straightforward to design for

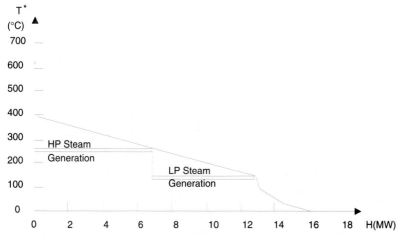

Figure 16.19 Grand composite curve for Example 16.3 showing two levels of steam generation.

maximum energy recovery in the minimum number of units by ticking off streams. Not all problems are so straightforward. Also, heat transfer area, number of shells when using 1-2 shells, capital cost, etc. should be considered when placing matches. Here, a more sophisticated approach is needed.[3]

When a match is placed, the duty needs to be chosen with some quantitative assessment of the match in the context of the whole network without having to complete the network. This can be done by exploiting the powers of targeting using a technique known as *remaining problem analysis.*[3]

Consider first the design for minimum energy in a more complex problem than seen so far. If the problem table analysis is performed on the stream data, $Q_{H\min}$ and $Q_{C\min}$ can be calculated. When the network is designed and a match placed, it would be useful to assess whether there will be any energy penalty caused by some feature of the match without having to complete the design. Whether there will be a penalty can be determined by performing a problem table analysis on the *remaining problem*. The problem table analysis is simply repeated on the stream data, leaving out those parts of the hot and cold stream satisfied by the match. One of two results would then occur:

1. The algorithm may calculate $Q_{H\min}$ and $Q_{C\min}$ to be unchanged. In this case, the designer knows that the match will not penalize the design in terms of increased utility usage.

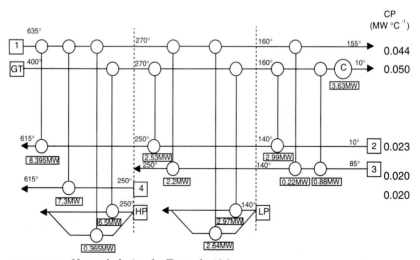

Figure 16.20 Network design for Example 16.3.

2. The algorithm may calculate an increase in $Q_{H\min}$ and $Q_{C\min}$. This means that the match is transferring heat across the pinch or that there is some feature of the design that will cause cross-pinch heat transfer if the design was completed. If the match is not transferring heat across the pinch directly, then the increase in utility will result from the match being too big as a result of the tick-off heuristic.

The remaining problem analysis technique can be applied to any feature of the network that can be targeted, such as minimum area. In Chap. 7 the approach to targeting for heat transfer area [Eq. (7.6)] was based on vertical heat transfer from the hot composite curve to the cold composite curve. If heat transfer coefficients do not vary significantly, this model predicts the minimum area requirements adequately for most purposes.[3] Thus, if heat transfer coefficients do not vary significantly, then the matches created in the design should come as close as possible to the conditions that would correspond with vertical transfer between the composite curves. Remaining problem analysis can be used to approach the area target, as closely as a practical design permits, using a minimum (or near-minimum) number of units. Suppose a match is placed, then its area requirement can be calculated. A remaining problem analysis can be carried out by calculating the area target for the stream data, leaving out those parts of the data satisfied by the match. The area of the match is now added to the area target for the remaining problem. Subtraction of the original area target for the whole-stream data A_{NETWORK} gives the area penalty incurred.

If heat transfer coefficients vary significantly, then the vertical heat transfer model adopted in Eq. (7.6) predicts a network area that is higher than the true minimum, as illustrated in Fig. 7.4. Under these circumstances, a careful pattern of nonvertical matching is required to approach the minimum network area. However, the remaining problem analysis approach can still be used to steer design toward a minimum area under these circumstances. When heat transfer coefficients vary significantly, the minimum network area can be predicted using linear programming.[4,5] The remaining problem analysis approach can then be applied using these more sophisticated area targeting methods. Under such circumstances, the design is likely to be difficult to steer toward the minimum area, and an automated design method based on the optimization of a reducible structure can be used, as will be discussed later.

Targets for number of shells, capital cost, and total cost also can be set. Thus remaining problem analysis can be used on these design parameters also.

Example 16.4   The stream data for a process are given in Table 16.5.[3] Steam is available condensing between 180 and 179°C and cooling water between 20 and 30°C. All film transfer coefficients are 200 W m^{-2}°C^{-1}. For $\Delta T_{min} = 10$°C, the minimum hot and cold utility duties are 7 and 4 MW, respectively, corresponding to a pinch at 90°C on the hot streams and 80°C on the cold streams.

a. Develop a maximum energy recovery design above the pinch which comes close to the area target in the minimum number of units.

b. Develop a maximum energy recovery design below the pinch which comes as close as possible to the minimum number of units.

Solution

a. The area target for the above-pinch problem shown in Fig. 16.21 is 8859 m^2. If the design is started at the pinch with stream 1, then Fig. 16.21a shows a feasible match which obeys the CP inequality. Maximizing its duty to 12 MW allows two streams to be ticked off simultaneously. This results from a coincidence in the stream data, the duties for streams 1 and 3 being equal above the pinch. The area of the match is 6592 m^2, and the target for the remaining problem above the pinch is 3419 m^2. Thus the match in Fig. 16.21a causes the overall target to be exceeded by 1152 m^2 (13 percent). This does not seem to be a good match.

Figure 16.21b shows an alternative match for stream 1 which also obeys the CP inequality. The tick-off heuristic also fixes its duty to be 12 MW. The area for this match is 5087 m^2, and the target for the remaining problem above the pinch is 3788 m^2. Thus the match in Fig. 16.21b causes the overall target to be exceeded by 16 m^2 (0.2 percent). This seems to be a better match and therefore is accepted.

Placing the next match above the pinch as shown in Fig. 16.21c also allows the CP inequality to be obeyed. The area for both matches in Fig. 16.21c is 7856 m^2, and the target for the remaining problem is 1020 m^2. Accepting both matches causes the overall area target to be exceeded by 17 m^2 (0.2 percent). This seems to be reasonable, and both matches are accepted. No further process-to-process matches are possible, and it remains to place hot utility.

b. The cold-utility target for the problem shown in Fig. 16.22 is 4 MW. If the design is started at the pinch with stream 3, then stream 3 must be split to satisfy the CP inequality (Fig. 16.22a). Matching one of the branches against stream 1 and ticking off stream 1 results in a duty of 8 MW.

This is a case in which the tick-off heuristic has caused problems. The match is infeasible, and its duty must be reduced to 6 MW to be feasible without either stream being ticked off (Fig. 16.22b).

TABLE 16.5   Stream Data

Stream		Supply temperature (°C)	Target temperature (°C)	Heat capacity flow rate (MW °C^{-1})
No.	Type			
1	Hot	150	50	0.2
2	Hot	170	40	0.1
3	Cold	50	120	0.3
4	Cold	80	110	0.5

Figure 16.22c shows an additional match placed on the other branch for stream 3 with its duty maximized to 3 MW to tick off stream 3. No further process-to-process matches are possible, and it remains to place cold utility.

Figure 16.23a shows the complete design, achieving maximum energy recovery in one more unit than the target minimum due to the inability to tick off streams below the pinch.

If the match in Fig. 16.21a had been accepted and the design completed, the design in Fig. 16.23b would have been obtained. This achieves the target for the minimum number of units of 7 (at the expense of excessive area). This results from the "coincidence of data" mentioned earlier in Fig. 16.21a, which allowed two streams to be ticked off simultaneously. The result is that the design above the pinch uses one fewer unit than target due to the formation of two components above the pinch (see Sec. 7.1). The design below the pinch uses one more than target, and the net result is that the overall design achieves the target for the minimum number of units.

16.6   Network Optimization

The design method used so far, the pinch design method, creates an irreducible structure in that no redundant features are added. However, the assumption is that no unit should have a temperature

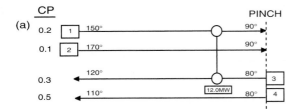

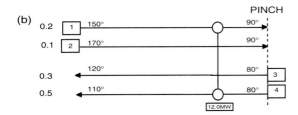

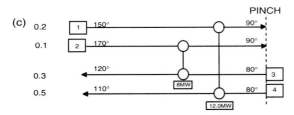

Figure 16.21   Above the pinch design for Example 16.4.

difference smaller than ΔT_{min}. A consequence of this is that there should be no heat transfer across the pinch. These constraints can now be relaxed and the heat exchanger network optimized. The optimization of heat exchanger networks is based on the redistribution of the exchanger duties. Some exchangers should perhaps be larger, some smaller, and some perhaps removed from the design altogether. Exchangers are removed from the design if the optimization sets their duty to zero.

Given a network structure, it is possible to identify loops and paths for it, as discussed in Chap. 7. Within the context of optimization, it is only necessary to consider those paths which connect two different utilities. This could be a path from steam to cooling water or a path from high-pressure steam used as a hot utility to low-pressure steam also used as a hot utility. These paths between two different utilities will be designated *utility paths*. Loops and utility paths both provide degrees of freedom in the optimization.[1,2]

Consider Fig. 16.24a, which shows the network design from Fig. 16.7 but with a loop highlighted. Heat can be shifted around loops. Figure 16.24a shows the effect of shifting heat duty U around the loop. In this loop, heat duty U is simply moved from unit E to unit B.

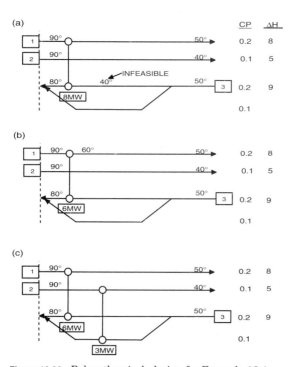

Figure 16.22 Below the pinch design for Example 16.4.

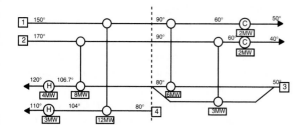

(a) The completed design.

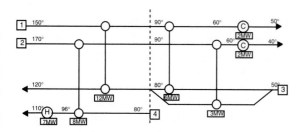

(b) An alternative design.

Figure 16.23 Alternative designs for Example 16.4.

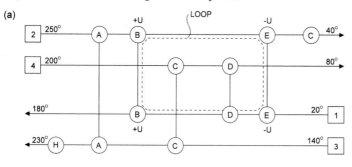

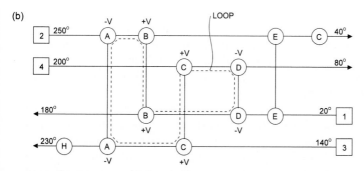

Figure 16.24 The loops which can be exploited for optimization of the design from Fig. 16.7.

The change in heat duties around the loop maintains the network heat balance and the supply and target temperatures of the streams. However, the temperatures around the loop change, and hence the temperature differences of the exchangers in the loop change in addition to their duties. The magnitude of U could be changed to different values and the network sized and costed at each value to find the optimal setting for U. If the optimal setting for U turns out to be 6.5 MW (the original duty on unit E), then the duty on unit E becomes zero, and this unit should be removed from the design.

Figure 16.24b shows the network with another loop marked. Figure 16.24b shows the effect of shifting heat duty V around the loop. Again, the heat balance is maintained, but the temperatures as well as the duties around the loop change. As before, the value of V can be optimized by costing the network at different settings of V. If V is optimized to 7.0 MW (the original duty on unit A), then the duty on unit A becomes zero, and this unit is removed from the design.

Figure 16.25a shows the network with a utility path highlighted. Heat duty can be shifted along utility paths in a similar way to that for loops. Figure 16.25a shows the effect of shifting heat duty W along the path. This time the heat balance changes because the loads imported from the hot utility and exported to the cold utility both change by W. The supply and target temperatures are maintained. If W is optimized to 7.0 MW, this will result in unit A being removed from the design. Different values of W can be taken and the network sized and costed at each value to find the optimal setting for W. Figure 16.25b through d shows other utility paths which can be exploited for optimization.

In fact, the optimization of the network requires that U, V, W, X, Y, and Z in Figs. 16.24 and 16.25 be optimized simultaneously. Furthermore, stream splits may exist in the design, and variations of their branch flow rates can be superimposed on the use of loops and paths in the optimization. During this optimization, the design is no longer constrained to have temperature differences larger than $\Delta T_{\min}$ (although very small values in individual exchangers should be avoided for practical reasons). Also, pinches no longer divide the design into independent thermodynamic regions, and there is no longer any concern about cross-pinch heat transfer. The objective now is simply to minimize cost.

Thus loops, utility paths, and stream splits offer the degrees of freedom for manipulating the network cost. The problem is one of multivariable nonlinear optimization.[6] The constraints are only those of feasible heat transfer: positive temperature difference and non-negative heat duty for each exchanger. Furthermore, if stream splits exist, then positive branch flow rates are additional constraints.

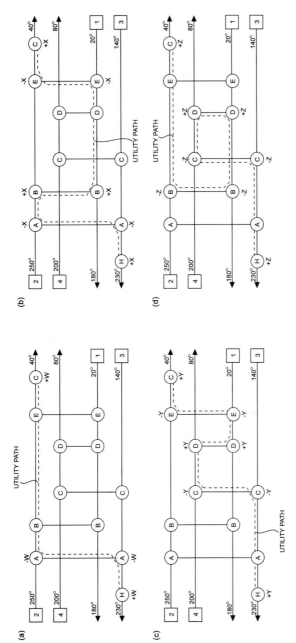

Figure 16.25 The utility paths which can be exploited for optimization of the design from Fig. 16.7.

If the network is optimized at fixed energy consumption, then only loops and stream splits are used. When energy consumption is allowed to vary, utility paths also must be included. As the network energy consumption increases, the overall capital cost decreases.

16.7   Heat Exchanger Network Design Based on the Optimization of a Reducible Structure

The approach to heat exchanger network design discussed so far is based on the creation of an irreducible structure. No redundant features were included. Of course, when the network is optimized, some of the features might be removed by the optimization. The scope for the optimization to remove features results from the assumptions made during the creation of the initial structure. However, no attempt was made to deliberately include redundant features.

An alternative approach is to create a reducible structure that deliberately includes redundant features and then subject this to optimization. Redundant features are then removed by the optimization.

Figure 16.26 shows a pair of composite curves divided into enthalpy intervals with a possible superstructure shown for one of the intervals.[3] The structure is created by splitting each hot stream

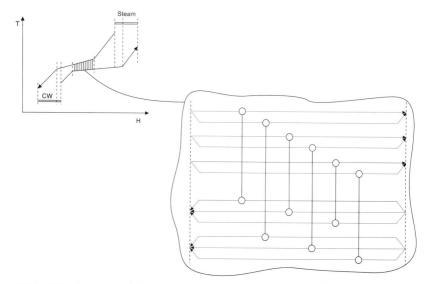

Figure 16.26 A superstructure can be developed from the enthalpy intervals of the composite curves for the design of the heat exchanger network.

into a number of branches equal to the number of cold streams in that interval and splitting each cold stream into a number of branches equal to the number of hot streams in that interval. In this way, a structure is created which allows each hot stream to be matched with each cold stream in that interval. By creating similar structures in each enthalpy interval and joining them together, a superstructure is created with many structural options. Not only are many options for alternative matches included but also the possibility of stream splitting, both at and away from the pinch. In some complex design problems, stream splitting is required away from the pinch. Whereas clear rules can be developed for stream splitting at the pinch, no such rules are available to determine necessary stream splits away from the pinch. Superstructures based on the one in Fig. 16.26 allow for the possibility of stream splitting away from the pinch. Of course, if it is not appropriate to stream split, then optimization of the superstructure would be expected to remove those features from the design.

While a superstructure based on the structure in Fig. 16.26 allows for many structural options, it is not comprehensive. Wood, Wilcox, and Grossmann[8] showed how direct contact heat transfer by mixing at unequal temperatures can be used to decrease the number of units in a heat exchanger network. Floudas, Ciric, and Grossman[9] showed how such features can be included in a heat exchanger network superstructure. Figure 16.27 shows the structure from Fig. 16.26 with possibilities for direct contact heat transfer included. In the

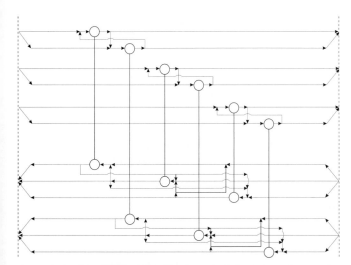

Figure 16.27 Possibilities for direct contact heat transfer can be added to the superstructure.

structure in Fig. 16.27, at the outlet of the units on the hot streams the branch is split into two parts, and one of the parts is directed to the inlet of the other unit on the same stream. At the outlet of the units on the cold streams the branch is split into three parts, and two of the parts are directed to the inlet of the other units on the same stream. It is possible that the final structure, after optimization of the superstructure in Fig. 16.27, will have fewer units than the final structure based on optimization of the superstructure in Fig. 16.26. In fact, reducing the number of units using direct contact heat transfer by features such as those included in Fig. 16.27 is rare. This is so because although it allows the number of units to be reduced compared with designs which do not feature direct contact heat transfer, those units which remain tend to suffer an excessive penalty in increased heat transfer area.[8]

Alternative superstructures to those in Figs. 16.26 and 16.27 can be developed.[10] On the one hand, it is desirable to include many structural options to ensure that all features which are candidates for an optimal solution have been included. On the other hand, including more and more structural features increases the computational load dramatically. Thus care should be taken not to include unnecessary features in the superstructure.

The major advantage of this approach to heat exchanger network design is that, in principle, it is capable of designing large networks with complex mixtures of constraints, mixed materials of construction, equipment types, etc. One significant disadvantage is that, because the optimization is carried out by a computer, the designer is removed from the decision making.

16.8   Heat Exchanger Network Design—Summary

A good initialization for heat exchanger network design is to assume that no individual exchanger should have a temperature difference smaller than $\Delta T_{\min}$. Having decided that no exchanger should have a temperature difference smaller than $\Delta T_{\min}$, two rules were deduced in Chap. 6. If the energy target set by the composite curves (or the problem table algorithm) is to be achieved, there must be no heat transfer across the pinch by

- process-to-process heat transfer
- inappropriate use of utilities

These rules are both necessary and sufficient for the design to achieve the energy target, given that no individual exchanger should have a temperature difference smaller than $\Delta T_{\min}$.

The design of heat exchanger networks can be summarized in five steps:

1. Divide the problem at the pinch into separate problems.

2. The design for the separate problems is started at the pinch, moving away.

3. Temperature feasibility requires constraints on the CP values to be satisfied for matches between the streams at the pinch.

4. The loads on individual units are determined using the tick-off heuristic to minimize the number of units. Occasionally, the heuristic causes problems.

5. Away from the pinch there is usually more freedom in the choice of matches. In this case, the designer can discriminate on the basis of judgment and process knowledge.

If the number of hot streams at the pinch above the pinch is greater than the number of cold streams, the cold streams must be split to satisfy the $\Delta T_{\min}$ constraint. If the number of cold streams at the pinch below the pinch is greater than the number of hot streams, the hot streams must be split to satisfy the $\Delta T_{\min}$ constraint. Not being able to satisfy the CP inequalities for all streams at the pinch also can necessitate stream splitting.

If the problem involves more than one pinch, then between pinches the design should be started from the most constrained pinch.

Remaining problem analysis can be used to make a quantitative assessment of matches in the context of the whole network without having to complete the network.

Once the initial network structure has been defined, then loops, utility paths, and stream splits offer the degrees of freedom for manipulating network cost in multivariable optimization. During the optimization, there is no constraint that temperature differences should be larger than $\Delta T_{\min}$ or that there should not be heat transfer across the pinch. The objective is simply to design for minimum total cost.

For more complex network designs, especially those involving many constraints, mixed equipment specifications, etc., design methods based on the optimization of a reducible structure can be used.

16.9 References

1. Linnhoff, B., and Hindmarsh, E., "The Pinch Design Method of Heat Exchanger Networks," *Chem. Eng. Sci.* 38: 745, 1983.
2. Linnhoff, B., Townsend, D. W., and Boland, D., et al., *A Users Guide on Process Integration for the Efficient Use of Energy,* IChemE, Rugby, U.K., 1982.

3. Linnhoff, B., and Ahmad, S., "Cost Optimum Heat Exchanger Networks: I. Minimum Energy and Capital Using Simple Models for Capital Cost," *Computers Chem. Eng.,* 14: 729, 1990.

4. Saboo, A. K., Morari, M., and Colberg, R. D., "RESHEX: An Interactive Software Package for the Synthesis and Analysis of Resilient Heat Exchanger Networks: II. Discussion of Area Targeting and Network Synthesis Algorithms," *Computers Chem. Eng.,* 10: 591, 1986.

5. Ahmad, S., Linnhoff, B., and Smith, R., "Cost Optimum Heat Exchanger Networks: II. Targets and Design for Detailed Capital Cost Models," *Computers Chem. Eng.,* 14: 751, 1990.

6. Gundersen, T., and Naess, L., "The Synthesis of Cost Optimal Heat Exchanger Networks: An Industrial Review of the State of the Art," *Computers Chem. Eng.,* 12: 503, 1988.

7. Yee, T. F., and Grossmann, I. E., "A Simultaneous Optimization Approach for the Synthesis of Heat Exchanger Networks," Paper 81d, Annual AIChE Meeting, Washington, 1988.

8. Wood, R. M., Wilcox, R. J., and Grossmann, I. E., "A Note on the Minimum Number of Units for Heat Exchanger Network Synthesis," *Chem. Eng. Commun.,* 39: 371, 1985.

9. Floudas, C. A., Ciric, A. R., and Grossmann, I. E., "Automatic Synthesis of Optimum Heat Exchanger Network Configurations," *AIChEJ,* 32: 276, 1986.

10. Grossmann, I. E., "Mixed-Integer Non-Linear Programming Techniques for the Synthesis of Engineering Systems," *Res. Eng. Design,* 1: 205, 1990.

17

Overall Strategy

17.1 Objectives

The purpose of chemical processes is not to make chemicals: The purpose is to make money. However, the profit must be made as part of a sustainable industrial activity which retains the capacity of ecosystems to support industrial activity and life. This means that process waste must be taken to its practical and economic minimum. Relying on methods of waste treatment is usually not adequate, since waste treatment processes tend not so much to solve the waste problem but simply to move it from one place to another. Sustainable industrial activity also means that energy consumption must be taken to its practical and economic minimum. Chemical processes also must not present significant short-term or long-term hazards, either to the operating personnel or to the community.

When developing a chemical process design, it helps if it is recognized that there is a hierarchy which is intrinsic to chemical processes.[1] Design starts at the reactor. The reactor design dictates the separation and recycle problem. The reactor design and separation problem together dictate the heating and cooling duties for the heat exchanger network. Those duties which cannot be satisfied by heat recovery dictate the need for external utilities. This hierarchy is represented by the layers in the onion diagram (see Fig. 1.6).

Following this hierarchy, all to often safety, health and environmental considerations are left to the final stages of design. This approach leaves much to be desired, since early decisions made purely for process reasons often can lead to problems of safety, health, and environment that require complex solutions. It is better to consider them as the design progresses. Designs which avoid the need for hazardous materials, or use less of them, or use them at lower temperatures and pressures, or dilute them with inert materials will be inherently safe and not require elaborate safety systems. Designs which minimize waste will not require elaborate treatment systems. These considerations need to be addressed as the design progresses, as each layer of the design is added.

17.2 The Hierarchy

1. *Choice of reactor.* The first and usually most important decisions to be made are those for the reactor type and its operating conditions. In choosing the reactor, the overriding consideration is usually raw materials efficiency (bearing in mind materials of construction, safety, etc.). Raw materials costs are usually the most important costs in the whole process. Also, any inefficiency in raw materials use is likely to create waste streams that become an environmental problem.

The design of the reactor usually interacts strongly with the rest of the flowsheet. Hence a return must be made to the reactor when the process design has progressed further.

2. *Choice of separator.* For a heterogeneous mixture, separation usually can be achieved by phase separation. Such phase separation normally should be carried out before any homogeneous separation. Phase separation tends to be easier and usually should be done first.

Distillation is by far the most commonly used method for the separation of homogeneous fluid mixtures. No attempt should be made to optimize pressure, reflux ratio, or feed condition of distillation in the early stages of design. The optimal values will almost certainly change later once heat integration with the overall process is considered.

The most common alternative to distillation for the separation of low-molecular-weight materials is absorption. Liquid flow rate, temperature, and pressure are important variables to be set, but no attempt should be made to carry out any optimization at this stage.

As with distillation and absorption, when evaporators and dryers are chosen, then no attempt should be made to carry out any optimization at this stage in the design.

3. *The synthesis of reaction-separation systems.* The recycling of material is an essential feature of most chemical processes. The use of excess reactants, diluents, or heat carriers in the reactor design has a significant effect on the flowsheet recycle structure. Sometimes

the recycling of unwanted byproduct to the reactor can inhibit its formation at the source.

Batch processes can be synthesized by first synthesizing a continuous process and then converting it to batch operation. A Gantt (time-event) diagram can be used to identify the scope for improved equipment utilization and the need for intermediate storage.

4. *Distillation sequencing.* Unless there are constraints severely restricting heat integration, sequencing of simple distillation columns can be carried out in two steps: (1) identify the best few nonintegrated sequences, and (2) study the heat integration. In most cases, there is no need to solve the problems simultaneously.[2]

The best few nonintegrated sequences can be identified most simply using the total vapor load as a criterion. If this is not satisfactory, then the alternative sequences can be sized and costed using shortcut techniques.

Complex column arrangements, such as the Petlyuk design,[3] offer large potential savings in energy compared with sequences of simple columns. The dividing-wall column also offers large potential savings in capital cost.[4] However, it is recommended that complex column arrangements should only be considered on a second pass through the design after first establishing a complete design with simple columns. Once this first complete design is established, then thermally coupled arrangements can be evaluated in the context of the overall design.[5]

5. *Heat exchanger network and utilities targets.* Having established a design for the two inner layers of the onion (reaction and separation and recycle), the material and energy balance is known. This allows the hot and cold streams for the heat recovery problem to be defined. Energy targets can then be calculated directly from the material and energy balance.[1] It is not necessary to design a heat exchanger network in order to establish the energy costs. Alternative utility scenarios and combined heat and power schemes can be screened quickly and conveniently using the grand composite curve.[1]

Targets also can be set for total heat exchange area, number of units, and number of shells for 1-2 shell-and-tube heat exchangers. These can be combined to establish a target for capital costs, taking into account mixed materials of construction, pressure rating, and equipment type. Furthermore, the targets for energy and capital cost can be optimized to produce an optimal setting for the capital/energy tradeoff before any network design is carried out.[6,7]

Once a design is known for the first two layers of the onion (i.e., reactors and separators only), the overall total cost of this design for all four layers of the onion (i.e., reactors, separators, heat exchanger network, and utilities) is simply the total cost of all reactors and separators (evaluated explicitly) plus the total cost target for heat exchanger network and utilities.[5]

Although it is not necessary to design the heat exchanger network in order for the design to progress, it is sometimes desirable to carry out a preliminary design to ensure that there are no significant features of the design which are unacceptable. If there are unacceptable features, the targets will have to be identified by inclusion of constraints, etc.

6. *Economic tradeoffs.* Interactions between the reactor and the rest of the process are extremely important. Reactor conversion is the most significant optimization variable because it tends to influence most operations through the process.[8,9] Also, when inerts are present in the recycle, the concentration of inerts is another important optimization variable, again influencing operations throughout the process.[8,9]

In carrying out these optimizations, targets should be used for the energy and capital cost of the heat exchanger network.[5] This is the only practical way to carry out these optimizations, since changes in reactor conversion and recycle inert concentration change the material and energy balance of the process, which in turn, changes the heat recovery problem. Each change in the material and energy balance, in principle, calls for a different heat exchanger network design. Furnishing a new heat exchanger network design for each setting of reactor conversion and recycle inert concentration is just not practical. On the other hand, targets for energy and capital cost of the heat exchanger network are by comparison easily generated.

7. *Effluent treatment.* It has been emphasized that waste minimization (along with safety and health considerations) should be considered as the design progresses. Inevitably, however, there will be some waste, and its treatment and disposal should be considered before the design is finalized. If treatment of the waste is particularly problematic, this might require fundamental design changes to reduce or change the nature of the waste.

8. *Process changes for improved heat integration.* Having minimized process waste, energy costs and utility waste can be reduced further by directing process changes to allow the energy targets to be reduced. The ultimate reference in guiding process changes is the plus/minus principle.[10]

The appropriate placement of the major items of equipment in relation to the heat recovery pinch[5,11,12] is as follows:

- Exothermic reactors should be above the pinch.

- Endothermic reactors should be below the pinch.

- Distillation columns, evaporators, and dryers should be above the pinch, below the pinch, but not across the pinch.

The grand composite curve can be used to quantitatively assess the appropriate placement of reactors and separators. If reactors and separators are not appropriately placed, then the plus/minus principle can be used to direct changes to bring about improvements through pressure change, etc. If a reactor is not appropriately placed, then it is more likely that the rest of the process would be changed to bring about appropriate placement rather than change the reactor design.

The sequence of distillation columns should be addressed again at this stage and the possibility of introducing complex configurations considered. Prefractionator arrangements (both with and without thermal coupling) can be used to replace direct or indirect distillation pairings. Alternatively, direct pairings can be replaced by side-rectifiers and indirect pairings replaced by side-strippers. The dividing-wall column can make significant reductions in both capital and operating costs when compared with a conventional distillation column pairing.

9. *Heat exchanger network design.* Having explored the major degrees of freedom, the material and energy balance is fixed, and the hot and cold streams which contribute to the heat exchanger network are firmly defined. The remaining task is to design the heat exchanger network.

The pinch design method is a step-by-step approach which allows the designer to interact as the design progresses.[13] For more complex network designs, especially those involving many constraints, mixed equipment specifications, etc., design methods based on the optimization of a reducible structure can be used.[14]

17.3 The Final Design

Although the sequence of the design follows the onion diagram in Fig. 1.6, the design rarely can be taken to a successful conclusion by a single pass. More often there is a flow in both directions. This follows from the fact that decisions are made for the inner layers on the basis of incomplete information. As more detail is added to the design in the outer layers with a more complete picture emerging, the decisions might need to be readdressed, moving back to the inner layers, and so on.

As the flowsheet becomes more firmly defined, the detailed process and mechanical design of the equipment can progress. The control scheme must be added and detailed hazard and operability studies carried out. All this is beyond the scope of the present text. However, all these considerations might require the flowsheet to be readdressed if problems are uncovered.

17.4 References

1. Linnhoff, B., Townsend, D. W., and Boland, D., et al., *A Users Guide on Process Integration for the Efficient Use of Energy,* IChemE, Rugby, U.K., 1982.
2. Stephanopoulos, G., Linnhoff, B., and Sophos, A., "Synthesis of Heat Integrated Distillation Sequences," *IChemE Symp. Ser.,* 74: 111, 1982.
3. Petlyuk, F. B., Plantonov, V. M., and Slavinskii, D. M., "Thermodynamically Optimal Method for Separating Multicomponent Mixtures," *Int. Chem. Eng.,* 5: 555, 1965.
4. Triantafyllou, C., and Smith, R., "The Design and Optimization of Fully Thermally Coupled Distillation Columns," *Trans. IChemE,* part A, 70: 118, 1992.
5. Smith, R., and Linnhoff, B., "The Design of Separators in the Context of Overall Processes," *Trans. IChemE ChERD,* 66: 195, 1988.
6. Linnhoff, B., and Ahmad, S., "Cost Optimum Heat Exchanger Networks: I. Minimum Energy and Capital Cost Using Simple Models for Capital Cost," *Computers Chem. Eng.,* 14: 729, 1990.
7. Ahmad, S., Linnhoff, B., and Smith, R., "Cost Optimum Heat Exchanger Networks: II. Targets and Design for Detailed Capital Cost Models," *Computers Chem. Eng.,* 14: 751, 1990.
8. Douglas, J. M., "A Hierarchical Decision Procedure for Process Synthesis," *AIChEJ,* 31: 353, 1985.
9. Douglas, J. M., *Conceptual Design of Chemical Processes,* McGraw-Hill, New York, 1988.
10. Linnhoff, B., and Vredeveld, D. R., "Pinch Technology Has Come of Age," *Chem. Eng. Prog.,* July, 80: 33, 1984.
11. Glavic, P., Kravanja, Z., and Homsak, M., "Heat Integration of Reactors: I. Criteria for the Placement of Reactors into Process Flowsheet," *Chem. Eng. Sci.,* 43(3): 593, 1988.
12. Linnhoff, B., Dunford, H., and Smith, R., "Heat Integration of Distillation Columns into Overall Processes," *Chem. Eng. Sci.,* 38: 1175, 1983.
13. Linnhoff, B., and Hindmarsh, E., "The Pinch Design Method for Heat Exchanger Networks," *Chem. Eng. Sci.,* 38: 745, 1983.
14. Grossmann, I. E., "Mixed-Integer Non-Linear Programming, Techniques for the Synthesis of Engineering Systems," *Res. Eng. Design,* 1: 205, 1990.

A

Preliminary Economic Evaluation

A.1 The Role of Process Economics

Process economics have three basic roles in preliminary process design:

1. *Evaluation of design options.* Costs are required to evaluate process design options; e.g., should unconverted raw material be recycled or disposed of?

2. *Preliminary process optimization.* Dominant process variables such as reactor conversion can have a major influence on the design. Preliminary optimization of these dominant variables is often required.

3. *Overall project profitability.* The economics of the overall project should be evaluated at different stages during the design to access whether the project is viable and whether major changes are needed.

A.2 Simple Economic Criteria

To evaluate design options and carry out preliminary process optimization, simple economic criteria are required. What happens to the revenue from product sales after the process has been commissioned? The sales revenue first pays for fixed costs which are independent of the rate of production. Variable costs, which do depend on the rate of production, also must be met. After this, taxes are deducted to leave the net profit.

Fixed costs include

- Capital cost repayments
- Routine maintenance
- Overheads (e.g., safety services, laboratories, personnel facilities, administrative services, etc.)
- Local taxes
- Labor
- Insurance, etc.

Variable costs include

- Raw materials
- Chemicals and catalysts consumed in manufacturing other than raw materials
- Utilities (fuel, steam, electricity, cooling water, process water, compressed air, inert gases, etc.)
- Maintenance costs incurred by operation
- Royalties
- Transport costs
- Quality control, etc.

There can be an element of maintenance costs that is fixed and an element which is variable. Fixed maintenance costs cover routine maintenance such as regular maintenance on safety valves which must be carried out irrespective of the rate of production. There also can be an element of maintenance costs which is variable. This arises from the fact that certain items of equipment can need more maintenance as the production rate increases. Also, royalties which cover the cost of purchasing another company's process technology may have different bases. Royalties may be a variable cost, since they can sometimes be paid in proportion to the rate of production. Alternatively, the royalty might be a single-sum payment at the beginning of the project. In this case, the single-sum payment will become part of the project's capital investment. As such, it will be included in the annual capital repayment, and this becomes part of the fixed cost.

Two simple economic criteria are useful in preliminary process design:

1. *Economic potential (EP):*

$$EP = \text{value of products} - \text{fixed costs}$$

$$- \text{variable costs} - \text{taxes} \qquad (A.1)$$

2. *Total annual cost (TAC):*

$$TAC = \text{fixed costs} + \text{variable costs} + \text{taxes} \qquad (A.2)$$

When synthesizing a flowsheet, these criteria are applied at various stages when there is an incomplete picture. Hence it is usually not possible to account for all the fixed and variable costs listed above. Also, there is little point in calculating taxes until a complete picture of operating costs and cash flows has been established.

The preceding definitions of economic potential and total annual cost can be simplified if it is accepted that they will be used to compare the relative merits of different structural options in the flowsheet and different settings of the operating parameters. Thus items which will be common to the options being compared can be neglected.

Let us now briefly review the most important costs which will be needed to compare options.

A.3 Operating Costs

A.3.1 Raw materials costs

In most processes, the largest individual cost is raw materials. Raw materials costs and product prices tend to have the largest influence on the economic performance of the process. The value of raw materials and products depends on whether the materials in question are being bought and sold under a contractual arrangement (either within or outside the company) or on the open market (the spot price). Open-market prices can fluctuate considerably with time. Products are normally sold at below open-market price when under a contractual arrangement.

The values of raw materials and products can be found in trade journals such as *Chemical Marketing Reporter* (Schnell Publishing Company) and *European Chemical News* (Reed Business Publishing

Group). This published information can be used to assess at what price a new product will sell or to assess the minimum allowable selling price for given raw materials costs.

A.3.2 Utility operating costs

Utility operating costs are usually the most significant variable operating cost after raw materials. Utility operating costs include

- Fuel
- Electricity
- Steam
- Cooling water
- Refrigeration
- Compressed air
- Inert gas

Utility costs vary enormously. This is especially true of fuel costs. Not only do costs vary considerably between different fuels (coal, oil, natural gas), but costs also tend to be sensitive to market fluctuations. Contractual relationships also have a significant effect on fuel costs. The price paid for fuel may depend very much on how much is purchased.

When electricity is bought from centralized power generation companies, the price tends to be more stable than fuel costs, since power generation companies tend to negotiate long-term contracts for fuel supply.

Steam costs vary with the price of fuel. If steam is only generated at low pressure and not used for power generation in steam turbines, then the cost can be estimated from local fuel costs assuming a boiler efficiency of around 75 percent (but can be significantly higher) and distribution losses of perhaps another 10 percent, giving an overall efficiency of around 65 percent.

If higher-pressure mains are used, then the cost of steam should be related in some way to its capacity to generate power in a steam turbine rather than simply to its additional heating value. The high-pressure steam is generated in the boiler and the low-pressure steam is generated by letting the pressure down through a steam turbine. One simple way to cost steam is to calculate the cost of the fuel required to generate the high-pressure steam (including any losses), and this fuel cost is then the cost of the high-pressure steam. Lower-pressure mains have a value equal to that of the high-pressure mains minus the value of power generated by letting the steam down to the lower pressure in a steam turbine. The following example illustrates the approach.

Example A.3.1 The pressures for three steam mains have been set to the conditions given in Table A.1. Medium- and low-pressure steam are generated by expanding high-pressure steam through a steam turbine with an isentropic efficiency of 80 percent. The cost of fuel is $4.00 GJ^{-1} and the cost of electricity is $0.07 kW^{-1}h^{-1}. Boiler feedwater is available at 100°C with a heat capacity of 4.2 kJ kg^{-1}°C^{-1}. Assuming a boiler efficiency of 75 percent and distribution losses of 10 percent, calculate the cost of steam for the three levels.

Solution
Cost of 41-barg steam. From steam tables, for 41-barg steam at 400°C,

$$\text{Enthalpy} = 3212 \text{ kJ kg}^{-1}$$

For boiler feedwater,

$$\text{Enthalpy} = 4.2(100 - 0)$$

$$= 420 \text{ kJ kg}^{-1}$$

To generate 41-barg steam at 400°C,

$$\text{Heat duty} = 3212 - 420$$

$$= 2792 \text{ kJ kg}^{-1}$$

For 41-barg steam,

$$\text{Cost} = 4.00 \times 10^{-6} \times 2792 \times \frac{1}{0.65}$$

$$= \$0.01718 \text{ kg}^{-1}$$

$$= \$17.18 \text{ t}^{-1}$$

TABLE A.1 Steam Mains Settings

Mains	Pressure (barg)
HP	41
MP	10
LP	3

Cost of 10-barg steam. Here, 41-barg steam is now expanded to 10 barg in a steam turbine. Details of steam turbine calculations were given in Sec. 6.7.
From steam tables, inlet conditions at 41 barg and 400°C are

$$h_1 = 3212 \text{ kJ kg}^{-1}$$

$$s_1 = 6.747 \text{ kJ kg}^{-1} \text{K}^{-1}$$

Turbine outlet conditions for isentropic expansion to 10 barg are

$$h_2 = 2873 \text{ kJ kg}^{-1}$$

$$s_2 = 6.747 \text{ kJ kg}^{-1} \text{K}^{-1}$$

For single-stage expansion with isentropic efficiency of 80 percent,

$$h_2' = h_1 - \eta_T(h_1 - h_2)$$

$$= 3212 - 0.8(3212 - 2873)$$

$$= 2941 \text{ kJ kg}^{-1}$$

From steam tables, the outlet temperature is 251°C, which is superheated by 67°C. Although steam for process heating is preferred at saturated conditions, it is not desirable in this case to desuperheat by boiler feedwater injection to bring to saturated conditions. If saturated steam is fed to the main, then the heat losses from the main will cause a large amount of condensation in the main, which is undesirable. Hence it is better to feed steam to the main with some superheat to avoid condensation in the main.

$$\text{Power generation} = 3212 - 2941$$

$$= 271 \text{ kJ kg}^{-1}$$

$$\text{Value of power generation} = 271 \times \frac{0.07}{3600}$$

$$= \$0.00527 \text{ kg}^{-1}$$

$$\text{Cost of 10-barg steam} = 0.01718 - 0.00527$$

$$= \$0.01191 \text{ kg}^{-1}$$

$$= \$11.91 \text{ t}^{-1}$$

Cost of 3-barg steam. Here, 10-barg steam from the exit of the first turbine is assumed to be expanded to 3 barg in another turbine.

From steam tables, inlet conditions of 10 barg and 251°C are

$$h_1 = 2941 \text{ kJ kg}^{-1}$$

$$s_1 = 6.880 \text{ kJ kg}^{-1} \text{K}^{-1}$$

Turbine outlet conditions for isentropic expansion to 3 barg are:

$$h_2 = 2732 \text{ kJ kg}^{-1}$$

$$s_2 = 6.880 \text{ kJ kg}^{-1} \text{K}^{-1}$$

For a single-stage expansion with isentropic efficiency of 80 percent,

$$h_2' = h_1 - \eta_T(h_1 - h_2)$$

$$= 2941 - 0.8(2941 - 2732)$$

$$= 2774 \text{ kJ kg}^{-1}$$

From steam tables, the outlet temperature is 160°C, which is superheated by 16°C. Again, it is desirable to have some superheat for the steam fed to the main to avoid condensation in the main.

$$\text{Power generation} = 2941 - 2774$$

$$= 167 \text{ kJ kg}^{-1}$$

$$\text{Value of power generation} = 167 \times \frac{0.07}{3600}$$

$$= \$0.00325 \text{ kg}^{-1}$$

$$\text{Cost of 3-barg steam} = 0.01191 - 0.00325$$

$$= \$0.00866 \text{ kg}^{-1}$$

$$= \$8.66 \text{ t}^{-1}$$

The problem with this approach is that if the steam generated in the boilers is at a very high pressure and/or the ratio of power to fuel costs is high, then the value of low-pressure steam can be extremely low or even negative. This is not sensible and discourages efficient use of low-pressure steam, since it leads to low-pressure steam with a value considerably less than its fuel value.

An alternative approach is to assume that the low-pressure steam is costed purely in terms of the heat input from fuel. The higher-pressure mains can then have a value over and above this in relation to their capacity to generate power. However, this approach has its drawbacks also. The high-pressure steam has a value far above the value of the fuel used for its generation.

These *marginal costing approaches* to costing steam are unsatisfactory. Within operating companies, few subjects generate more controversy than the value placed on different steam levels.

The only reliable way to cost steam is to create a simulation and costing model of the steam system. Figure A.1 shows a typical steam system. Raw water is usually treated by filtration to remove suspended solids and some form of ion exchange to remove dissolved salts. The water is then steam stripped in the deaerator to remove dissolved gases and treated chemically before being fed to the boilers. The chemical treatment usually involves oxygen scavengers (since oxygen causes corrosion in the boilers), phosphates to precipitate any

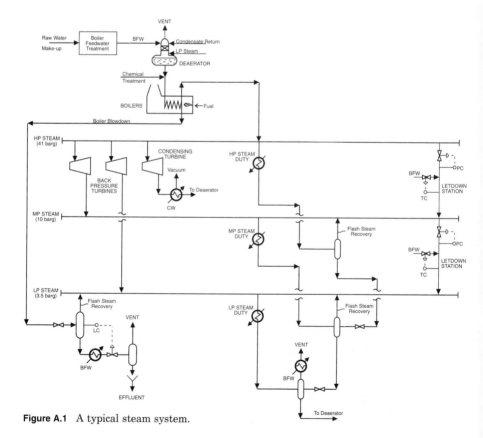

Figure A.1 A typical steam system.

calcium, magnesium, etc., and polymer dispersants to keep precipit-ates dispersed. The water fed to the boilers still contains dissolved solids not removed in the treatment, which accumulate in the boilers. These are removed by taking a purge, i.e., boiler blowdown.

In Fig. A.1, high-pressure steam is generated and fed to the high-pressure mains. The medium- and low-pressure mains are fed by expansion through steam turbines to generate power. Figure A.1 shows three mains with typical mains pressures, but these vary in both number and pressure from site to site. Figure A.1 also shows the possibility of using a condensing turbine, which is used when there is a desire to generate power but the exhaust steam from a back-pressure turbine is not needed. Letdown stations are used to control the mains' pressures. Because the letdown from high-pressure to lower-pressure mains creates steam with a large superheat, boiler feedwater is injected directly to reduce the superheat. As discussed in Example A.3.1, although steam for process heating is preferred saturated, if it is fed through the mains saturated, this leads to excessive condensation in the mains due to heat losses, which is undesirable. Hence steam is fed to the mains with some superheat. Another feature shown in Fig. A.1 is that when water (blowdown or condensate) is reduced in pressure, flash steam is recovered. Altho-ugh as much condensate as practicable and economical should be returned to the deaerator, levels of condensate return tend to be on the order of 50 percent but can be significantly higher.

Thus steam systems typified by the one in Fig. A.1 exhibit many complex interactions. It is extremely difficult to say exactly what the cost of any of the steam levels is. The only truly reliable way to determine the economic benefit of saving, say, low-pressure steam is to simulate the effect of that saving on the steam system and cost the change. Although this low-pressure steam saving will bring a fuel saving and water treatment saving (and hence cost saving), it will lead to lower power generation, and any shortfall in power will need to be generated by the condensing turbine (if there is one and it has spare capacity) or imported from centralized power generation (giving a cost penalty). Thus these various changes can be costed to evaluate the true economic benefit from the low-pressure saving.

In preliminary process design it might be necessary to use marginal costs for steam, but the designer should keep in mind the dangers of using such costs.

Cooling water costs tend to be low relative to the value of both fuel and electricity. The cost of cooling duty provided by cooling water is on the order of 1 percent that of the cost of power. For example, if power costs $0.07\,kW^{-1}\,h^{-1}$, then cooling water will typically cost $0.07 \times 0.01/3600 = 0.19 \times 10^{-6}\,kJ^{-1}$ or $0.19\,GJ^{-1}$.

The cost of power required to run a refrigeration system can be estimated approximately as a multiple of the power required for an ideal system. Thus, for an ideal system,

$$\frac{W_{\text{IDEAL}}}{Q_C} = \frac{T_H - T_C}{T_C} \qquad (A.3)$$

where W_{IDEAL} = ideal shaftwork required for the refrigeration cycle
$\quad Q_C$ = the cooling duty
$\quad T_C$ = temperature at which heat is taken into the refrigeration cycle (K)
$\quad T_H$ = temperature at which heat is rejected from the refrigeration cycle (K)

The ratio of ideal to actual shaftwork is often around 0.6. Thus,

$$W = \frac{Q_C}{0.6}\left(\frac{T_H - T_C}{T_C}\right) \qquad (A.4)$$

where W is the actual shaftwork required for the refrigeration cycle.

Example A.3.2. A process requires 0.5 MW of cooling at $-20°C$. A refrigeration cycle is required to remove this heat and reject it to cooling water supplied at $25°C$ and returned at $30°C$. Assuming $\Delta T_{\text{min}} = 5°C$ and both vaporization and condensation of the refrigerant occur isothermally, estimate the annual operating cost of refrigeration for an electrically driven system operating 8000 hours per year. The cost of electricity is $0.07 kW^{-1} h^{-1}$.

Solution

$$W = \frac{Q_C}{0.6}\left(\frac{T_H - T_C}{T_C}\right)$$

$$T_H = \quad 30 + 5 = \quad 35°C = 308\,\text{K}$$

$$T_C = -20 - 5 = -25°C = 248\,\text{K}$$

$$W = \frac{0.5}{0.6}\left(\frac{308 - 248}{248}\right)$$

$$= 0.202\,\text{MW}$$

$$\text{Cost of electricity} = 0.202 \times 10^3 \times 0.07 \times 8000$$

$$= \$113,120\,\text{yr}^{-1}$$

A.3.3   Labor costs

Labor costs are extremely difficult to estimate. They depend on whether the process is batch or continuous, the level of automation, the number of processing steps, and the level of production. When synthesizing a flowsheet, it is usually only necessary to screen process options which are of the same nature (e.g., continuous), have the same level of automation, have a similar number of processing steps, and the same level of production. In this case, labor costs will be common to all options and hence will not affect the comparison.

If, however, alternatives are to be compared which are very different in nature, such as a comparison between batch and continuous operation, some allowance for the difference in labor costs must be made.

A.3.4 Maintenance costs

Maintenance costs depend on whether processing fluids are solids on the one hand or gas and liquid on the other. Solids handling tends to increase maintenance costs. Highly corrosive process fluids increase maintenance costs. Average maintenance costs tend to be around 6 percent of the fixed capital investment.[1]

A.4 Capital Costs

The total investment required for a project can be broken down into four parts:

- Battery limits investment
- Off-site investment
- Engineering fees
- Working capital

A.4.1 Battery limits investment

The *battery limit* is a geographic boundary which defines the manufacturing area of the process. This includes process equipment and buildings or structures to house it but excludes boilerhouse facilities, pollution control, site infrastructure, etc.

The battery limits investment may be estimated by applying installation factors to the cost of individual items of equipment:[2]

$$I_F = f_I I_E \qquad\qquad (A.6)$$

where I_F = fixed capital investment in complete system
I_E = equipment cost
f_I = installation factor

The installation factor allows for

- Cost of installation
- Insulation
- Piping
- Foundations
- Structures
- Fireproofing
- Electrical
- Painting
- Engineering fees
- Contingency

Typical installation factors[3] are as follows:

Type of equipment	f_I
Distillation columns	4
Pressure vessels	4
Heat exchangers	3.5
Furnaces	2
Pumps	4
Compressors	2.5
Instruments	4
Miscellaneous equipment	2.5

Equipment costs may be obtained from equipment vendors or published cost data. Published cost data are usually presented as cost versus capacity charts or expressed as a power law of capacity:

$$I_E = I_B \left(\frac{Q}{Q_B} \right)^M \tag{A.7}$$

where I_E = required equipment cost of capacity Q
I_B = known base investment for equipment with capacity Q_B
M = constant depending on equipment type

A number of sources of such data are available in the open literature.[1,4–9] Unfortunately, the data to be used are often old, sometimes from a variety of sources, with different ages. Such data can be brought up to date and put on a common basis using cost indexes:

$$\frac{I_1}{I_2} = \frac{\text{INDEX}_1}{\text{INDEX}_2} \tag{A.8}$$

where I_1 = investment in year 1
  I_2 = investment in year 2
INDEX$_1$ = cost index in year 1
INDEX$_2$ = cost index in year 2

The most commonly used index is the chemical engineering plant cost index published in *Chemical Engineering Magazine.*

Factors should be included to allow for variations in design pressure and material of construction:

$$I_F = f_I I_E + f_P f_M I_E \qquad (A.9)$$

where f_P = design pressure cost factor
  f_M = material-of-construction factor

The factors f_P and f_M have not been applied to installation costs because installation costs are not a simple function of purchase cost.[6] Although process piping and fittings made for the same unusual conditions are proportionally more expensive, labor, foundations, insulation, etc. are not. Furthermore, only about 70 percent of piping is directly exposed to process fluid. The balance is auxiliary or utility piping made of conventional materials.

It should be emphasized that capital cost estimates using installation factors are at best crude and at worst highly misleading. When preparing such an estimate, the designer spends most of the time on the equipment costs, which represent typically 20 to 40 percent of the total installed cost. The bulk costs (civil engineering, labor, etc.) are factored costs which lack definition. At best, this type of estimate can be expected to be accurate to ±30 percent.

A.4.2 Off-site investment

In addition to battery limits investment, off-site investment is required. This includes all structures, equipment, and services that do not enter directly into the chemical process. Within this broad category there are two major classifications, namely, utilities and service facilities.

1. *Utility plant:*

- Electricity generation
- Electricity distribution
- Steam generation
- Steam distribution

- Process water
- Cooling water
- Water treatment
- Refrigeration
- Compressed air
- Inert gas

Costs of utilities are considered from their sources within the site to the limits of the chemical process served.

2. *Service facilities:*

- Auxiliary buildings such as offices, medical, personnel, locker rooms, guardhouses, warehouses, and maintenance shops
- Roads and paths
- Railroads
- Fire protection systems
- Communication systems
- Waste disposal systems
- Storage facilities for end product, water, and fuel not directly connected with the process
- Plant service vehicles, loading and weighing devices

The cost of off-sites ranges typically from 20 to 40 percent of the total installed cost of the plant.[10] In general terms, the larger the plant, the larger will be the fraction of the total project cost which goes to off-sites. In other words, a small project will require typically 20 percent of the total installed cost as off-sites. For a large project, the figure will be typically up to 40 percent.

A.4.3 Working capital

Working capital is what must be invested to get the process into productive operation. This is money invested before there is a product to sell and includes

- Raw materials for plant start-up
- Raw materials and product inventories
- Cost of handling and transportation of materials
- Money to carry accounts receivable (i.e., credit extended to customers) less accounts payable (i.e., credit extended by suppliers)

- Money to meet payroll when starting up

Theoretically, in contrast with fixed investment, this money is not lost but can be recovered when the process is closed down.
   For an estimate, take either[11]

1. 30 percent of annual sales, or
2. 15 percent of fixed capital investment.

A.4.4   Annualized capital costs

Capital for new installations may be obtained from

1. Loans from banks
2. Issue of company stock and bonds (or debentures)
3. Net cash flow arising from profit.

   The cost of the capital depends on its source. The source of the capital often will not be known during the early stages of a project, and yet there is a need to select between process options and carry out preliminary optimization on the basis of both capital and operating costs. This is difficult to do unless both capital and operating costs can be expressed on a common basis. Capital costs can be expressed on an annual basis if it is assumed that the capital has been borrowed over a fixed period (usually 5 to 10 years) at a fixed rate of interest, in which case the capital costs can be annualized according to

$$\text{Annualized capital cost} = \text{capital cost} \times \frac{i(1+i)^n}{(1+i)^n - 1} \qquad \text{(A.10)}$$

where i = fractional interest rate per year
   n = number of years

   Derivation of Eq. (A.10) is as follows:[11] Let

$$P = \text{present worth of estimated capital cost}$$

$$F = \text{future worth of estimated capital cost}$$

After the first year, the future worth F of the capital cost present value P is given by

$$F(1) = P + Pi = P(1+i) \qquad \text{(A.11)}$$

After the second year, the worth is

$$F(2) = P(1 + i) + P(1 + i)i$$
$$= P(1 + i)^2 \qquad\qquad\text{(A.12)}$$

After the third year, the worth is

$$F(3) = P(1 + i)^2 + P(1 + i)^2 i$$
$$= P(1 + i)^3 \qquad\qquad\text{(A.13)}$$

After the nth year, the worth is

$$F(n) = P(1 + i)^n$$

which is normally written as

$$F = P(1 + i)^n \qquad\qquad\text{(A.14)}$$

Take the capital cost and spread it as a series of equal annual payments A made at the end of each year over n years. The first payment gains interest over $(n - 1)$ years, and its future value after $(n - 1)$ years is

$$F = A(1 + i)^{n-1} \qquad\qquad\text{(A.15)}$$

The future worth of the second annual payment after $(n - 2)$ years is

$$F = A(1 + i)^{n-2} \qquad\qquad\text{(A.16)}$$

The combined worth of all the annual payments is

$$F = A[(1 + i)^{n-1} + (1 + i)^{n-2} + (1 - i)^{n-3} + \cdots + (1 + i)^{n-n}] \quad\text{(A.17)}$$

Multiplying both sides of this equation by $(1 + i)$ gives

$$F(1 + i) = A[(1 + i)^n + (1 + i)^{n-1} + (1 + i)^{n-2} + \cdots + (1 + i)] \quad\text{(A.18)}$$

Subtracting the Eqs. (A.18) and (A.17) gives

$$F(1 + i) - F = A[(1 + i)^n - 1] \qquad\qquad\text{(A.19)}$$

which on rearranging gives

$$F = \frac{A[(1 + i)^n - 1]}{i} \qquad\qquad\text{(A.20)}$$

Combining Eq. (A.20) with Eq. (A.14) gives

$$A = \frac{P[i(1+i)^n]}{(1+i)^n - 1} \tag{A.21}$$

Thus Eq. (A.10) is obtained.

As stated previously, the source of capital is often not known, and hence it is not known whether or not Eq. (A.10) is appropriate to represent the cost of capital. Equation (A.10) is, strictly speaking, only appropriate if the money for capital expenditure is to be borrowed over a fixed period at a fixed rate of interest. Moreover, if Eq. (A.10) is accepted, then the number of years over which the capital is to be annualized is unknown, as is the rate of interest. However, the most important thing is that even if the source of capital is not known, etc., and uncertain assumptions are necessary, Eq. (A.10) provides a common basis for the comparison of competing projects.

Example A.4.1 The purchased cost of a distillation column is $1 million, and the reboiler and condenser are $100,000. Calculate the annual cost of installed capital if the capital is to be annualized over a 5-year period at a fixed rate of interest of 5 percent.

Solution First, calculate the installed capital cost from Eq. (A.6):

$$I_F = f_I I_E$$

$$= 4 \times 1,000,000 + 3.5 \times 100,000$$

$$= \$4,350,000$$

$$\text{Annualization factor} = \frac{i(1+i)^n}{(1+i)^n - 1}$$

$$= \frac{0.05(1+0.05)^5}{(1+0.05)^5 - 1}$$

$$= 0.2310$$

$$\text{Annualized capital cost} = 4,350,000 \times 0.2310$$

$$= \$1,004,850 \text{ yr}^{-1}$$

When using annualized capital cost to carry out optimization, the designer should not lose sight of the uncertainties involved in the capital annualization. In particular, changing the annualization period can lead to very different results when, for example, carrying out a capital/energy tradeoff. When carrying out optimization, the sensitivity of the result to changes in the assumptions should be tested.

A.4.5 Project cash flow and economic evaluation

As the design progresses, more information is accumulated. The best methods of assessing the profitability of alternatives are based on projections of the cash flows during the project life.[12]

Figure A.2 shows the cash-flow pattern for a typical project. The cash flow is a cumulative cash flow. Consider curve 1 in Fig. A.2. From the start of the project at A, cash is spent without any

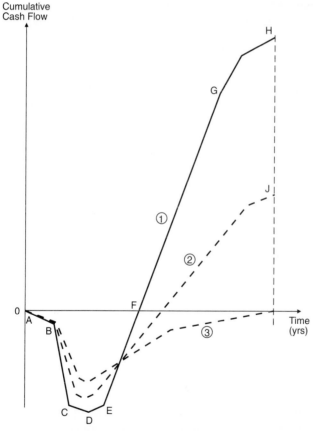

Figure A.2 Cash-flow pattern for a typical project. *(From Allen, IChemE, 1980; reproduced by permission of the Institution of Chemical Engineers.)*

immediate return. The early stages of the project consist of development, design, and other preliminary work, which causes the cumulative curve to dip to B. This is followed by the main phase of capital investment in buildings, plant, and equipment, and the curve drops more steeply to C. Working capital is spent to commission the process between C and D. Production starts at D, where revenue from sales begins. Initially, the rate of production is likely to be below design conditions until full production is achieved at E. At F, the cumulative cash flow is again zero. This is the *project break-even point*. Toward the end of the project life, at G, the net rate of cash flow may decrease due, for example, to increasing maintenance costs, a fall in the market price for the product, etc.

Ultimately, the process might be permanently shut down or given a major revamp. This marks the end of the project, H. If the process is shut down, working capital is recovered, and there may be salvage value, which would create a final cash inflow at the end of the project.

The predicted cumulative cash-flow curve for a project throughout its life forms the basis for more detailed evaluation. Many quantitative measures or indices have been proposed. In each case, important features of the cumulative cash-flow curve are identified and transformed into a single numerical measure as an index.

1. *Payback time.* Payback time is the time that elapses from the start of the project, A, to the break-even point, F. The shorter the payback time, the more attractive is the project.

2. *Net present value (NPV).* Since money can be invested to earn interest, money received now has a greater present value than money received at some time in the future. The net present value of a project is the sum of the present values of each individual cash flow. In this case, the *present* is taken to be the start of a project.

Time is taken into account by discounting the annual cash flow A_{CF} with the rate of interest to obtain the annual discounted cash flow A_{DCF}. Thus, at the end of year 1,

$$A_{DCF1} = \frac{A_{CF1}}{(1+i)}$$

At the end of year 2,

$$A_{DCF2} = \frac{A_{CF2}}{(1+i)^2}$$

And at the end of year n,

$$A_{DCF_n} = \frac{A_{CF_n}}{(1+i)^n} \tag{A.22}$$

The sum of the annual discounted cash flows over n years $\sum A_{\text{DCF}}$ is known as the *net present value* (NPV) of the project:

$$\text{NPV} = \sum A_{\text{DCF}} \qquad (A.23)$$

The value of NPV is, of course, directly dependent on the choice of the fractional interest rate i.

Returning to the cumulative cash-flow curve for a project, the effect of discounting is also shown in Fig. A.2. Curve 1 is the original curve with no discounting, i.e., $i = 0$, and the project NPV is equal to the final net cash position given by H. Curve 2 shows the effect of discounting at a fixed rate of interest, and the corresponding project NPV is given by J. Curve 3 in Fig. A.2 shows a larger rate of interest but it is chosen such that the NPV is zero at the end of the project.

The greater the positive NPV for a project, the more economically attractive it is. A project with a negative NPV is not a profitable proposition.

3. *Discounted cash-flow rate of return.* Discounted cash-flow rate of return is defined as the discount rate i which makes the NPV of a project zero (curve 3 in Fig. A.2):

$$\text{NPV} = \sum A_{\text{DCF}} = 0 \qquad (A.24)$$

The value of i given by this equation is known as the *discounted cash-flow rate of return* (DCFRR). It may be found graphically or by trial and error.

The higher the value of the DCFRR for a project, the more attractive it is. The minimum acceptable value of the DCFRR is the market interest rate. If the DCFRR is lower than market interest rate, it would be better to put money in the bank. For a DCFRR value greater than this, the project will show a profit; for a lesser value, it will show a loss.

4. *Comparison between NPV and DCFRR*
a. *Net present value.* Net present value measures profit but does not indicate how efficiently capital is being used.
b. *Discounted cash-flow rate of return.* DCFRR measures how efficiently capital is being used but gives no indication of how large the profits will be.
If the goal is to maximize profit, NPV is used. If the supply of capital is restricted (which is usual), DCFRR is used to decide which projects will use the capital most efficiently.

Example A.4.2 A company has the alternative of investing in one of two projects, A or B. The capital cost of both projects if $10 million. The predicted annual cash flows for both projects are shown in Table A.2. Capital is restricted and a choice is to be made based on discounted cash flow rate of return.

Project A
Start with an initial guess for DCFRR of 20 percent and increase:

Year	A_{CF}	DCF 20%		DCF 30%		DCF 25%	
		A_{DCF}	ΣA_{DCF}	A_{DCF}	ΣA_{DCF}	A_{DCF}	ΣA_{DCF}
0	−10	−10	−10	−10	−10	−10	−10
1	1.6	1.33	−8.67	1.23	−8.77	1.28	−8.72
2	2.8	1.94	−6.73	1.66	−7.11	1.79	−6.93
3	4.0	2.31	−4.42	1.82	−5.29	2.05	−4.88
4	5.2	2.51	−1.91	1.82	−3.47	2.13	−2.75
5	6.4	2.57	0.66	1.72	−1.75	2.10	−0.65

Twenty percent is too low, since ΣA_{DCF} is positive at the end of year 5. Thirty percent is too large, since ΣA_{DCF} is negative at the end of year 5, and is the case with 25 percent. The answer must be between 20 and 25 percent. Interpolating on the basis of ΣA_{DCF}, the DCFRR ≈ 23 percent.

Project B
Again, start with an initial guess for DCFRR of 20 percent and increase:

Year	A_{CF}	DCF 20%		DCF 40%		DCF 35%	
		A_{DCF}	ΣA_{DCF}	A_{DCF}	ΣA_{DCF}	A_{DCF}	ΣA_{DCF}
0	−10	−10	−10	−10	−10	−10	−10
1	6.5	5.42	−4.58	4.64	−5.36	4.81	−5.19
2	5.2	3.61	−0.97	2.65	−2.71	2.85	−2.34
3	4.0	2.31	1.34	1.46	−1.25	1.63	−0.71
4	2.8	1.35	2.69	0.729	−0.521	0.843	0.133
5	1.6	0.643	3.33	0.297	−0.224	0.357	0.490

From ΣA_{DCF} at the end of year 5, 20 percent is too low, 40 percent too high,

TABLE A.2 Predicted Annual Cash Flows

Year	Cash flows ($1,000,000)	
	Project A	Project B
0	−10	−10
1	1.6	6.5
2	2.8	5.2
3	4.0	4.0
4	5.2	2.8
5	6.4	1.6

and 35 percent also too low. Interpolating on the basis of ΣA_{DCF}, the DCFRR $\simeq 38$ percent. Project B is therefore chosen.

Predicting future cash flows for a project is extremely difficult with many uncertainties, including the project life. However, providing that consistent assumptions are made, projections of cash flows can be used to choose between competing projects.

A.5   References

1. Peters, M. S., and Timmerhaus, K. B., *Plant Design and Economics for Chemical Engineers,* McGraw-Hill, New York, 1980.
2. Lang, H. J., "Cost Relationships in Preliminary Cost Estimation," *Chem. Engg.,* 54: 117, 1947.
3. Hand, W. E., "From Flowsheet to Cost Estimate," *Petrol. Refiner,* 37: 331, 1958.
4. Guthrie, K. M., "Data and Techniques for Preliminary Capital Cost Estimating," *Chem. Engg.,* 76: 114, 1969.
5. Hall, R. S., Matley, J., and McNaughton, K. J., "Current Costs of Process Equipment," *Chem. Engg.,* 89: 80, 1982.
6. Ulrich, G. D., *A Guide to Chemical Engineering Process Design and Economics,* Wiley, New York, 1984.
7. Hall, R. S., Vatavuk, W. M., and Matley, J., "Estimating Process Equipment Costs," *Chem. Engg.,* 95: 66, 1988.
8. *A Guide to Capital Cost Estimating,* 3d ed., IChemE, Rugby, U.K., 1988.
9. Remer, D. S., and Chai, L. H., "Design Cost Factors for Scaling-up Engineering Equipment," *Chem. Engg. Progr.,* 86 Aug.: 77, 1990.
10. Bauman, H. C., "Estimating Cost of Process Auxiliaries," *Chem. Engg. Progr.,* 51: 45, 1955.
11. Holland, F. A., Watson, F. A., and Wilkinson, J. K., *Introduction to Process Economics,* 2d ed., Wiley, New York, 1983.
12. Allen, D. H., *A Guide to the Economic Evaluation of Projects,* IChemE, Rugby, U.K., 1980.

Algorithm
for the Heat Exchange
Area Target

Figure B.1 shows a pair of composite curves divided into vertical enthalpy intervals. Also shown in Fig. B.1 is a heat exchanger network for one of the enthalpy intervals which will satisfy all the heating and cooling requirements. The network shown in Fig. B.1 for the enthalpy interval is in grid diagram form. The network arrangement in Fig. B.1 has been placed such that each match experiences the ΔT_{LM} of the interval. The network also uses the minimum number of matches $(S - 1)$. Such a network can be developed for any interval, providing each match within the interval (1) satisfies completely the enthalpy change of a stream in the interval and (2) achieves the same ratio of CP values as exists between the composite curves (by stream splitting if necessary).

As each match is successively placed in the interval, the minimum number of matches can be achieved because there is one fewer stream to match *and* the CP ratio of the remaining streams (i.e., ratio of ΣCP_H and ΣCP_C of the remaining streams) in the interval still satifies the CP ratio between the composite curves.

It is thus always possible to achieve the interval design with $(S - 1)$ matches and each match operating with the log mean temperature difference of the interval.[1]

Now consider the heat transfer area required by enthalpy interval k, in which the overall heat transfer coefficient is allowed to vary between individual matches.[2]

$$A_k = \frac{1}{\Delta T_{\text{LM}k}} \sum_{ij} \frac{Q_{ij}}{U_{ij}} \qquad (\text{B.1})$$

where A_k = network area based on vertical heat exchange in enthalpy interval k

$\Delta T_{\text{LM}k}$ = log mean temperature difference for enthalpy interval k

Q_{ij} = heat load on match between hot stream i and cold stream j

U_{ij} = overall heat transfer coefficient between hot stream i and cold stream j

Introducing individual film transfer coefficients,

$$A_k = \frac{1}{\Delta T_{\text{LM}k}} \sum_{ij} Q_{ij}\left(\frac{1}{h_i} + \frac{1}{h_j}\right) \qquad (\text{B.2})$$

where h_i and h_j are film transfer coefficients for hot stream i and cold stream j (including wall and fouling resistances).

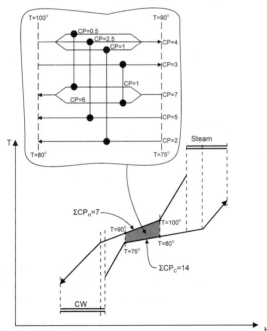

Figure B.1 Within each enthalpy interval it is possible to design a network in $(S - 1)$ matches. *(From Ahmad and Smith, IChemE, ChERD, 67: 481, 1989; reproduced by permission of the Institution of Chemical Engineers.)*

From Eq. (B.2),

$$A_k = \frac{1}{\Delta T_{\mathrm{LM}k}} \left(\sum_{ij} \frac{Q_{ij}}{h_i} + \sum_{ij} \frac{Q_{ij}}{h_j} \right) \tag{B.3}$$

Since enthalpy interval k is in heat balance, then summing over all cold stream matches with hot stream i gives the stream duty on hot stream i:

$$\sum_{j}^{J} Q_{ij} = q_i \tag{B.4}$$

where q_i = stream duty on hot stream i in enthalpy interval k
$\quad J$ = total number of cold streams in enthalpy interval k

Similarly, summing over all hot stream matches with cold stream j gives the stream duty on cold stream j:

$$\sum_{i}^{I} Q_{ij} = q_j \tag{B.5}$$

where q_j = stream duty on cold stream j in enthalpy interval k
$\quad I$ = total number of hot streams in enthalpy interval k

Thus, from Eq. (B.4),

$$\sum_{ij} \frac{Q_{ij}}{h_i} = \sum_{i}^{I} \frac{q_i}{h_i} \tag{B.6}$$

and from Eq. (B.5),

$$\sum_{ij} \frac{Q_{ij}}{h_j} = \sum_{j}^{J} \frac{q_j}{h_j} \tag{B.7}$$

Substituting these expressions in Eq. (B.3) gives

$$A_k = \frac{1}{\Delta T_{\mathrm{LM}k}} \left(\sum_{i}^{I} \frac{q_i}{h_i} + \sum_{j}^{J} \frac{q_j}{h_j} \right) \tag{B.8}$$

Extending this equation to all enthalpy intervals in the composite curves gives[2,3]

$$A_{\mathrm{NETWORK}} = \sum_{k}^{\mathrm{INTERVALS}\,K} \frac{1}{\Delta T_{\mathrm{LM}k}} \left(\sum_{i}^{\mathrm{HOT\,STREAMS}\,I} \frac{q_i}{h_i} + \sum_{j}^{\mathrm{COLD\,STREAMS}\,J} \frac{q_j}{h_j} \right) \tag{B.9}$$

References

1. Ahmad, S., and Smith, R., "Targets and Design for Minimum Number of Shells in Heat Exchanger Networks," *IChemE, ChERD,* 67: 481, 1989.
2. Linnhoff, B., and Ahmad, S., "Cost Optimum Heat Exchanger Networks: I. Minimum Energy and Capital Using Simple Models for Capital Cost," *Computers Chem. Eng.,* 14: 729, 1990.
3. Townsend, D. W., and Linnhoff, B., "Surface Area Targets for Heat Exchanger Networks," IChemE Annual Research Meeting, Bath. U.K., 1984.

C

Maximum Thermal Effectiveness for 1-2 Shell-and-Tube Heat Exchangers

The derivation of Eq. (7.12) is as follows.[1] From Bowman et al.[2] when $R \neq 1$,

$$F_T = \frac{\sqrt{R^2 + 1} \ln \left[\dfrac{(1 - P)}{(1 - RP)} \right]}{(R - 1) \ln \left\{ \dfrac{[2 - P(R + 1 - \sqrt{R^2 + 1})]}{[2 - P(R + 1 + \sqrt{R^2 + 1})]} \right\}} \tag{C.1}$$

When $R = 1$,

$$F_T = \frac{\left[\dfrac{\sqrt{2} P}{(1 - P)} \right]}{\ln \left\{ \dfrac{[2 - P(2 - \sqrt{2})]}{[2 - P(2 + \sqrt{2})]} \right\}} \tag{C.2}$$

The maximum value of P, for any R, occurs as F_T tends to $-\infty$. Study of the F_T functions above reveals that for F_T to be determinate,

1. $P < 1$
2. $RP < 1$
3. $\dfrac{2 - P(R + 1 - \sqrt{R^2 + 1})}{2 - P(R + 1 + \sqrt{R^2 + 1})} > 0$

Condition (3) applies to Eq. (C.2) when $R = 1$. Both conditions (1) and (2) are always true for a feasible heat exchange with positive temperature differences.

Taking condition (3), then

(a) $P < \dfrac{2}{R + 1 - \sqrt{R^2 + 1}}$ and $P < \dfrac{2}{R + 1 + \sqrt{R^2 + 1}}$ (C.3)

or

(b) $P > \dfrac{2}{R + 1 - \sqrt{R^2 + 1}}$ and $P > \dfrac{2}{R + 1 + \sqrt{R^2 + 1}}$ (C.4)

but not both. Consider condition (b) in more detail. For positive values of R, $R + 1 - \sqrt{R^2 + 1}$ is a continuously increasing function of R, and

- As R tends to 0, $R + 1 - \sqrt{(R^2 + 1)}$ tends to 0.
- As R tends to ∞, $R + 1 - \sqrt{(R^2 + 1)}$ tends to 1.

For condition b to apply, for positive values of R, $P > 2$. However, $P < 1$ for feasible heat exchange. Thus, condition (b) does not apply.

Consider now condition (a). Because

$$R + 1 + \sqrt{R^2 + 1} > R + 1 - \sqrt{R^2 + 1} \qquad (C.5)$$

then both inequalities for condition (a) are satisfied when

$$P < \frac{2}{R + 1 + \sqrt{R^2 + 1}} \qquad (C.6)$$

Thus the maximum value of P for any value of R, P_{max}, is given by

$$P_{max} = \frac{2}{R + 1 + \sqrt{R^2 + 1}} \qquad (C.7)$$

References

1. Ahmad, S., "Heat Exchanger Networks: Cost Trade-offs in Energy and Capital," Ph.D. thesis, UMIST, U.K., 1985.
2. Bowman, R. A., Mueller, A. C., and Nagle, W. M., "Mean Temperature Differences in Design," *Trans. ASME*, 62: 283, 1940.

Expression for the Minimum Number of 1-2 Shell-and-Tube Heat Exchangers for a Given Unit

The derivation of Eqs. (7.14) to (7.16) is as follows.[1] From Bowman,[2] the value of P over N number of 1-2 shells in series P_{N-2N} can be related to the value of P for each 1-2 shell P_{1-2} according to

$R \neq 1$:

$$P_{N-2N} = \frac{1 - \left(\dfrac{1 - P_{1-2}R}{1 - P_{1-2}}\right)^N}{R - \left(\dfrac{1 - P_{1-2}R}{1 - P_{1-2}}\right)^N} \tag{D.1}$$

$R = 1$:

$$P_{N-2N} = \frac{P_{1-2}N}{P_{1-2}N - P_{1-2} + 1} \tag{D.2}$$

Now, the maximum possible value of P_{1-2} in a 1-2 shell is (see App. C)

$$P_{\mathrm{max}1-2} = \frac{2}{R + 1 + \sqrt{R^2 + 1}} \tag{D.3}$$

The value of P_{1-2} required in each 1-2 shell to satisfy a chosen value of X_P is defined by

$$P_{1-2} = X_P P_{1-2\mathrm{max}} \tag{D.4}$$

This therefore requires that

$R \neq 1$:

$$P_{N-2N} = \frac{1 - \left(\dfrac{1 - \dfrac{2X_P R}{R + 1 + \sqrt{R^2 + 1}}}{1 - \dfrac{2X_P}{R + 1 + \sqrt{R^2 + 1}}} \right)^N}{R - \left(\dfrac{1 - \dfrac{2X_P R}{R + 1 + \sqrt{R^2 + 1}}}{1 - \dfrac{2X_P}{R + 1 + \sqrt{R^2 + 1}}} \right)^N} \tag{D.5}$$

$R = 1$:

$$P_{N-2N} = \frac{\dfrac{2X_P N}{2 + \sqrt{2}}}{\dfrac{2X_P N}{2 + \sqrt{2}} - \dfrac{2X_P}{2 + \sqrt{2}} + 1} \tag{D.6}$$

These expressions define P_{N-2N} for N number of 1-2 shells in series in terms of R and X_P in each shell. The expressions can be used to define the number of 1-2 shells in series required to satisfy a specified value of X_P in each shell for a given R and P_{N-2N}. Hence the relationship can be inverted to find the value of N which satisfies X_P exactly in each 1-2 shell in the series:

$R \neq 1$:

$$N = \frac{\ln \left(\dfrac{1 - RP_{N-2N}}{1 - P_{N-2N}} \right)}{\ln W} \tag{D.7}$$

where
$$W = \frac{R + 1 + \sqrt{R^2 + 1} - 2X_P R}{R + 1 + \sqrt{R^2 + 1} - 2X_P} \tag{D.8}$$

$R = 1$:

$$N = \left(\frac{P_{N-2N}}{1 - P_{N-2N}} \right) \left(\frac{1 + \dfrac{\sqrt{2}}{2} - X_P}{X_P} \right) \tag{D.9}$$

Choosing the number of 1-2 shells in series to be the next largest integer above N ensures a practical exchanger design satisfying X_P.

References

1. Ahmad, S., "Heat Exchanger Networks: Cost Trade-offs in Energy and Capital," Ph.D. thesis, UMIST, U.K., 1985.
2. Bowman, R. A., "Mean Temperature Difference Correction in Multipass Exchangers," *Ind. Eng. Chem.,* 28: 541, 1936.

Algorithm for the Number-of-Shells Target

One particularly important property of the relationships for multipass exchangers is illustrated by the two streams shown in Fig. E.1. The problem overall is predicted to require 3.889 shells (4 shells in practice). If the problem is divided arbitrarily into two parts S and T as shown in Fig. E1, then part S requires 2.899 and Part T requires 0.990, giving a total of precisely 3.889. It does not matter how many vertical sections the problem is divided into or how big the sections are, the same identical result is obtained, provided fractional (noninteger) numbers of shells are used. When the problem is divided into four arbitrary parts A, B, C, and D (Fig. E.1), adding up the individual shell requirements gives precisely 3.889 again.

This additive property takes on fundamental practical significance when the problem of setting a target for the number of shells in a network from the composite curves is considered.[1]

To establish the shells target, the composite curves are first divided into vertical enthalpy intervals as done for the area target algorithm. It was shown in App. B that it is always possible to design a network for an enthalpy interval with $(S_k - 1)$ matches, with each match having the same temperature profile as the enthalpy interval.

If such a design is established within an interval, then the number of shells for each match in interval k will be the same. This is so because the number of shells in Eqs. (7.14) to (7.16) depends only on the temperatures of the streams being matched, and since each match within an interval operates with the same temperatures, each match will require the same number of shells.

It is important to work in fractional numbers of shells (rather than integer) in order to use the additive property for 1-2 shells from one interval to another. If each match in enthalpy interval k requires N_k shells using the temperatures of interval k in Eqs. (7.14) to (7.16), then the minimum shells count for the interval is

$$N_k(S_k - 1) \tag{E.1}$$

since the temperatures defining N_k are achieved by a minimum of $(S_k - 1)$ matches.

The real (noninteger) number of shells target is then simply the sum of the real number of shells from all the enthalpy intervals:

$$N_{\text{SHELLS}} = \sum_{k=1}^{K} N_k(S_k - 1) \tag{E.2}$$

where K is the total number of enthalpy intervals on the composite curves.

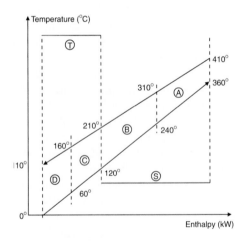

N_{overall}	$= 3.889$	
$N_{\textcircled{S}} = N_{\textcircled{A}} + N_{\textcircled{B}} = 2.899$		3.889
$N_{\textcircled{T}} = N_{\textcircled{C}} + N_{\textcircled{D}} = 0.990$		
$N_{\textcircled{A}}$	$= 1.660$	
$N_{\textcircled{B}}$	$= 1.239$	3.889
$N_{\textcircled{C}}$	$= 0.520$	
$N_{\textcircled{D}}$	$= 0.470$	

Figure E.1. If the real (noninteger) number of shells is calculated, the heat exchange profiles can be divided in any way and the sum is always the same. *(From Ahmad, Linnhoff, and Smith, Trans. ASME, J. Heat Transfer, 110: 304, 1988; reproduced by permission of the American Society of Mechanical Engineers.)*

The large number of matches assumed in Eq. (E.2) is not a complication in establishing the target. This is so because the additive property shows that the total fractional number of shells is independent of how many vertical sections are used to divide a given heat exchange profile.

Equation (E.2) can be considered further by using the *contribution* of the total fractional number of shells from each stream. The shells contribution of each stream i per match is

$$N(i) = \sum_{k=k_1}^{k_2} N_k \tag{E.3}$$

where k_1 is the starting interval of stream i and k_2 is the interval where stream i ends. Then,

$$\begin{aligned}
N_{\text{SHELLS}} &= \sum_{k=1}^{K} N_k(S_k - 1) \\
&= \sum_{k=1}^{K} N_k S_k - \sum_{k=1}^{K} N_k \\
&= \sum_{i=1}^{S} N(i) - \sum_{k=1}^{K} N_k
\end{aligned} \tag{E.4}$$

The reason why

$$\sum_{k=1}^{K} N_k S_k = \sum_{i=1}^{S} N(i) \tag{E.5}$$

is that for the whole problem the shells contributions can either be added on an interval basis or on a stream basis. Equation (E.4) shows that the fractional number-of-shells target is the sum of the contributions from all the streams less the contribution of one stream running across all intervals.

This result is useful in making the following observations: Ultimately, N_{SHELLS} will have to be converted into an integer, and streams contributing a small fractional number of shells, much fewer than 1.0 shells, will nevertheless eventually require at least one whole shell in the design. Therefore, if many such small streams exist in a problem, they only make a small contribution to N_{SHELLS} but will actually require several shells between them in the design. There is thus the possibility of a significant underestimate if N_{SHELLS} is simply made integer. To overcome the problem, each stream i with $N(i) < 1.0$ should have $N(i)$ reset to 1.0 in Eq. (E.4) before converting N_{SHELLS} to an integer.

Furthermore, actual designs will normally observe the pinch division. Hence N_{SHELLS} should be evaluated and taken as the next largest integer for each side of the pinch. The number-of-shells target is then

$$[N_{\text{SHELLS}}] = [(N_{\text{SHELLS}})_{\text{ABOVE PINCH}}] + [(N_{\text{SHELLS}})_{\text{BELOW PINCH}}] \tag{E.6}$$

where the symbol $[N]$ represents the next largest integer to the fractional number N. The approach naturally progresses toward the units target N_{UNITS} in cases where the temperatures are such that all exchangers require only a single shell each.

E.1 Minimum Area Target for Networks of 1-2 Shells

In practice, any design of 1-2 shells using the maximum available driving force in interval k can never be achieved with $N_k(S_k - 1)$ shells because a fractional number of shells is implied. The nearest actual approach to this is with $[N_k](S_k - 1)$ shells, where $[N_k]$ is the next largest integer to N_k. Hence the area target for a network of 1-2 exchangers is calculated by applying the F_T factor given by $[N_k]$ in the area target formula:

$$A_{1-2} = \sum_{k=1}^{\text{INTERVALS}} \frac{1}{\Delta T_{\text{LM}k} F_{Tk}} \sum_{i=1}^{\text{STREAMS}} \left(\frac{q_i}{h_i}\right)_k \tag{E.7}$$

F_{Tk} for $[N_k]$ shells in series can be calculated from the well-established relationships for $F_T{}^2$. First, P_{N-2N} is calculated for interval k, from which P_{1-2} is calculated for the $[N_k]$ shells from

$$P_{1-2} = \frac{1 - \left(\dfrac{1 - P_{N-2N}R}{1 - P_{N-2N}}\right)^{1/[N_k]}}{R - \left(\dfrac{1 - P_{N-2N}R}{1 - P_{N-2N}}\right)^{1/[N_k]}} \qquad \text{for } R \neq 1 \tag{E.8}$$

$$P_{1-2} = \frac{P_{N-2N}}{P_{N-2N} - P_{N-2N}[N_k] + [N_k]} \qquad \text{for } R = 1 \tag{E.9}$$

Equations (E.8) and (E.9) are simple inversion of Eqs. (D.1) and (D.2). F_T can then be calculated from

$$F_T = \frac{\sqrt{R^2 + 1} \ln\left[\dfrac{(1 - P_{1-2})}{(1 - RP_{1-2})}\right]}{(R - 1) \ln\left\{\dfrac{[2 - P_{1-2}(R + 1 - \sqrt{R^2 + 1})]}{[2 - P_{1-2}(R + 1 + \sqrt{R^2 + 1})]}\right\}} \qquad \text{for } R \neq 1 \tag{E.10}$$

$$F_T = \frac{\left[\dfrac{\sqrt{2}\,P_{1-2}}{(1 - P_{1-2})}\right]}{\ln\left\{\dfrac{[2 - P_{1-2}(2 - \sqrt{2})]}{[2 - P_{1-2}(2 + \sqrt{2})]}\right\}} \qquad \text{for } R = 1 \tag{E.11}$$

At the targeting stage, it is now possible to predict the minimum average area per shell in the network as $A_{1-2}/[N_{\text{SHELLS}}]$. This may be larger than the allowable maximum area per shell, $A_{\text{SHELL,max}}$. If so, then the number of shells targets should be increased to the integer $[A_{1-2}/A_{\text{SHELL,max}}]$. However, if $A_{1-2}/[N_{\text{SHELLS}}] \le A_{\text{SHELL,max}}$, this will actually have an area less than $A_{\text{SHELL,max}}$. The true outcome is largely design-dependent on the distribution of exchanger areas and cannot be fully evaluated as a target. Despite this, the error incurred by using $[A_{1-2}/A_{\text{SHELL,max}}]$, when required, will usually be small and is adequate for targeting purposes. This approach can be used on each side of the pinch if required.

An algorithm to target the number of shells and network area for networks consisting of 1-2 shells can now be formulated:

1. The composite curves (including utilities) are divided into enthalpy intervals. The minimum (fractional) number of shells N_k for the temperatures of each interval k is evaluated using Eqs. (D.7) to (D.9).

2. The streams (including utilities) are drawn in a grid diagram which shows the intervals, their N_k values, and the pinch location.

3. The total number of shells for each side of the pinch is calculated by summing the number of shells from all the intervals for each side of the pinch:

$$(N_{\text{SHELLS}})_{\text{ABOVE PINCH}} = \sum_{k=1}^{K} N_k(S_k - 1) \qquad (E.12)$$

where K is the number of the enthalpy intervals above the pinch, and

$$(N_{\text{SHELLS}})_{\text{BELOW PINCH}} = \sum_{k=K+1}^{M} N_k(S_k - 1) \qquad (E.13)$$

where $(M - K)$ is the number of enthalpy intervals below the pinch. M is the number of enthalpy intervals on the composite curves.

4. Find the total real number of shells contributed by each stream i on each side of the pinch:

$$N(i)_{\text{ABOVE PINCH}} = \left(\sum_{k=\alpha_i}^{\beta_i} N_k \right)_{\text{ABOVE PINCH}} \qquad (E.14)$$

$$N(i)_{\text{BELOW PINCH}} = \left(\sum_{k=\alpha_i}^{\beta_i} N_k \right)_{\text{BELOW PINCH}} \qquad (E.15)$$

where α_i and β_i are the start and end intervals for stream i on each side of the pinch.

- If $N(i)_{\text{ABOVE PINCH}} < 1.0$, set $N(i)_{\text{ABOVE PINCH}} = 1.0$.
- If $N(i)_{\text{BELOW PINCH}} < 1.0$, set $N(i)_{\text{BELOW PINCH}} = 1.0$.

Reevaluate:

$$(N_{\text{SHELLS}})_{\text{ABOVE PINCH}} = \sum_i N(i)_{\text{ABOVE PINCH}} - \sum_{k=1}^{K} N_k \qquad \text{(E.16)}$$

$$(N_{\text{SHELLS}})_{\text{BELOW PINCH}} = \sum_i N(i)_{\text{BELOW PINCH}} - \sum_{k=K+1}^{M} N_k \qquad \text{(E.17)}$$

5. The 1-2 area target above the pinch is

$$(A_{1-2})_{\text{ABOVE PINCH}} = \sum_{k=1}^{K} \frac{1}{\Delta T_{\text{LM}k} F_{Tk}} \sum_{i=1}^{S_k} \left(\frac{q_i}{h_i} \right)_k \qquad \text{(E.18)}$$

If the average area per shell above the pinch

$$\frac{(A_{1-2})_{\text{ABOVE PINCH}}}{[N_{\text{SHELLS}}]_{\text{ABOVE PINCH}}}$$

is greater than maximum allowable area per shell, $A_{\text{SHELL.max}}$, then set

$$(N_{\text{SHELLS}})_{\text{ABOVE PINCH}} = [(A_{1-2})_{\text{ABOVE PINCH}}/A_{\text{SHELL.max}}]$$

The 1-2 area target below the pinch is

$$(A_{1-2})_{\text{BELOW PINCH}} = \sum_{k=K+1}^{M} \frac{1}{\Delta T_{\text{LM}k} F_{Tk}} \sum_{i=1}^{S_k} \left(\frac{q_i}{h_i} \right)_k \qquad \text{(E.19)}$$

If the average area per shell below the pinch

$$\frac{(A_{1-2})_{\text{BELOW PINCH}}}{[N_{\text{SHELLS}}]_{\text{BELOW PINCH}}}$$

is greater than maximum allowable area per shell, $A_{\text{SHELL.max}}$, then set

$$(N_{\text{SHELLS}})_{\text{BELOW PINCH}} = [(A_{1-2})_{\text{BELOW PINCH}}/A_{\text{SHELL.max}}]$$

Then

$$[N_{\text{SHELLS}}] = [(N_{\text{SHELLS}})_{\text{ABOVE PINCH}}] + [(N_{\text{SHELLS}})_{\text{BELOW PINCH}}]$$
$$(E.20)$$

Example E.1 Target the number of shells for the process shown in Fig. 6.2 for which the stream data are given in Table 7.1. Assume $\Delta T_{\min} = 10°C$, $X_p = 0.9$, and the maximum area per shell is 500 m².

Solution

a. The composite curves for this problem have already been divided into enthalpy intervals in Fig. 7.5.

b. The streams (including utilities) are shown in grid form in Fig. E.2. Also shown in Fig. E.2 are the values of R, P, and N_k determined from Eqs. (7.14) to (7.16).

c. The total (fractional) number of shells for each side of the pinch is given by

$$(N_{\text{SHELLS}})_{\text{ABOVE PINCH}} = \sum_{k=1}^{K} N_k(S_k - 1)$$
$$= 0.03412 \times 1 + 0.6343 \times 2 + 0.9172 \times 1 + 2.7980 \times 3$$
$$= 10.61$$

$$(N_{\text{SHELLS}})_{\text{BELOW PINCH}} = \sum_{k=K+1}^{M} N_k(S_k - 1)$$
$$= 2.5225 \times 2 + 0.1501 \times 3 + 0.5793 \times 2$$
$$= 6.65$$

d. Check the number of shells for each stream:

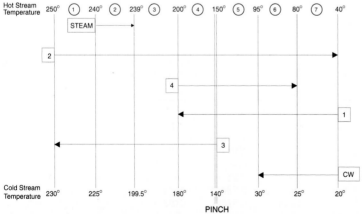

Figure E.2. Stream population for targeting the number of shells for the data from Table 7.1.

Above the pinch

$$N_{\text{SHELLS}} \text{ (STEAM)} = 0.6343$$

$$N_{\text{SHELLS}} (2) = 4.3836$$

$$N_{\text{SHELLS}} (4) = 2.7980$$

$$N_{\text{SHELLS}} (1) = 2.7980$$

$$N_{\text{SHELLS}} (3) = 4.3836$$

Below the pinch

$$N_{\text{SHELLS}} (2) = 3.2519$$

$$N_{\text{SHELLS}} (4) = 2.6726$$

$$N_{\text{SHELLS}} (1) = 3.2519$$

$$N_{\text{SHELLS}} \text{ (CW)} = 0.7294$$

N_{SHELLS} is greater than 1 for all streams except steam and cooling water. The contribution of these is now increased to one. Thus

$$(N_{\text{SHELLS}})_{\text{ABOVE PINCH}} = 10.61 + (1 - 0.6343)$$

$$= 10.98$$

$$= 11$$

$$(N_{\text{SHELLS}})_{\text{BELOW PINCH}} = 6.65 + (1 - 0.7294)$$

$$= 6.92$$

$$= 7$$

e. The 1-2 network area target can be calculated by applying the F_T correction factors in Fig. E.2 to the interval areas from Table 7.2:

Enthalpy interval	A_{k1-1}	F_{Tk}	A_{k1-2}
1	194.2	0.9717	199.9
2	482.7	0.9930	486.1
3	459.4	0.8140	564.4
4	3566.1	0.7847	4544.5
5	2113.8	0.8493	2488.9
6	227.3	0.9965	228.1
7	366.1	0.9699	377.5

Average area per shell above the pinch $= \dfrac{199.9 + 486.1 + 564.4 + 4544.5}{11}$

$$= 527 \text{ m}^2$$

This is greater than the maximum area per shell. Thus

$$(N_{\text{SHELLS}})_{\text{ABOVE PINCH}} = \frac{199.9 + 486.1 + 564.4 + 4544.5}{500}$$

$$= 11.6$$

Average area per shell below pinch $= \dfrac{2488.9 + 228.1 + 377.5}{7}$

$$= 442.1 \text{ m}^2$$

This is lower than the maximum area per shell.

f. The target for the number of shells is given by

$$[N_{\text{SHELLS}}] = [(N_{\text{SHELLS}})_{\text{ABOVE PINCH}}] + [(N_{\text{SHELLS}})_{\text{BELOW PINCH}}]$$

$$= 12 + 7$$

$$= 19$$

Thus the target is 19 shells.

E.2   References

1. Ahmad, S., and Smith, R., "Targets and Design for Minimum Number of Shells in Heat Exchanger Networks," *IChemE, ChERD,* 67: 481, 1989.
2. Bowman, R. A., Mueller, A. C., and Nagle, W. M., "Mean Temperature Differences in Design," *Trans. ASME,* 62: 283, 1940.

Algorithm for Heat Exchanger Capital Cost Target

The area target, Eq. (7.6), sums the area contributions from each enthalpy interval. This equation can be rearranged to an equivalent expression which sums the area contribution of each stream[1,2]

$$
\begin{aligned}
A_{\text{NETWORK}} &= \sum_{k}^{\text{INTERVALS } K} \frac{1}{\Delta T_{\text{LM}k}} \left(\sum_{i}^{\text{HOT STREAMS } I} \frac{q_i}{h_i} + \sum_{j}^{\text{COLD STREAMS } J} \frac{q_j}{h_j} \right) \\
&= \sum_{i}^{\text{HOT STREAMS } I} \left(\sum_{k}^{\text{INTERVALS } K} \frac{q_i}{\Delta T_{\text{LM}k} h_i} \right) + \sum_{j}^{\text{COLD STREAMS } J} \\
&\qquad\qquad\qquad\qquad\qquad \left(\sum_{k}^{\text{INTERVALS } K} \frac{q_j}{\Delta T_{\text{LM}k} h_j} \right) \\
&= \sum_{i}^{\text{HOT STREAMS } I} A_i + \sum_{j}^{\text{COLD STREAMS } J} A_j
\end{aligned} \tag{F.1}
$$

where A_i = contribution to area target from hot stream i
A_j = contribution to area target from cold stream j

Equation (F.1) shows that each stream makes a contribution to total heat transfer area defined only by its duty, position in the composite curves, and its h value. This contribution to area means also a contribution to capital cost. If, for example, a corrosive stream requires special materials of construction, it will have a greater contribution to capital cost than a similar noncorrosive stream. If only one cost law is to be used for a network comprising mixed materials of construction, the area contribution of streams requiring special materials must somehow increase. One way this may be done is by weighting the heat transfer coefficients to reflect the cost of the material the stream requires.

To develop the approach, first consider a single exchanger whose cost may be representetd as

$$\text{Installed capital cost of reference exchanger} = a_1 + b_1 A^{c_1} \quad \text{(F.2)}$$

where a_1, b_1, and c_1 are cost law coefficients for the reference exchanger.

If, instead, the heat exchanger is made to a different specification, its cost may be represented as

$$\text{Installed capital cost of special exchanger} = a_2 + b_2 A^{c_2} \quad \text{(F.3)}$$

where a_2, b_2, and c_2 are cost law coefficients for a special exchanger.

In the approach to be developed, the cost of the special exchanger can be determined from the reference cost law by using a modified area A^*:

$$\text{Installed capital cost of special exchanger} = a_1 + b_1 A^{*c_1} \quad \text{(F.4)}$$

Heat exchanger cost data can usually be manipulated such that fixed costs, represented by the coefficient a in Eqs. (F.2) to (F.4), do not vary with exchanger specification.[2] Equations (F.3) and (F.4) can now be rearranged to give the modified exchanger area A^* as a function of actual area A and the cost law coefficients:

$$A^* = \left(\frac{b_2}{b_1}\right)^{1/c_1} A^{(c_2/c_1)-1} A \quad \text{(F.5)}$$

The relationship between heat exchanger area and overall heat transfer coefficient U is given by

$$A = \frac{Q}{\Delta T_{\text{LM}} U} \quad \text{(F.6)}$$

where Q is exchanger heat load. The ratio $Q/\Delta T_{\text{LM}}$ is constant for a given heat exchanger, and hence the modified U value U^* can be related to actual U value:

$$\frac{1}{U^*} = \left(\frac{b_2}{b_1}\right)^{1/c_1} A^{(c_2/c_1)-1} \frac{1}{U} \quad \text{(F.7)}$$

The overall heat transfer coefficient in a single exchanger comprises resistance contributions from both streams. Each contribution contains allowances for film, wall, and fouling resistances. In practice, the overall heat transfer coefficient will depend to some extent on the exchanger flow arrangement. It is not possible to specify such details at the targeting stage; hence the overall heat transfer coefficient must be assumed independent of the flow arrangement:

$$\frac{1}{U} = \frac{1}{h_H} + \frac{1}{h_C} \quad \text{(F.8)}$$

Equation (F.7) may be split stream-wise to obtain an expression for the modified h value h_j^* of either stream in the match:

$$h_j^* = \left(\frac{b_1}{b_2}\right)^{1/c_1} A^{1-(c_2/c_1)} h_j \qquad (F.9)$$

A stream-specific cost-weighting factor ϕ_j to apply to the h value of a special stream j can now be defined. This is the ratio of weighted to actual stream h values:

$$\phi_j = \frac{h_j^*}{h_j} = \left(\frac{b_1}{b_2}\right)^{1/c_1} A^{1-(c_2/c_1)} \qquad (F.10)$$

The same philosophy of weighting area contributions in a single exchanger can be extended to weighting stream area contributions for a whole network. Some additional error in the targets is incurred by this extension, resulting from the fact that a stream may pass through several exchangers all with different areas. At the targeting stage, exchangers are assumed to be all the same size. The special stream cost-weighting factor is then expressed as

$$\phi_j = \left(\frac{b_1}{b_2}\right)^{1/c_1} \left(\frac{A_{\text{NETWORK}}}{N}\right)^{1-(c_2/c_1)} \qquad (F.11)$$

Once the ϕ factor has been evaluated for each stream, a weighted network area target A_{NETWORK}^* can be calculated:

$$A_{\text{NETWORK}}^* = \sum_k^{\text{INTERVALS } K} \frac{1}{\Delta T_{\text{LM}k}}$$

$$\times \left(\sum_i^{\text{HOT STREAMS } I} \frac{q_i}{\phi_i h_i} + \sum_j^{\text{COLD STREAMS } J} \frac{q_j}{\phi_j h_j}\right) \qquad (F.12)$$

References

1. Ahmad, S., "Heat Exchanger Networks: Cost Trade-Offs in Energy and Capital," Ph.D. thesis, UMIST, U.K., 1985.
2. Hall, S. G., Ahmad, S., and Smith, R., "Capital Cost Targets for Heat Exchanger Networks Comprising Mixed Materials of Construction, Pressure Ratings and Equipment Types," *Computers Chem. Eng.*, 14: 319, 1990.

Index

ABOUT THE AUTHOR

Robin Smith is a Senior Lecturer in Chemical Engineering at the University of Manchester Institute of Science and Technology in the United Kingdom, as well as a consultant for major companies worldwide. He previously worked in process intvestigation and design with Rohm & Haas, and in process modeling and the development of process integration techniques with ICI. The author of many technical papers, Dr. Smith is a Fellow of the Institution of Chemical Engineers.